INTRODUCTION TO MINIMAX

V. F. Dem'yanov and V. N. Malozemov

INTRODUCTION TO MINIMAX

Translated from Russian by D. Louvish

A HALSTED PRESS BOOK

JOHN WILEY & SONS
New York · Toronto

ISRAEL PROGRAM FOR SCIENTIFIC TRANSLATIONS
Jerusalem · London

Sole distributors for the Western Hemisphere and Japan

HALSTED PRESS, a division of
JOHN WILEY & SONS, INC., NEW YORK

Library of Congress Cataloging in Publication Data

Dem'ĭanov, Vladimir Fedorovich.
Introduction to minimax.

Translation of *Vvedenie v minimaks.*
"A Halsted Press book."
Bibliography: p.
1. Maxima and minima. 2. Mathematical optimization. I. Malozemov, Vasiliĭ Nikolaevich, joint author. II. Title.
QA402.5.D3913 515'.7 74-8156
ISBN 0-470-20850-3

Distributors for the U.K., Europe, Africa and the Middle East

JOHN WILEY & SONS, LTD., CHICHESTER

Distributed in the rest of the world by

KETER PUBLISHING HOUSE JERUSALEM LTD.
ISBN 0 7065 1400 9
IPST cat. no. 22100

This book is a translation from Russian of
VVEDENIE V MINIMAKS
Izdatel'stvo "Nauka"
Moscow, 1972

Printed in Israel

CONTENTS

PREFACE

1. Minimax is an important principle in optimal selection of parameters. The gist of this principle is easily illustrated in regard to the processing of empirical data. Suppose we are given a table of values of some function:

$$y_k = y(t_k), \qquad k \in [0:N]. \tag{1}$$

Of all algebraic polynomials $P_n(A, t) = \sum_{i=0}^{n} a_i t^i$, $A = (a_0, \ldots, a_n)$, we wish to select a polynomial $P_n(A^*, t)$, which provides a good approximation of table (1). The phrase "good approximation" may be made rigorous as follows: we seek a polynomial $P_n(A^*, t)$ for which the maximum deviation

$$\max_{k \in [0:N]} |y_k - P_n(A^*, t_k)|$$

is minimal with respect to the maximum deviations of all other algebraic polynomials $P_n(A, t)$:

$$\max_{k \in [0:N]} |y_k - P_n(A^*, t_k)| = \min_{\{A\}} \max_{k \in [0:N]} |y_k - P_n(A, t_k)|.$$

The polynomial $P_n(A^*, t)$ is known as a polynomial of best approximation for table (1) in Chebyshev's sense. In other words, A^* is a solution to the following minimax problem:

$$\max_{k \in [0:N]} |y_k - P_n(A, t_k)| \to \min_{\{A\}}.$$

Of course, $P_n(A, t)$ may be replaced by any other parametric family of functions $Q(X, t)$, where $X = (x_1, \ldots, x_n)$; the parameter vector X may be subjected to a constraint $X \in \Omega$, where Ω is a subset of E_n. Solving the minimax problem

$$\max_{k \in [0:N]} |y_k - Q(X, t_k)| \to \min_{X \in \Omega},$$

one then finds a vector $X^* \in \Omega$ such that $Q(X^*, t)$ is a best approximation (in Chebyshev's sense) to the table (1) with respect to all other functions $Q(X, t)$, $X \in \Omega$.

Minimax is not the only principle for optimal selection of parameters. In the same problem — processing of empirical data — a "best polynomial" may be defined as a polynomial $P_n(\tilde{A}, t)$ which minimizes the sum of squared deviations

$$\sum_{k=0}^{N}(y_k - P_n(\tilde{A}, t_k))^2 = \min_{\{A\}} \sum_{k=0}^{N}(y_k - P_n(A, t_k))^2.$$

The question of what optimization principle to choose depends on the nature of the problem and on the requirements its solution must meet. Leaving this question aside, we only remark that in some cases the minimax solution has a remarkable property, known as alternation. In the problem of approximation, this means that there are $n+2$ abscissas $t_{i_0} < t_{i_1} < \ldots < t_{i_{n+1}}$ (one more than the dimensionality of the parameter vector A) at which the difference $y_{i_k} - P_n(A^*, t_{i_k})$ attains its maximum absolute value, and its sign alternates at consecutive abscissas.

Two more examples of minimax problems will now be presented.

Mandel'shtam's problem. Let $X = (x_1, \ldots, x_n)$ and

$$F(X, t) = \left| \sum_{k=1}^{n} \cos(kt + x_k) \right|.$$

Of all $X \in E_n$, we wish to select a vector X^* such that

$$\max_{t \in [0, 2\pi]} F(X^*, t) = \min_{\{X\}} \max_{t \in [0, 2\pi]} F(X, t).$$

This problem occurs in electrical circuit theory.

Matrix games. Let $A = \{a_{ij}\}, i \in [1:m], j \in [1:n]$, be a given real $m \times n$ matrix, and let

$$T_n = \left\{ X = (x_1, \ldots, x_n) \mid x_i \geqslant 0, \sum_{i=1}^{n} x_i = 1 \right\},$$

$$G_m = \left\{ Y = (y_1, \ldots, y_m) \mid y_j \geqslant 0, \sum_{j=1}^{m} y_j = 1 \right\}.$$

We wish to find vectors $X^* \in T_n$ and $Y^* \in G_m$ for which

$$\max_{Y \in G_m} (AX^*, Y) = \min_{X \in T_n} \max_{Y \in G_m} (AX, Y),$$

$$\min_{X \in T_n} (AX, Y^*) = \max_{Y \in G_m} \min_{X \in T_n} (AX, Y).$$

In this book we consider the minimax problem in its general form

$$\max_{Y \in G} F(X, Y) \to \min_{X \in \Omega}, \tag{2}$$

where Ω is a convex closed subset of E_n and G a bounded closed subset of E_m.

If $F(X, Y)$ is a linear function of X for every fixed $Y \in G$ and the set Ω is defined by linear equalities and inequalities, problem (2) is known as a linear minimax problem. Otherwise the problem is said to be nonlinear. According as $\Omega \neq E_n$ or $\Omega = E_n$, we shall speak of a minimax problem with or without constraints. The variable X is known as the parameter.

There are three basic ideas that can be applied to solving minimax problems.

I. The search for an extremal basis. Let the function $F(X, Y)$ be convex in X on Ω for every fixed $Y \in G$. Then there are r points $Y_1, \ldots, Y_r$ in $G, 1 \leqslant r \leqslant n+1$, such that the minimax problem (2) is equivalent to the following problem:

$$\max_{Y \in G_r} F(X, Y) \to \min_{X \in \Omega}, \qquad (3)$$

where $G_r = \{Y_1, \ldots, Y_r\}$. The set G_r is known as an extremal basis. Once an extremal basis is available, a solution to problem (2) is obtained by solving problem (3), which is usually simpler.

II. Minimization of the maximum function. Set

$$\varphi(X) = \max_{Y \in G} F(X, Y).$$

The original problem (2) is then equivalent to minimizing the function $\varphi(X)$ on Ω.

III. Determination of a saddle point. A point $[X^*, Y^*]$ is known as a saddle point of $F(X, Y)$ on the set $\Omega \times G$ if $F(X^*, Y) \leqslant F(X^*, Y^*) \leqslant F(X, Y^*)$ for all $X \in \Omega$, $Y \in G$. Assuming that the function $F(X, Y)$ has a saddle point $[X^*, Y^*]$ on $\Omega \times G$, we have

$$\min_{X \in \Omega} \max_{Y \in G} F(X, Y) = F(X^*, Y^*) = \max_{Y \in G} \min_{X \in \Omega} F(X, Y).$$

In the case considered, therefore, the minimax problem (2) and its dual

$$\min_{X \in \Omega} F(X, Y) \to \max_{Y \in G}$$

both reduce to determination of a saddle point.

The first idea is due to Chebyshev. Its most sophisticated application to date is in one-dimensional approximation problems.

The second idea is very simple. It has developed considerably over the past few years, thanks to the proof that, under certain natural assumptions, the maximum function is differentiable in all directions.

The third concept is naturally associated with the name of von Neumann, and is used in solving problems of game theory.

2. We now outline the book's contents chapter by chapter.

The first two chapters are devoted to the problem of the best approximation (in Chebyshev's sense) of functions by algebraic polynomials (Chapter I — the discrete case, Chapter II — the continuous case). The exposition is based on Chebyshev interpolation. This yields a rapid derivation of all the basic results of Chebyshev's theory and, as by-products, methods for determination of polynomials of best approximation. No serious difficulties are involved in going from the discrete to the continuous case. The problem of best uniform approximation of a function is the simplest (and historically the earliest) linear minimax problem.

In the third chapter we discuss the linear minimax problem without constraints (discrete case). In the simplest situation we establish the basic results of the minimax theory: directional differentiability of the maximum function, necessary conditions for a minimax, sufficient conditions for a local minimax. Much attention is paid to methods of successive approximation for finding stationary points of the maximum function $\varphi(X)$.

The same problems, with convex constraints added, are examined in Chapters IV and V (discrete case).

The transition from the discrete to the continuous case is made in Chapter VI.

The pivotal chapter of the book is Chapter III.

The book has four appendices, containing all the requisite auxiliary material.

There are no references to literature in the text. The concluding section of the book, "Bibliographic notes," present historical information and bibliographic references.

3. The authors use the inductive method of exposition (particular cases of the minimax problem are discussed first, then more general ones), for two reasons. First, our goal is to acquaint the unversed reader with the subject. Second, the specific features of certain particular cases require more space for their exposition than would be expected in the light of the general theorems. This is the case with polynomial approximation and methods for solving discrete minimax problems. We therefore elected first to examine particular cases in full detail, and to go over only then to the general theory.

4. The book consists of three almost independent parts: Chapters I and II — polynomial approximation; Chapters III through V — discrete minimax problems and nonlinear programming; Chapter VI — the main results of minimax theory in its general formulation and their applications.

Starred sections and subsections may be skipped in a first reading.

The sections are numbered independently in each chapter. The chapter number is not cited in references to formulas, lemmas and theorems within the chapter.

Two special symbols used are:

$\stackrel{df}{=}$: equal by definition,

$[0:N]$: the set of all integers from zero to N inclusive.

5. This book is the result of four years of work. All chapters and sections were written by both authors in collaboration.

Now that our work on the book is completed, we express our gratitude to all those who helped and supported us. First and foremost, we thank N. N. Moiseev, who supported the idea of writing the book, manifested exceptional interest in our work, and was highly instrumental in speeding up publication. Yu. B. Germeier, I. K. Daugavet, and B. N. Pshenichnyi read the manuscript carefully and offered several highly pertinent remarks which were gratefully accepted and taken into consideration. Most fruitful for us were our years of collaboration with E. F. Voiton, who proposed a number of minimax nonlinear problems in radio-engineering applications and in many ways stimulated our work.

We had the happy opportunity to discuss various results with A. M. Vershik, M. K. Gavurin, E. G. Gilbert, D. M. Danskin, V. I. Zubov, I. P. Mysovskikh, L. W. Neustadt, E. Polak, I. V. Romanovskii, A. M. Rubinov and V. A. Yakubovich. We are indebted to all those mentioned.

The authors have now gained some practical experience in solving minimax problems. This is due to a great extent to the help of M. S. Al'tmark, V. M. Belykh, T. K. Vinogradova, V. A. Daugavet, B. F. Mitchell, A. B. Pevnyi, A. L. Fradkov and V. I. Shushkov.

We are sincerely indebted to E. V. Artem'eva, L. K. Nikitina, V. I. Savchenko and V. S. Filimonova, whose help in the shaping of the manuscript cannot be overestimated.

The main body of the book was presented in lectures at the Third All-Union Summer School of Mathematics on Optimization Methods at Tiraspol, August 1969.

The Authors.

Leningrad,
November 1970

Chapter I

BEST APPROXIMATION BY ALGEBRAIC POLYNOMIALS — DISCRETE CASE

§1. STATEMENT OF THE PROBLEM

Suppose we are given a table of values for some function:

$$y_k = y(t_k), \qquad k \in [0:N].$$

t	t_0	t_1	t_2	...	t_N
y	y_0	y_1	y_2	...	y_N

We shall assume throughout this chapter that the values of the independent variable (abscissas) are arranged in increasing order:

$$t_0 < t_1 < t_2 < \ldots < t_N.$$

The goodness of fit of an algebraic polynomial $P_n(A, t)$ of degree at most n $(n \leqslant N)$,

$$P_n(A, t) = a_0 + a_1 t + \ldots + a_n t^n, \qquad A = (a_0, a_1, \ldots, a_n),$$

to this table is most naturally characterized by the maximum deviation:

$$\max_{k \in [0:N]} | y_k - P_n(A, t_k) |.$$

For fixed n, set

$$\rho = \inf_{\{A\}} \max_{k \in [0:N]} | y_k - P_n(A, t_k) |.$$

A polynomial $P_n(A^*, t)$ for which the maximum deviation

$$\max_{k \in [0:N]} | y_k - P_n(A^*, t_k) | = \rho$$

is minimal is known as a polynomial of best approximation to the original table [and the corresponding deviation (the number ρ) will be called the b e s t f i t].

Our subject in this chapter is the actual construction of polynomials of best approximation. Success in this project depends essentially on the relation between n and N.

Thus, if $n=N$ (the degree of the required polynomial equals the number of abscissas in the table), the solution to our problem is clearly given by an interpolating polynomial (see Appendix I, §2). In this case, $\rho=0$.

The first nontrivial (and, as we shall see, fundamental) case is $N=n+1$, which leads to what is known as Chebyshev interpolation. Chebyshev interpolation ultimately provides a polynomial of best approximation in the general case $N>n+1$.

§2. CHEBYSHEV INTERPOLATION

1. Retaining the previous notation and setting $N=n+1$, we state and prove the following theorem.

Theorem 2.1. Let

$$\rho=\inf_{\{A\}} \max_{k\in[0:n+1]} |y_k - P_n(A, t_k)|.$$

There exists a polynomial of best approximation, and it is unique. A polynomial $P_n(A^*, t)$ is the polynomial of best approximation if and only if, for some h,

$$(-1)^k h + P_n(A^*, t_k) = y_k, \quad k\in[0:n+1]. \tag{2.1}$$

In that case, $\rho=|h|$.

The proof will be divided into steps as follows: sufficiency, existence, necessity, uniqueness.

Sufficiency. Let $P_n(A^*, t)$ be a polynomial satisfying (2.1) for some h. We must show that $P_n(A^*, t)$ is a polynomial of best approximation.

Suppose, on the contrary, that

$$\rho< \max_{k\in[0:n+1]} |y_k - P_n(A^*, t_k)| \overset{df}{=} |h|.$$

Then, by the definition of ρ, there is a polynomial $P_n(A_1, t)$ such that

$$\max_{k\in[0:n+1]} |y_k - P_n(A_1, t_k)| < |h|. \tag{2.2}$$

Consider the polynomial $Q_n(t) = P_n(A_1, t) - P_n(A^*, t)$; its values at the abscissas t_k, $k\in[0:n+1]$, are

$$Q_n(t_k) = \{P_n(A_1, t_k) - y_k\} + \{y_k - P_n(A^*, t_k)\} = \\ = (-1)^k h - \{y_k - P_n(A_1, t_k)\}.$$

By (2.2), $Q_n(t_k)$ has the same sign as $(-1)^k h$.

Since the abscissas t_k, $k\in[0:n+1]$, are arranged in increasing order, we see that $Q_n(t)$ has $n+1$ consecutive sign changes and therefore

$n+1$ zeros. But a nontrivial polynomial of degree n cannot have $n+1$ zeros, and so $Q_n(t)\equiv 0$, i. e., $P_n(A_1, t)\equiv P_n(A^*, t)$, contradicting (2.2). This proves sufficiency, and also that $\rho=|h|$.

Existence. The existence problem for a polynomial of best approximation is now reduced to solvability of system (2.1), which is a system of $(n+2)$ linear equations in $(n+2)$ unknowns $h, a_0, a_1, \ldots, a_n$:

$$\left.\begin{aligned} h+a_0+a_1t_0+\ldots+a_nt_0^n&=y_0,\\ -h+a_0+a_1t_1+\ldots+a_nt_1^n&=y_1,\\ \cdots\cdots\cdots\cdots&\cdots\\ (-1)^{n+1}h+a_0+a_1t_{n+1}+\ldots+a_nt_{n+1}^n&=y_{n+1}.\end{aligned}\right\} \qquad (2.1')$$

The determinant Δ of this system is

$$\Delta=\begin{vmatrix} 1 & 1 & t_0 & \ldots & t_0^n \\ -1 & 1 & t_1 & \ldots & t_1^n \\ \cdot & \cdot & \cdot & \cdot & \cdot \\ (-1)^{n+1} & 1 & t_{n+1} & \ldots & t_{n+1}^n \end{vmatrix}.$$

Set

$$v(z_1, z_2, \ldots, z_n)\overset{\text{df}}{=}\begin{vmatrix} 1 & z_1 & \ldots & z_1^n \\ 1 & z_2 & \ldots & z_2^n \\ \cdot & \cdot & \cdot & \cdot \\ 1 & z_n & \ldots & z_n^n \end{vmatrix}=\prod_{n\geqslant k>i\geqslant 0}(z_k-z_i).$$

Expanding the determinant Δ in terms of its first column, we get

$$\Delta=\sum_{\nu=0}^{n+1} v(t_0, t_1, \ldots, t_{\nu-1}, t_{\nu+1}, \ldots, t_{n+1}).$$

Since the t_k are arranged in increasing order, all the terms in this sum are positive. Thus $\Delta>0$.

It follows that system (2.1') is always solvable and the solution is unique. Hence the existence of a polynomial of best approximation.

Necessity. We first develop an explicit expression for h. For all $k\in[0:n+1]$, set

$$\alpha_k=\frac{\dfrac{(-1)^k}{(t_k-t_0)\ldots(\ \ )\ldots(t_k-t_{n+1})}}{\displaystyle\sum_{\nu=0}^{n+1}\frac{(-1)^\nu}{(t_\nu-t_0)\ldots(\ \ )\ldots(t_\nu-t_{n+1})}},$$

where

$$(t_k-t_0)\ldots(\ \ )\ldots(t_k-t_{n+1})=\prod_{i=0,\ i\neq k}^{n+1}(t_k-t_i).$$

The numbers α_k possess the following properties:

a) $\alpha_k > 0, \ k \in [0:n+1]$;

b) $\sum_{k=0}^{n+1} \alpha_k = 1$;

c) for any polynomial $P_n(A, t)$ of degree at most n,

$$\sum_{k=0}^{n+1} (-1)^k \alpha_k P_n(A, t_k) = 0.$$

(Concerning these properties, see Appendix I, §1.)

Using (2.1), we get the equality

$$\sum_{k=0}^{n+1} (-1)^k \alpha_k [(-1)^k h + P_n(A^*, t_k)] = \sum_{k=0}^{n+1} (-1)^k \alpha_k y_k.$$

Hence it follows that

$$h = \sum_{k=0}^{n+1} (-1)^k \alpha_k y_k. \tag{2.3}$$

We can now take up the necessity proof. Let $P_n(A^*, t)$ be some polynomial of best approximation.

We have to prove that it satisfies (2.1), with h defined by (2.3). Since $\rho = |h|$, we have

$$\max_{k \in [0:n+1]} |y_k - P_n(A^*, t_k)| = |h|. \tag{2.4}$$

We claim that in fact

$$y_k - P_n(A^*, t_k) = (-1)^k h, \quad k \in [0:n+1]. \tag{2.5}$$

We distinguish between the cases $h = 0$, $h > 0$, $h < 0$. If $h = 0$, the validity of (2.5) is evident from (2.4). Let $h < 0$ (the treatment of the case $h > 0$ is analogous), and suppose that at least one of the equalities (2.5) fails to hold. Then, first, by (2.4),

$$(-1)^k [y_k - P_n(A^*, t_k)] \geqslant h, \quad k \in [0:n+1],$$

and, second, for some k_0,

$$(-1)^{k_0} [y_{k_0} - P_n(A^*, t_{k_0})] > h.$$

Using (2.3) and properties (a)–(c) of the numbers α_k, we obtain

$$h = \sum_{k=0}^{n+1} (-1)^k \alpha_k y_k = \sum_{k=0}^{n+1} (-1)^k \alpha_k [y_k - P_n(A^*, t_k)] =$$
$$= \sum_{k=0}^{n+1} \alpha_k \{(-1)^k [y_k - P_n(A^*, t_k)]\} > h \sum_{k=0}^{n+1} \alpha_k = h,$$

which is absurd. This proves necessity.

Uniqueness. That the polynomial of best approximation is unique now follows from the unique solvability of system (2.1)–(2.1'). The proof is complete.

2. Definition. The construction of a polynomial satisfying (2.1) is known as Chebyshev interpolation; the polynomial itself is called a Chebyshev interpolating polynomial.

Thus, the content of the preceding subsection may be summed up thus: we have shown that when $N=n+1$ construction of a polynomial of best approximation reduces to Chebyshev interpolation. Moreover, the value of h is found in advance by (2.3).

We shall need another expression for h. Let χ be a function defined at the abscissas t_k by

$$\chi_k=\chi(t_k)=(-1)^k, \quad k\in[0:n+1].$$

Recalling the formula for divided differences (Appendix I, §1) and the definition of α_k, we rewrite (2.3) as

$$h=\frac{y[t_0, t_1, \ldots, t_{n+1}]}{\chi[t_0, t_1, \ldots, t_{n+1}]}. \tag{2.6}$$

This is the required expression.

Proceeding now to construction of the polynomial of best approximation, we first introduce an interpolating polynomial $L_{n+1}(t)$ of degree at most $n+1$, uniquely determined by

$$L_{n+1}(t_k)=w_k; \quad w_k=y_k-h\chi_k, \quad k\in[0:n+1]. \tag{2.1''}$$

Using Newton's formula, we express $L_{n+1}(t)$ as

$$L_{n+1}(t)=w_0+\sum_{k=1}^{n+1} w[t_0, t_1, \ldots, t_k](t-t_0)(t-t_1)\ldots(t-t_{k-1})$$

and consider the last coefficient

$$w[t_0, t_1, \ldots, t_{n+1}]=y[t_0, \ldots, t_{n+1}]-h\chi[t_0, \ldots, t_{n+1}].$$

By (2.6), $w[t_0, t_1, \ldots, t_{n+1}]=0$, so that $L_{n+1}(t)$ is indeed a polynomial of degree at most n, satisfying (2.1)–(2.1''). Thus $L_{n+1}(t)\equiv P_n(A^*, t)$, and we finally obtain the following formula:

$$P_n(A^*, t)=y_0-h+\sum_{k=1}^{n}\{y[t_0, t_1, \ldots, t_k]-$$
$$-h\chi[t_0, t_1, \ldots, t_k]\}(t-t_0)(t-t_1)\ldots(t-t_{k-1}). \tag{2.7}$$

Practical use of this formula involves the following steps:

I. Construct a table of divided differences for the functions y and χ.

II. Compute h from (2.6).

III. Compute y_0-h and

$$y[t_0, t_1, \ldots, t_k] - h\chi[t_0, t_1, \ldots, t_k], \quad k \in [1:n],$$

and substitute these values into formula (2.7) to obtain $P_n(A^*, t)$.

IV. Check for the validity of the equalities

$$y_k - P_n(A^*, t_k) = (-1)^k h, \quad k \in [0:n+1].$$

Example. Consider the function $y(t) = |t|$, and the following two systems of abscissas in $[-1, 1]$:

1) $t_0 = -1,\ t_1 = -\frac{1}{2},\ t_2 = \frac{1}{2},\ t_3 = 1;$

2) $t_0 = -1,\ t_1 = -\frac{1}{2},\ t_2 = 0,\ t_3 = \frac{1}{2},\ t_4 = 1.$

We shall construct polynomials of best approximation for $|t|$ with these abscissas, supposing that $N = n + 1$; in the first case we shall get a polynomial of degree at most 2, in the second — of degree at most 3.

The computations are arranged in accordance with steps I through IV as recommended above.

1)

t	y	Divided differences		
		I	II	III
-1	1			
		-1		
$-\frac{1}{2}$	$\frac{1}{2}$		$\frac{2}{3}$	
		0		0
$\frac{1}{2}$	$\frac{1}{2}$		$\frac{2}{3}$	
		1		
1	1			

Here there is no need to set up a table of differences for χ, because $y[t_0, t_1, t_2, t_3] = 0$ and so $h = 0$. We at once find the polynomial of best approximation

$$P_2(t) = 1 - (t+1) + \frac{2}{3}(t+1)\left(t + \frac{1}{2}\right) = \frac{1}{3}(2t^2 + 1).$$

It is readily checked that

$$|t_k| - P_2(t_k) = 0, \quad k \in [0:3].$$

2)

t	y	Divided differences I	II	III	IV
-1	1				
		-1			
$-\frac{1}{2}$	$\frac{1}{2}$		0		
		-1		$\frac{4}{3}$	
0	0		2		$-\frac{4}{3}$
		1		$-\frac{4}{3}$	
$\frac{1}{2}$	$\frac{1}{2}$		0		
		1			
1	1				

t	χ	Divided differences I	II	III	IV
-1	1				
		-4			
$-\frac{1}{2}$	-1		8		
		4		$-\frac{32}{3}$	
0	1		-8		$\frac{32}{3}$
		-4		$\frac{32}{3}$	
$\frac{1}{2}$	-1		8		
		4			
1	1				

$$h = \frac{-4/3}{32/3} = -\frac{1}{8};$$

$$P_3(t) = \frac{9}{8} - \frac{3}{2}(t+1) + (t+1)\left(t+\frac{1}{2}\right) + \\ + 0\,(t+1)\left(t+\frac{1}{2}\right)t = t^2 + \frac{1}{8}.$$

A direct check shows that

$$|t_k| - P_3(t_k) = (-1)^{k+1}\cdot\frac{1}{8}, \qquad k \in [0:4].$$

3*. To conclude this section, we consider a special choice of abscissas:

$$t_k = \cos\frac{(n+1-k)\pi}{n+1}, \qquad k \in [0:n+1];$$
$$-1 = t_0 < t_1 < \ldots < t_{n+1} = 1.$$

In this case we shall be able to derive simpler formulas than (2.6) and (2.7) for h and $P_n(A^*, t)$.

Theorem 2.2. Let $t_k = \cos\frac{(n+1-k)\pi}{n+1}$, $y_k = y(t_k)$, $k \in [0:n+1]$. *Then*

$$h = \frac{1}{2n+2}\left\{ y_0 + (-1)^{n+1} y_{n+1} + 2\sum_{s=1}^{n} (-1)^s y_s \right\} \tag{2.8}$$

and

$$P_n(A^*, t) = \frac{1}{n+1}\sum_{s=0}^{n+1} y_s \tau_s(t), \tag{2.9}$$

where

$$\tau_s(t)=\begin{cases} D_n(x-\pi)=\frac{1}{2}+\sum_{\nu=1}^{n}(-1)^{\nu}T_{\nu}(t) & \textit{if} \quad s=0; \\ D_n(x+x_s)+D_n(x-x_s)= & \\ =1+2\sum_{\nu=1}^{n}(-1)^{\nu}\cos\frac{\nu s\pi}{n+1}T_{\nu}(t) & \textit{if} \quad s\in[1:n]; \\ D_n(x)=\frac{1}{2}+\sum_{\nu=1}^{n}T_{\nu}(t) & \textit{if} \quad s=n+1. \end{cases}$$

Here $D_n(x)=\frac{1}{2}+\cos x+\cos 2x+\ldots+\cos nx$ *is the Dirichlet kernel;*

$$x=\arccos t; \quad x_s=\frac{(n+1-s)\pi}{n+1}, \quad s\in[1:n];$$

$$T_{\nu}(t)=\cos\nu\arccos t, \quad \nu=0,\ 1,\ 2,\ \ldots$$

Proof. It suffices to check the relations

$$(-1)^k h+P_n(A^*,\ t_k)=y_k, \quad k\in[0:n+1].$$

We first prove two lemmas.

Lemma 2.1. The function $T_{\nu}(t)=\cos\nu\arccos t$, $-1\leqslant t\leqslant 1$, *is an algebraic polynomial of degree* ν *with leading coefficient* $2^{\nu-1}$, $\nu=1,\ 2,\ \ldots$

Proof. We first note that

$$T_0(t)\equiv 1, \quad T_1(t)=t.$$

Next, since

$$\cos(\nu+2)x+\cos\nu x=2\cos(\nu+1)x\cos x,$$

we see, setting $x=$ arccost t in this identity, that the functions $T_{\nu}(t)$ obey the following recurrence relations:

$$T_{\nu+2}(t)=2tT_{\nu+1}(t)-T_{\nu}(t), \quad \nu=0,\ 1,\ 2,\ \ldots,$$

and the statement of our lemma follows immediately. In particular,

$$T_2(t)=2t^2-1, \quad T_3(t)=4t^3-3t, \text{ etc.}$$

The polynomials $T_{\nu}(t)$ are known as the Chebyshev polynomials.

Corollary. The functions $\tau_s(t)$, $s\in[0:n+1]$ *defined above are algebraïc polynomials of degree* n.

Thus, all the expressions on the right of (2.9) are algebraic polynomials of degree $\leqslant n$.

Lemma 2.2. The Dirichlet kernel $D_n(x)$ ($n=0,1,2,\ldots$) satisfies the equalities

$$D_n\left(\frac{s\pi}{n+1}\right)=\begin{cases} n+\frac{1}{2}, & \text{if } s=0; \\ \frac{1}{2}(-1)^{s+1}, & \text{if } s=\pm 1,\ldots,\pm(2n+1). \end{cases}$$

Proof. By definition, $D_n(0)=n+\frac{1}{2}$. To determine the values of $D_n(x)$ at the points $x=\frac{s\pi}{n+1}$, $s=\pm 1,\ldots,\pm(2n+1)$, we observe that

$$\frac{1}{2}+\cos x+\cos 2x+\ldots+\cos nx=\frac{\sin\frac{2n+1}{2}x}{2\sin\frac{x}{2}}.$$

Therefore

$$D_n\left(\frac{s\pi}{n+1}\right)=\frac{\sin\frac{2n+1}{2}\left(\frac{s\pi}{n+1}\right)}{2\sin\frac{s\pi}{2n+2}}=\frac{\sin\left(s\pi-\frac{s\pi}{2n+2}\right)}{2\sin\frac{s\pi}{2n+2}}=$$

$$=\frac{(-1)^{s+1}\sin\frac{s\pi}{2n+2}}{2\sin\frac{s\pi}{2n+2}}=\frac{1}{2}(-1)^{s+1},\quad s=\pm 1,\ldots,\pm(2n+1).$$

Q. E. D.

We now return to the proof of the theorem, writing the following expression for $P_n(A^*,t)$ at the abscissas t_k, $k\in[0:n+1]$:

$$P_n(A^*,t_k)=\frac{1}{n+1}\sum_{s=0}^{n+1}y_s\tau_s(t_k)=$$

$$=\frac{1}{n+1}\left\{y_0D_n\left(\frac{k\pi}{n+1}\right)+y_{n+1}D_n\left(\frac{(n+1-k)\pi}{n+1}\right)+\right.$$

$$\left.+\sum_{s=1}^{n}y_s\left[D_n\left(\frac{(s+k)\pi}{n+1}\right)+D_n\left(\frac{(k-s)\pi}{n+1}\right)\right]\right\}.$$

Let $k=0$. Then, since $(-1)^{\nu}=(-1)^{-\nu}$ for any integer ν, we get

$$P_n(A^*,t_0)=\frac{1}{n+1}\left\{y_0\left(n+\frac{1}{2}\right)+y_{n+1}\frac{1}{2}(-1)^n+\right.$$

$$\left.+\sum_{s=1}^{n}y_s\frac{1}{2}[(-1)^{s+1}+(-1)^{s+1}]\right\}=$$

$$=y_0-\frac{1}{2n+2}\left\{y_0+(-1)^{n+1}y_{n+1}+2\sum_{s=1}^{n}(-1)^s y_s\right\}=y_0-h.$$

If $k=n+1$, then

$$P_n(A^*, t_{n+1})=$$
$$=\frac{1}{n+1}\left\{y_0\frac{1}{2}(-1)^n+y_{n+1}\left(n+\frac{1}{2}\right)+\sum_{s=1}^{n}(-1)^{n+s}y_s\right\}=$$
$$=y_{n+1}-\frac{(-1)^{n+1}}{2n+2}\left\{y_0+(-1)^{n+1}y_{n+1}+2\sum_{s=1}^{n}(-1)^s y_s\right\}=$$
$$=y_{n+1}-(-1)^{n+1}h.$$

Finally, let $k\in[1:n]$. Then

$$P_n(A^*, t_k)=\frac{1}{n+1}\left\{y_0\frac{1}{2}(-1)^{k+1}+y_{n+1}\frac{1}{2}(-1)^{n+k}+\right.$$
$$\left.+y_k n+\sum_{\substack{s=1\\ s\neq k}}^{n}y_s(-1)^{s+k+1}\right\}=$$
$$=y_k-\frac{(-1)^k}{2n+2}\left\{y_0+(-1)^{n+1}y_{n+1}+2\sum_{s=1}^{n}(-1)^s y_s\right\}=$$
$$=y_k-(-1)^k h.$$

Q. E. D.

Corollary. Let $t_k=\cos\frac{(n+1-k)\pi}{n+1}$, $k\in[0:n+1]$. *Then, for any algebraic polynomial* $Q_n(t)$ *of degree* $\leqslant n$,

$$Q_n(t_0)+(-1)^{n+1}Q_n(t_{n+1})+2\sum_{s=1}^{n}(-1)^s Q_n(t_s)=0. \qquad (2.10)$$

Proof. Set $y_k=Q_n(t_k)$, $k\in[0:n+1]$. Carrying out Chebyshev interpolation, we see via (2.6) and property 4 of divided differences (see Appendix I, §1), that

$$h=\frac{y[t_0, t_1, \ldots, t_{n+1}]}{\chi[t_0, t_1, \ldots, t_{n+1}]}=0.$$

Formula (2.10) now follows from (2.8).

§3. GENERAL DISCRETE CASE; DE LA VALLÉE-POUSSIN ALGORITHM

1. We now consider the general case of the discrete problem, $N>n+1$. Set

$$y_k=y(t_k),\quad k\in[0:N];\quad t_0<t_1<\ldots<t_N;$$
$$\rho=\inf_{\{A\}}\max_{k\in[0:N]}|y_k-P_n(A, t_k)|.$$

We wish to construct a polynomial of best approximation $P_n(A^*, t)$:

$$\max_{k \in [0:N]} |y_k - P_n(A^*, t_k)| = \rho.$$

We shall show in this section that there is an exact solution to this problem, which is obtained by carrying out finitely many Chebyshev interpolations.

2. Definition. A basis is any subsystem of $n+2$ abscissas

$$\sigma = \{t_{i_0} < t_{i_1} < \dots < t_{i_{n+1}}\}. \tag{3.1}$$

The set of bases $\{\sigma\}$ will be denoted by Ξ.

For each basis σ of type (3.1), we can perform Chebyshev interpolation, i. e., construct the (unique) algebraic polynomial $P_n(A(\sigma), t)$ such that

$$(-1)^k h(\sigma) + P_n(A(\sigma), t_{i_k}) = y_{i_k}, \qquad k \in [0 : n+1],$$

where

$$h(\sigma) = \sum_{k=0}^{n+1} (-1)^k \alpha_{i_k}(\sigma)\, y_{i_k} \tag{3.2}$$

and

$$\alpha_{i_k}(\sigma) = \frac{\dfrac{(-1)^k}{(t_{i_k} - t_{i_0}) \dots (\;) \dots (t_{i_k} - t_{i_{n+1}})}}{\sum\limits_{\nu=0}^{n+1} \dfrac{(-1)^\nu}{(t_{i_\nu} - t_{i_0}) \dots (\;) \dots (t_{i_\nu} - t_{i_{n+1}})}},$$
$$k \in [0 : n+1].$$

Let $\rho(\sigma)$ be the best fit on the basis σ:

$$\rho(\sigma) = \inf_{\{A\}} \max_{k \in [0:n+1]} |y_{i_k} - P_n(A, t_{i_k})|.$$

As we have seen,

$$\rho(\sigma) = \max_{k \in [0:n+1]} |y_{i_k} - P_n(A(\sigma), t_{i_k})| = |h(\sigma)|.$$

Finally, we set

$$\varphi(A(\sigma)) = \max_{k \in [0:N]} |y_k - P_n(A(\sigma), t_k)|,$$

so that $\varphi(A(\sigma))$ is the maximum deviation of the polynomial $P_n(A(\sigma), t)$ for the whole system of abscissas.

It is obvious that for any basis $\sigma \in \Xi$

$$\varphi(A(\sigma)) \geqslant \rho(\sigma).$$

Lemma 3.1. Suppose that for some basis σ^*,

$$\sigma^* = \{t^*_{i_0} < t^*_{i_1} < \ldots < t^*_{i_{n+1}}\}$$

we have $\varphi(A(\sigma^*)) = \rho(\sigma^*)$; *then*

$$\varphi(A(\sigma^*)) \stackrel{df}{=} \max_{k \in [0:N]} |y_k - P_n(A(\sigma^*), t_k)| = \rho,$$

i. e., $P_n(A(\sigma^*), t)$ *is the required polynomial of best approximation.*

Proof. By the definition of $\rho(\sigma^*)$, we have for any polynomial $P_n(A, t)$

$$\max_{k \in [0:n+1]} \left| y^*_{i_k} - P_n\left(A, t^*_{i_k}\right) \right| \geqslant \rho(\sigma^*).$$

A fortiori,

$$\max_{k \in [0:N]} |y_k - P_n(A, t_k)| \geqslant \rho(\sigma^*). \tag{3.3}$$

Since by assumption

$$\rho(\sigma^*) = \varphi(A(\sigma^*)) = \max_{k \in [0:N]} |y_k - P_n(A(\sigma^*), t_k)|, \tag{3.4}$$

it follows via (3.3) and (3.4) that for any polynomial $P_n(A, t)$:

$$\max_{k \in [0:N]} |y_k - P_n(A, t_k)| \geqslant \max_{k \in [0:N]} |y_k - P_n(A(\sigma^*), t_k)|.$$

Thus (by the definition) $P_n(A(\sigma^*), t)$ is indeed a polynomial of best approximation for the whole system of abscissas.

3. To summarize: if it turns out that $\varphi(A(\sigma)) = \rho(\sigma)$ for some basis σ, then $P_n(A(\sigma), t)$ is a polynomial of best approximation in the general discrete case.

We now consider bases $\sigma \in \Xi$ of type (3.1) for which

$$\varphi(A(\sigma)) > \rho(\sigma). \tag{3.5}$$

We shall introduce a standard transformation S that converts any basis σ satisfying (3.5) into a new basis $\sigma_1 = S\sigma$,

$$\sigma_1 = \{t^{(1)}_{i_0} < t^{(1)}_{i_1} < \ldots < t^{(1)}_{i_{n+1}}\},$$

such that

$$\rho(\sigma_1) > \rho(\sigma). \tag{3.6}$$

This type of transformation will have a major role to play in the sequel.

Prior to describing the transformation S, we set

$$\Delta_\sigma(t_k) = y_k - P_n(A(\sigma), t_k), \quad k \in [0:N],$$

and let t_{k_0} be an abscissa at which

$$|\Delta_\sigma(t_{k_0})| = \varphi(A(\sigma))$$

(if there are several, we choose any one).

There are three possibilities:

1) $t_{i_0} < t_{k_0} < t_{i_{n+1}}$;
2) $t_{k_0} < t_{i_0}$;
3) $t_{k_0} > t_{i_{n+1}}$.

The transformation S is defined differently in each case.

Case 1. A typical situation is illustrated in Figure 1.

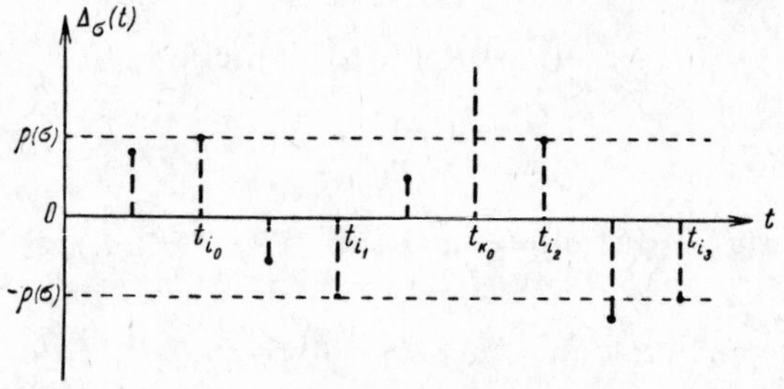

FIGURE 1.

Let ν be an integer such that

$$t_{i_\nu} < t_{k_0} < t_{i_{\nu+1}}.$$

We first set

$$t_{i_l}^{(1)} = t_{i_l}, \quad l \neq \nu, \ \nu+1; \quad l \in [0:n+1].$$

If $\operatorname{sign} \Delta_\sigma(t_{k_0}) = \operatorname{sign} \Delta_\sigma(t_{i_\nu})$, then

$$t_{i_\nu}^{(1)} = t_{k_0}, \quad t_{i_{\nu+1}}^{(1)} = t_{i_{\nu+1}}.$$

But if this condition fails to hold, then

$$t_{i_\nu}^{(1)} = t_{i_\nu}, \quad t_{i_{\nu+1}}^{(1)} = t_{k_0}.$$

In this case, then the basis $\sigma_1 = S\sigma$ is completely defined. In the situation of Figure 1, the basis σ_1 is obtained from σ by replacing t_{i_2} by t_{k_0}.

Case 2. If $\operatorname{sign} \Delta_\sigma(t_{k_0}) = \operatorname{sign} \Delta_\sigma(t_{i_0})$, we set

$$t_{i_0}^{(1)} = t_{k_0}; \quad t_{i_l}^{(1)} = t_{i_l}, \quad l \in [1:n+1].$$

If the above condition fails to hold, then

$$t_{i_0}^{(1)} = t_{k_0}; \quad t_{i_l}^{(1)} = t_{i_{l-1}}, \quad l \in [1 : n+1]$$

(the abscissa $t_{i_{n+1}}$ is discarded).

Case 3. If $\operatorname{sign} \Delta_\sigma(t_{k_0}) = \operatorname{sign} \Delta_\sigma(t_{i_{n+1}})$, then

$$t_{i_l}^{(1)} = t_{i_l}, \quad l \in [0 : n]; \quad t_{i_{n+1}}^{(1)} = t_{k_0}.$$

Otherwise,

$$t_{i_l}^{(1)} = t_{i_{l+1}}, \quad l \in [0 : n]; \quad t_{i_{n+1}}^{(1)} = t_{k_0}$$

(the abscissa t_{i_0} is discarded).

This completes our description of the transformation S.

The construction and inequality (3.5) imply the following properties of the basis $\sigma_1 = S\sigma$ if $\rho(\sigma) > 0$:

I. $\operatorname{sign} \Delta_\sigma(t_{i_k}^{(1)}) = -\operatorname{sign} \Delta_\sigma(t_{i_{k+1}}^{(1)})$, $k \in [0 : n]$.

II. If $t_{i_s}^{(1)}$ denotes the abscissa in the basis σ_1 corresponding to t_{k_0}, then

$$\left|\Delta_\sigma(t_{i_k}^{(1)})\right| = \rho(\sigma) \quad \text{for all} \quad k \neq s,$$
$$\left|\Delta_\sigma(t_{i_s}^{(1)})\right| > \rho(\sigma).$$

We can now prove the fundamental property of the basis σ_1:

$$\rho(\sigma_1) > \rho(\sigma).$$

Proof. By (3.2) and property 4 of divided differences (see Appendix I, §1),

$$\rho(\sigma_1) = \left|\sum_{k=0}^{n+1} (-1)^k \alpha_{i_k}(\sigma_1)\, y_{i_k}^{(1)}\right| =$$
$$= \left|\sum_{k=0}^{n+1} \alpha_{i_k}(\sigma_1) \left\{(-1)^k \left[y_{i_k}^{(1)} - P_n\left(A(\sigma), t_{i_k}^{(1)}\right)\right]\right\}\right| =$$
$$= \left|\sum_{k=0}^{n+1} \alpha_{i_k}(\sigma_1) \left\{(-1)^k \Delta_\sigma\left(t_{i_k}^{(1)}\right)\right\}\right|.$$

First let $\rho(\sigma) > 0$.

Since $\alpha_{i_k}(\sigma_1) > 0$, $k \in [0 : n+1]$, and $\sum_{k=0}^{n+1} \alpha_{i_k}(\sigma_1) = 1$, it follows via I and II that

$$\rho(\sigma_1) = \sum_{k=0}^{n+1} \alpha_{i_k}(\sigma_1) \left|\Delta_\sigma\left(t_{i_k}^{(1)}\right)\right| > \rho(\sigma) \sum_{k=0}^{n+1} \alpha_{i_k}(\sigma_1) = \rho(\sigma).$$

But if $\rho(\sigma) = 0$, then

$$\rho(\sigma_1) = \alpha_{i_s}(\sigma_1) \left|\Delta_\sigma\left(t_{i_s}^{(1)}\right)\right| > 0.$$

In all cases, then, $\rho(\sigma_1) > \rho(\sigma)$.

This proves our assertion.

The result is summarized in the following lemma.

Lemma 3.2. Let $\sigma \in \Xi$ *be a basis for which* $\varphi(A(\sigma)) > \rho(\sigma)$. *Then there exists a new basis* $\sigma_1 = S\sigma$ *such that* $\rho(\sigma_1) > \rho(\sigma)$.

4. Set

$$\bar{\rho} = \max_{\sigma \in \Xi} \rho(\sigma).$$

Definition. A basis σ^* for which

$$\rho(\sigma^*) = \bar{\rho},$$

is called an extremal basis.

Since the set Ξ is finite, it is clear that an extremal basis exists, though in general it need not be unique.

Theorem 3.1. In the general discrete case, there exists a unique polynomial of best approximation. A polynomial $P_n(A^*, t)$ *is the polynomial of best approximation if and only if it is a Chebyshev interpolating polynomial for some extremal basis* σ^*, *i. e.,*

$$P_n(A^*, t) \equiv P_n(A(\sigma^*), t).$$

Finally, $\rho = \bar{\rho}$, *so that*

$$\inf_{\{A\}} \max_{k \in [0:N]} |y_k - P_n(A, t_k)| = \max_{\sigma \in \Xi} \rho(\sigma). \tag{3.7}$$

Proof. Sufficiency. Let σ^* be some extremal basis. This means that for any $\sigma \in \Xi$

$$\rho(\sigma^*) \geqslant \rho(\sigma). \tag{3.8}$$

We must show that $P_n(A(\sigma^*), t)$ is a polynomial of best approximation, i. e.,

$$\varphi(A(\sigma^*)) \stackrel{df}{=} \max_{k \in [0:N]} |y_k - P_n(A(\sigma^*), t_k)| = \rho.$$

By Lemma 3.1, it will suffice to verify that $\varphi(A(\sigma^*)) = \rho(\sigma^*)$.

If this is false, then $\varphi(A(\sigma^*)) > \rho(\sigma^*)$. By Lemma 3.2, there is a new basis $\sigma_1 = S\sigma^*$ such that

$$\rho(\sigma_1) > \rho(\sigma^*),$$

and this contradicts (3.8), proving sufficiency.

We have at the same time proved (3.7), for

$$\rho = \varphi(A(\sigma^*)) = \rho(\sigma^*) = \bar{\rho}. \tag{3.9}$$

The existence of a polynomial of best approximation follows from the existence of an extremal basis.

Necessity. Fix some extremal basis σ^*:

$$\sigma^* = \{t^*_{i_0} < t^*_{i_1} < \ldots < t^*_{i_{n+1}}\},$$

and let $P_n(A^*, t)$ be some polynomial of best approximation. Since

$$\rho(\sigma^*) \leqslant \max_{k\in[0:n+1]} \left| y^*_{i_k} - P_n\left(A^*, t^*_{i_k}\right)\right| \leqslant$$
$$\leqslant \max_{i\in[0:N]} |y_i - P_n(A^*, t_i)| = \rho = \bar{\rho} = \rho(\sigma^*),$$

it follows that

$$\max_{k\in[0:n+1]} \left| y^*_{i_k} - P_n\left(A^*, t^*_{i_k}\right)\right| = \rho(\sigma^*).$$

Thus $P_n(A^*, t)$ is a polynomial of best approximation for σ^*. Since the σ^* latter is unique, we conclude that

$$P_n(A^*, t) \equiv P_n(A(\sigma^*), t).$$

This proves both necessity and uniqueness, completing the proof.

Corollary. A basis

$$\sigma = \{t_{i_0} < t_{i_1} < \ldots < t_{i_{n+1}}\}$$

is extremal if and only if

$$\varphi(A(\sigma)) = \rho(\sigma),$$

or, equivalently,

$$\max_{i\in[0:N]} |\Delta_\sigma(t_i)| = \max_{k\in[0:n+1]} \left|\Delta_\sigma\left(t_{i_k}\right)\right|.$$

5. The fundamental theorem yields a procedure for constructing polynomials of best approximation in the general (discrete) case. One first examines all bases, and selects a basis σ^* for which $\rho(\sigma^*)$ is maximal. The Chebyshev interpolating polynomial $P_n(A(\sigma^*), t)$ will then be the required polynomial.

The difficulty here is how to organize examination of the bases in the most rational manner. This is particularly important when N is large.

Since we are looking for a basis giving the maximal best fit, it is reasonable to demand that we discard a basis σ_k and adopt a new one σ_{k+1} only when

$$\rho(\sigma_{k+1}) > \rho(\sigma_k). \tag{3.10}$$

Fulfillment of this condition also guarantees that no basis can come up more than once in any examination.

The above idea is implemented in the de la Vallée-Poussin algorithm for constructing polynomials of best approximation. This algorithm employs the transformation S described in subsection 3.

Select an initial basis σ_1 arbitrarily, and construct the polynomial $P_n(A(\sigma_1), t)$. If

$$\varphi(A(\sigma_1)) = \rho(\sigma_1),$$

then (Lemma 3.1) $P_n(A(\sigma_1), t)$ is the polynomial of best approximation. On the other hand, if

$$\varphi(A(\sigma_1)) > \rho(\sigma_1),$$

we proceed to a new basis $\sigma_2 = S\sigma_1$. By Lemma 3.2,

$$\rho(\sigma_2) > \rho(\sigma_1).$$

Suppose we have already constructed a basis σ_k. Determine $P_n(A(\sigma_k), t)$. If

$$\varphi(A(\sigma_k)) = \rho(\sigma_k),$$

then $P_n(A(\sigma_k), t)$ is the required polynomial. Otherwise, i. e., if

$$\varphi(A(\sigma_k)) > \rho(\sigma_k),$$

we go on to the basis $\sigma_{k+1} = S\sigma_k$.

This procedure will yield an extremal basis, and hence a polynomial of best approximation, in finitely many trials. That the procedure is indeed finite follows from the fact that the number of bases is finite and by (3.10) no basis can be examined twice.

§4. R-ALGORITHM

1. The above algorithm of de la Vallée-Poussin utilizes the simple S-transformation of bases, which essentially involves replacing one point of the old basis σ. The new basis $\sigma_1 = S\sigma$ then contains a point at which $|\Delta_\sigma(t_k)|$ is a maximum.

When using the S-transformation, one must compute all the $\Delta_\sigma(t_k)$, $k \in [0:N]$, at each step. A shortcoming of this algorithm is that the information given by the computation of the deviations $\Delta_\sigma(t_k)$ is poorly utilized; in fact it is used only to determine the deviation of maximum absolute value.

In this section we describe a new transformation R of bases, which generally involves replacing several (sometimes all) of the abscissas of the old basis and leads to a higher increase of $\rho(R\sigma)$ in comparison with $\rho(S\sigma)$.

2. The R-transformation is defined only for bases σ such that $\rho(\sigma)>0$. Set

$$\sigma=\{t_{i_0}<t_{i_1}<\ldots<t_{i_{n+1}}\};$$
$$\Delta_\sigma(t_k)=y_k-P_n(A(\sigma),\,t_k),\qquad k\in[0:N].$$

We shall construct a basis $\sigma_1=R\sigma$,

$$\sigma_1=\{t_{i_0}^{(1)}<t_{i_1}^{(1)}<\ldots<t_{i_{n+1}}^{(1)}\},$$

taking care throughout to fulfill the following conditions:

I. sign $\Delta_\sigma\left(t_{i_k}^{(1)}\right)=-$ sign $\Delta_\sigma\left(t_{i_{k+1}}^{(1)}\right)$, $\quad k\in[0:n]$.

II. $\left|\Delta_\sigma\left(t_{i_k}^{(1)}\right)\right|\geqslant\rho(\sigma),\quad k\in[0:n+1]$.

III. The basis σ_1 includes at least one abscissa at which $\max\limits_{k\in[0:N]}|\Delta_\sigma(t_k)|\overset{df}{=}\varphi(A(\sigma))$ is attained.

As in the case of the S-transformation, these conditions guarantee that

$$\rho(\sigma_1)>\rho(\sigma)\tag{4.1}$$

provided $\varphi(A(\sigma))>\rho(\sigma)$.

The basis transformation R involves $n+1$ steps.

Step 1. Consider the abscissas in the interval $[t_0,\,t_{i_1}]$. Select two of them, t_{v1} and t_{w1}, such that $t_{v1}<t_{w1}$ and one of them corresponds to the maximum $\Delta_\sigma(t_k)$, $k\in[0:i_1]$, the other to the minimum.

Set

$$t_{i_0}^{(1)}=t_{v1},\qquad t_{i_1}^{(1)}=t_{w1}.$$

Step 2. Consider the abscissas in $[t_{i_1},\,t_{i_2}]$. Select two of them, t_{v2} and t_{w2}, such that $t_{v2}<t_{w2}$ and one of them corresponds to maximum $\Delta_\sigma(t_k)$, $k\in[i_1:i_2]$, the other to the minimum.

Let

$$\operatorname{sign}\Delta_\sigma\left(t_{i_1}^{(1)}\right)=\operatorname{sign}\Delta_\sigma(t_{v2}).$$

Now set $t_{i_1}^{(1)}=t_{w2}$. If we then have $\left|\Delta_\sigma\left(t_{i_1}^{(1)}\right)\right|<|\Delta_\sigma(t_{v2})|$, we redefine $t_{i_1}^{(1)}$: $t_{i_1}^{(1)}\overset{df}{=}t_{v2}$. Store $z_2=0$.

Let sign $\Delta_\sigma\left(t_{i_1}^{(1)}\right)=-$ sign $\Delta_\sigma(t_{v2})$. In that case, we set $t_{i_2}^{(1)}=t_{v2}$. Store $t_{z2}=t_{w2}$ and $z_2=\Delta_\sigma(t_{z2})$.

Step $\nu+1$. Suppose we have already selected abscissas

$$t_{i_0}^{(1)}<t_{i_1}^{(1)}<\ldots<t_{i_\nu}^{(1)}.$$

The value of z_ν remains from the previous step, and if $z_\nu\neq0$ we also have an abscissa $t_{z\nu}$.

Consider the abscissas in $[t_{i_\nu},\,t_{i_{\nu+1}}]$. Select two of them, $t_{v,\nu+1}$ and $t_{w,\nu+1}$, such that $t_{v,\nu+1}<t_{w,\nu+1}$ and one of them corresponds to maximum $\Delta_\sigma(t_k)$, $k\in[i_\nu:i_{\nu+1}]$, the other to the minimum.

Let $\operatorname{sign}\Delta_\sigma(t^{(1)}_{i_\nu}) = \operatorname{sign}\Delta_\sigma(t_{v,\nu+1})$. Then, first, we store $z_{\nu+1}=0$. If $|\Delta_\sigma(t^{(1)}_{i_\nu})| \geqslant |\Delta_\sigma(t_{v,\nu+1})|$, we set

$$t^{(1)}_{i_{\nu+1}} = \begin{cases} t_{w,\nu+1}, & \text{if } \quad |\Delta_\sigma(t_{w,\nu+1})| \geqslant |z_\nu|; \\ t_{z\nu} & \text{otherwise.} \end{cases}$$

If $|\Delta_\sigma(t^{(1)}_{i_\nu})| < |\Delta_\sigma(t_{v,\nu+1})|$, we set $t^{(1)}_{i_{\nu+1}} = t_{w,\nu+1}$. We then redefine $t^{(1)}_{i_\nu} : t^{(1)}_{i_\nu} \overset{df}{=} t_{v,\nu+1}$. Finally, if $|z_\nu| > |\Delta_\sigma(t^{(1)}_{i_{\nu-1}})|$, we also redefine $t^{(1)}_{i_{\nu-1}} : t^{(1)}_{i_{\nu-1}} \overset{df}{=} t_{z\nu}$.

Now let $\operatorname{sign}\Delta_\sigma(t^{(1)}_{i_\nu}) = -\operatorname{sign}\Delta_\sigma(t_{v,\nu+1})$. Then we set

$$t^{(1)}_{i_{\nu+1}} = \begin{cases} t_{v,\nu+1}, & \text{if } \quad |\Delta_\sigma(t_{v,\nu+1})| \geqslant |z_\nu|, \\ t_{z\nu} & \text{otherwise.} \end{cases}$$

Store $t_{z,\nu+1} = t_{w,\nu+1}$, $z_{\nu+1} = \Delta_\sigma(t_{z,\nu+1})$.

Two remarks will complete our description of the transformation R.

First, at the last step (Step $n+1$), one considers the abscissas not in $[t_{i_n}, t_{i_{n+1}}]$ but in $[t_{i_n}, t_N]$.

Second, at the last step we are left with z_{n+1} and, if $z_{n+1} \neq 0$, also $t_{z,n+1}$. If it turns out that $|z_{n+1}| > |\Delta_\sigma(t^{(1)}_{i_0})|$, we "shift" the abscissas, assigning $t^{(1)}_{i_k}$ the value of $t^{(1)}_{i_{k+1}}$, $k \in [0:n]$; $t^{(1)}_{i_{n+1}} \overset{df}{=} t_{z,n+1}$ Otherwise, no changes are necessary.

This completes the description of the R-transformation. We must prove that conditions I, II and III are fulfilled.

Note that at each step the abscissas under consideration always include two abscissas of the old basis. Therefore, for $k \in [1:n+1]$,

$$\operatorname{sign}\Delta_\sigma(t_{vk}) = -\operatorname{sign}\Delta_\sigma(t_{wk})$$

and

$$|\Delta_\sigma(t_{vk})| \geqslant \rho(\sigma), \qquad |\Delta_\sigma(t_{wk})| \geqslant \rho(\sigma).$$

Together with the details of the construction, this implies the validity of conditions I and II. It is also readily shown that the new basis contains an abscissa at which $\max\limits_{k \in [0:N]} \Delta_\sigma(t_k)$ is attained.

Thus the R-transformation answers all our demands.

3. We now describe the R-algorithm for constructing a polynomial of best approximation, utilizing the R-transformation.

Select an initial basis $\sigma_1, \rho(\sigma_1) > 0$, arbitrarily; construct the polynomial $P_n(A(\sigma_1), t)$ and the basis $\sigma_2 = R\sigma_1$. If

$$\rho(\sigma_2) = \rho(\sigma_1), \tag{4.2}$$

the procedure terminates and $P_n(A(\sigma_1), t)$ is the polynomial of best approximation. Indeed, by (4.1), equality (4.2) may hold only if $\varphi(A(\sigma_1)) = \rho(\sigma_1)$, and this implies that $P_n(A(\sigma_1), t)$ is a polynomial of best approximation.

If $\rho(\sigma_2) > \rho(\sigma_1)$, we construct $P_n(A(\sigma_2), t)$ and the basis $\sigma_3 = R\sigma_2$.

Suppose we have already constructed the polynomial $P_n(A(\sigma_\nu), t)$ and the basis $\sigma_{\nu+1} = R\sigma_\nu$. If it turns out that $\rho(\sigma_{\nu+1}) = \rho(\sigma_\nu)$, the procedure terminates and $P_n(A(\sigma_\nu), t)$ is the required polynomial. Otherwise, we construct $P_n(A(\sigma_{\nu+1}), t)$ and $\sigma_{\nu+2} = R\sigma_{\nu+1}$.

After finitely many steps, we ultimately obtain an extremal basis and hence also a polynomial of best approximation. That the procedure is indeed finite follows from the fact that the number of bases is finite and condition (4.1) is observed throughout the process, so that no basis may come up twice.

Once the extremal basis σ^* has been constructed, the result at the next step is $\rho(R\sigma^*) = \rho(\sigma^*)$, and this is the condition for termination of the procedure.

Remark. The initial basis σ_1 in the R-algorithm must satisfy the condition $\rho(\sigma_1) > 0$, whereas the initial basis in the de la Vallée-Poussin algorithm is arbitrary.

4. Example. Considering the interval $[-1, 1]$, introduce a grid T_{1000}, whose abscissas are

$$t_k = -1 + \frac{2k}{1000}, \quad k \in [0:1000].$$

We wish to determine a 7-th degree best approximation polynomial $P_7(t)$ for the function $|t|$ on the grid T_{1000}.

The problem is solved using the R-algorithm. The results are shown in Table 1.

TABLE 1.

	I	II	III	IV
t_{i_0}	−1.000	−1.000	−1.000	−1.000
t_{i_1}	−0.750	−0.876	−0.884	−0.884
t_{i_2}	−0.500	−0.538	−0.572	−0.572
t_{i_3}	−0.250	−0.184	−0.194	−0.196
t_{i_4}	0.000	0.000	0.000	0.000
t_{i_5}	0.250	0.184	0.194	0.196
t_{i_6}	0.500	0.538	0.572	0.572
t_{i_7}	0.750	0.876	0.884	0.884
t_{i_8}	1.000	1.000	1.000	1.000
$\rho(\sigma)$	0.0390625	0.0455908	0.0459271	0.0459284

Column I lists the components of the initial basis and the best fit for this basis. Column IV lists the components of the extremal basis and the corresponding best fit; the extremal basis is obtained here after three steps. The polynomial $P_7(t)$ of best approximation for $|t|$ on T_{1000} is

$$P_7(t) = 2.31038t^6 - 4.17842t^4 + 2,86804t^2 + 0.0459284.$$

§5. REDUCTION TO LINEAR PROGRAMMING

1. The discrete problem of best approximation by algebraic polynomials involves finding the number ρ and the coefficient vector A^* of the polynomial of best approximation.

In this section we formulate two problems equivalent to best approximation.

2. We shall use the previous notation:

$$A=(a_0, a_1, \ldots, a_n);$$

$$\rho=\inf_{\{A\}} \max_{k\in[0:N]} \left| y_k - \sum_{i=0}^{n} a_i t_k^i \right|.$$

Set

$$\gamma_{ki}=t_k^i, \quad i\in[0:n], \quad k\in[0:N];$$

$$\gamma_{N+1+s,i}=-\gamma_{si}, \quad y_{N+1+s}=-y_s \quad \text{for} \quad s\in[0:N].$$

In this case, since $|\xi|=\max\{\xi, -\xi\}$, we get

$$\max_{k\in[0:N]} |y_k - P_n(A, t_k)| = \max_{k\in[0:N]} \left| \sum_{i=0}^{n} \gamma_{ki}a_i - y_k \right| =$$
$$= \max_{k\in[0:2N+1]} \left(\sum_{i=0}^{n} \gamma_{ki}a_i - y_k \right) = \max_{k\in[0:2N+1]} \left(y_k - \sum_{i=0}^{n} \gamma_{ki}a_i \right). \tag{5.1}$$

In the euclidean $(n+2)$-space whose elements are vectors

$$V=(a_0, a_1, \ldots, a_n, z),$$

we consider the half-space

$$H_k=\left\{ V \mid z \geqslant \sum_{i=0}^{n} \gamma_{ki}a_i - y_k \right\}, \quad k\in[0:2N+1].$$

Let G be the intersection of these half-spaces (Figure 2):

$$G=\bigcap_{k=0}^{2N+1} H_k.$$

Theorem 5.1. *$A^*=(a_0^*, a_1^*, \ldots, a_n^*)$ is the coefficient vector of a polynomial of best approximation and ρ is the best fit if and only if the point*

$$V^*=(a_0^*, a_1^*, \ldots, a_n^*, \rho)$$

is in G and has the smallest last coordinate of any point of G.

Proof. Necessity. Since

$$\rho=\max_{k\in[0:N]} |y_k - P_n(A^*, t_k)|,$$

it follows from (5.1) that V^* is in each H_k, $k \in [0:2N+1]$, and therefore in G.

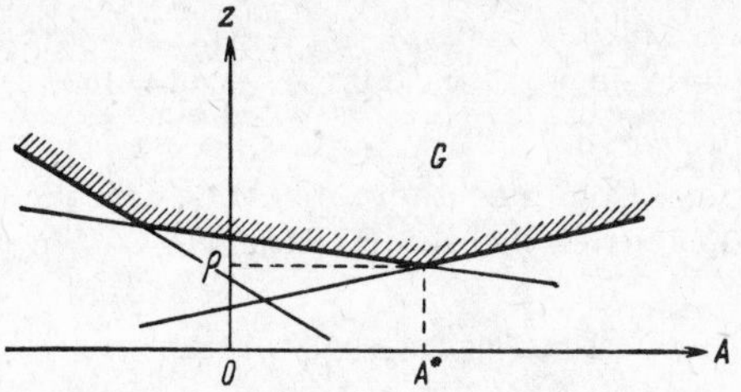

FIGURE 2.

Suppose now that there exists a point $V_0 \in G$,

$$V_0 = (a_0^{(0)}, a_1^{(0)}, \ldots, a_n^{(0)}, z^{(0)}),$$

such that $z^{(0)} < \rho$. Then, setting $A_0 = (a_0^{(0)}, a_1^{(0)}, \ldots, a_n^{(0)})$, and using the definition of G and (5.1), we get

$$\max_{k \in [0:N]} |y_k - P_n(A_0, t_k)| \leqslant z^{(0)} < \rho,$$

contradicting the definition of ρ. This proves necessity.

Sufficiency. Let $V_0 = (a_0^{(0)}, a_1^{(0)}, \ldots, a_n^{(0)}, z^{(0)})$ be a point of G with smallest last coordinate. Then obviously

$$\max_{k \in [0:N]} |y_k - P_n(A_0, t_k)| = z^{(0)}, \qquad (5.2)$$

where $A_0 = (a_0^{(0)}, a_1^{(0)}, \ldots, a_n^{(0)})$. We claim that A_0 is the coefficient vector of a polynomial of best approximation and $z^{(0)}$ is the best approximation.

Let A be an arbitrary $(n+1)$-vector. Then, by the definition of V_0,

$$\max_{k \in [0:N]} |y_k - P_n(A, t_k)| \geqslant z^{(0)}.$$

In view of (5.2), this completes the proof.

Remark. A point $(a_0^*, a_1^*, \ldots, a_n^*, z^*)$ in G with smallest last coordinate may be defined as the solution of the following linear programming problem: find min z under the constraints

$$\sum_{i=0}^{n} \gamma_{ki} a_i - z \leqslant y_k, \qquad k \in [0:2N+1].$$

Thus, solution of this linear programming problem yields the solution of the discrete problem of best approximation.

3. In the euclidean $(n+2)$-space whose elements are vectors $W = (\xi_0, \xi_1, \ldots, \xi_n, y)$, we consider the points

$$W_k = (\gamma_{k0}, \ldots, \gamma_{kn}, y_k), \qquad k \in [0:2N+1].$$

The set $\{W_k\}$ consists of pairs of points $W_{N+1+k} = -W_k$, $k \in [0:N]$, symmetric with respect to the origin. Let D denote the convex hull of $\{W_k\}$, i. e., the set of all vectors

$$W = \sum_{k=0}^{2N+1} \alpha_k W_k; \quad \alpha_k \geqslant 0, \quad \sum_{k=0}^{2N+1} \alpha_k = 1. \tag{5.3}$$

D contains the origin. Consider the point at which the ray $W = (0, \ldots, 0, y)$, $y \geqslant 0$, leaves the set D (Figure 3). We denote this point by W^*:

$$W^* = (0, \ldots, 0, y^*).$$

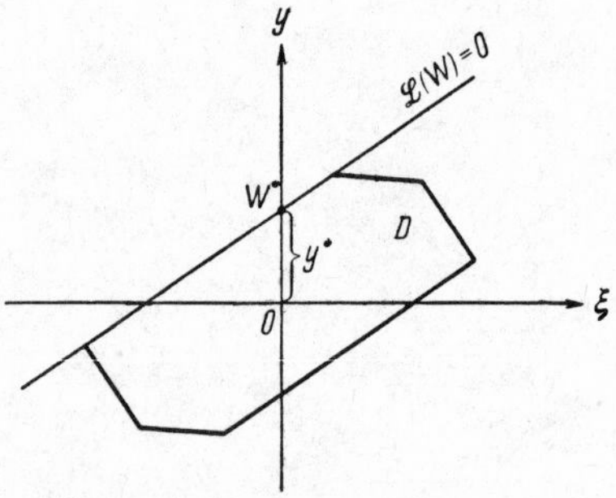

FIGURE 3.

Lemma 5.1. $y^* = \rho$.

Proof. W^* is the point of D with first $n+1$ coordinates zero and last coordinate maximal. Thus, by (5.3),

$$y^* = \max \sum_{k=0}^{2N+1} \alpha_k y_k$$

provided that the α_k satisfy the relations

$$\left.\begin{aligned} \sum_{k=0}^{2N+1} \alpha_k \gamma_{ki} = 0, \quad i \in [0:n], \\ \alpha_k \geqslant 0, \quad \sum_{k=0}^{2N+1} \alpha_k = 1. \end{aligned}\right\} \tag{5.4}$$

We claim that $y^* \leqslant \rho$.

Indeed, for any vector $A = (a_0, a_1, \ldots a_n)$, and any numbers α_k, $k \in [0:2N+1]$, satisfying the constraints (5.4), we have

$$\sum_{k=0}^{2N+1} \alpha_k y_k = \sum_{k=0}^{2N+1} \alpha_k \left(y_k - \sum_{i=0}^{n} \gamma_{ki} a_i\right) \leqslant \max_{k \in [0:2N+1]} \left(y_k - \sum_{i=0}^{n} \gamma_{ki} a_i\right).$$

Hence

$$y^* \leqslant \min_{\{A\}} \max_{k \in [0:2N+1]} \left(y_k - \sum_{i=0}^{n} \gamma_{ki} a_i \right) \overset{df}{=} \rho.$$

To establish the converse inequality it will suffice to prove the existence of numbers $\bar{a}_k$, $k \in [0:2N+1]$, satisfying (5.4) such that

$$\sum_{k=0}^{2N+1} \bar{a}_k y_k = \rho.$$

To this end, we refer to the fundamental Theorem 3.1. Let σ be an extremal basis,

$$\sigma = \{t_{k_0} < t_{k_1} < \ldots < t_{k_{n+1}}\}.$$

Then, by (3.7) and (3.2),

$$\rho = | h(\sigma) | = \left| \sum_{s=0}^{n+1} (-1)^s \alpha_{k_s}(\sigma) y_{k_s} \right|,$$

with $\alpha_{k_s}(\sigma) > 0$, $\sum_{s=0}^{n+1} \alpha_{k_s}(\sigma) = 1$,

$$\sum_{s=0}^{n+1} (-1)^s \alpha_{k_s}(\sigma) \gamma_{k_s i} \overset{df}{=} \sum_{s=0}^{n+1} (-1)^s \alpha_{k_s}(\sigma) t_{k_s}^i = 0, \quad i \in [0:n].$$

If $h(\sigma) \geqslant 0$, we set:
$\bar{a}_{k_s} = \alpha_{k_s}(\sigma)$ for even s,
$\bar{a}_{N+1+k_s} = \alpha_{k_s}(\sigma)$ for odd s,
$\bar{a}_k = 0$ for all other $k \in [0:2N+1]$.
If $h(\sigma) < 0$, we set:
$\bar{a}_{k_s} = \alpha_{k_s}(\sigma)$ for odd s,
$\bar{a}_{N+1+k_s} = \alpha_{k_s}(\sigma)$ for even s,
$\bar{a}_k = 0$ for all other $k \in [0:2N+1]$.
It is now readily seen that

$$\bar{a}_k \geqslant 0, \quad \sum_{k=0}^{2N+1} \bar{a}_k = 1; \quad \sum_{k=0}^{2N+1} \bar{a}_k \gamma_{ki} = 0, \quad i \in [0:n];$$

$$\sum_{k=0}^{2N+1} \bar{a}_k y_k = \rho.$$

This proves the lemma.

Theorem 5.2. $A^* = (a_0^*, a_1^*, \ldots, a_n^*)$ *is the coefficient vector of a polynomial of best approximation and* ρ *is the best fit if and only if the hyperplane*

$$\mathscr{L}(W) \overset{df}{=} -\sum_{i=0}^{n} a_i^* \xi_i + y - \rho = 0$$

supports the set D at the point W^* (see Figure 3), i.e., $\mathscr{L}(W^*)=0$ and $\mathscr{L}(W_k)\leqslant 0,\ k\in[0:2N+1]$.

Proof. Necessity. By (5.1) and the definition of the polynomial of best approximation, we obtain

$$-\sum_{i=0}^{n} a_i^*\gamma_{ki} + y_k \leqslant \rho,$$

i. e., $\mathscr{L}(W_k)\leqslant 0,\ k\in[0:2N+1]$. It remains to note that by Lemma 5.1 $\mathscr{L}(W^*)=0$.

Sufficiency. Suppose that the hyperplane

$$\mathscr{L}_0(W)\overset{df}{=} -\sum_{i=0}^{n} a_i^{(0)}\xi_i + y - \rho^{(0)} = 0$$

supports D at W^*. Then it follows from the equality $\mathscr{L}_0(W^*)=0$ and from Lemma 5.1 that $\rho^{(0)}=\rho$, i. e., $\rho^{(0)}$ is the best fit.

Now, since $\mathscr{L}_0(W_k)\leqslant 0, k\in[0:2N+1]$, we see from (5.1) that

$$\max_{k\in[0:\ N]} |y_k - P_n(A_0,\ t_k)| \leqslant \rho,$$

where $A_0=(a_0^{(0)},\ a_1^{(0)},\ \ldots,\ a_n^{(0)})$. The converse inequality is obvious from the definition of ρ. Thus A_0 is the coefficient vector of a polynomial of best approximation. This completes the proof.

Chapter II

BEST APPROXIMATION BY ALGEBRAIC POLYNOMIALS — CONTINUOUS CASE

§1. STATEMENT OF THE PROBLEM

Let $f(t)$ be a continuous function defined on an interval $[c, d]$. Set

$$\rho_n(f; c, d) = \rho_n(f) = \inf_{\{A\}} \max_{c \leqslant t \leqslant d} |f(t) - P_n(A, t)|.$$

A polynomial $P_n(A^*, t)$ such that

$$\max_{c \leqslant t \leqslant d} |f(t) - P_n(A^*, t)| = \rho_n(f),$$

will be called a polynomial of best approximation for $f(t)$ on $[c, d]$, or simply a polynomial of best approximation.

Our object in this chapter is the actual construction of polynomials of best approximation.

As before, a basis σ is defined as any system of $n+2$ points in $[c, d]$, arranged in increasing order:

$$\sigma = \{t_0 < t_1 < \ldots < t_{n+1}\}.$$

The set of bases $\{\sigma\}$, again denoted by Ξ, is infinite (even uncountable). Nevertheless, as in the discrete case, it can be shown that the fundamental equality

$$\inf_{\{A\}} \max_{c \leqslant t \leqslant d} |f(t) - P_n(A, t)| = \sup_{\sigma \in \Xi} \rho(\sigma),$$

where $\rho(\sigma)$ is the best fit of $f(t)$ for the basis σ, remains valid.

In the sequel, basing ourselves on this equality, we shall elaborate a method of successive Chebyshev interpolations, due to Remez, for constructing polynomials of best approximation.

We shall also consider another approach to construction of polynomials of best approximation, using grids.

To this end, we introduce on $[c, d]$ a uniform grid

$$z_k = c + k\frac{d-c}{N}; \quad w_k = f(z_k), \quad k \in [0:N].$$

Using the standard methods, we can solve the discrete analog of our problem on this grid. It turns out that as $N \to \infty$ the polynomials of best approximation for the discrete case tend to a polynomial of best approximation for the continuous problem.

At the end of the chapter we shall study the behavior of the coefficients of the polynomials of best approximation as $n \to \infty$.

§2. CHEBYSHEV THEOREM. CHEBYSHEV POLYNOMIALS

1. Set

$$\bar{\rho} = \sup_{\sigma \in \Xi} \rho(\sigma).$$

We claim that $\bar{\rho}$ is finite. Indeed, if $M = \max_{c \leqslant t \leqslant d} |f(t)|$, then for any basis $\sigma = \{t_0 < t_1 < \ldots < t_{n+1}\}$ we have

$$\rho(\sigma) = \inf_{\{A\}} \max_{k \in [0:n+1]} |f(t_k) - P_n(A, t_k)| \leqslant$$
$$\leqslant \max_{k \in [0:n+1]} |f(t_k) - P_n(0, t_k)| = \max_{k \in [0:n+1]} |f(t_k)| \leqslant M.$$

Thus $\bar{\rho} \leqslant M$.

A basis $\sigma^* \in \Xi$ for which $\rho(\sigma^*) = \bar{\rho}$ is known as an extremal basis.

Our immediate goal is to prove the existence of an extremal basis.

2. Definition. We shall say that a sequence of bases $\{\sigma_s\}$, $s = 1, 2, \ldots$,

$$\sigma_s = \{t_0^{(s)} < t_1^{(s)} < \ldots < t_{n+1}^{(s)}\},$$

converges to a basis σ,

$$\sigma = \{t_0 < t_1 < \ldots < t_{n+1}\},$$

writing $\sigma_s \xrightarrow[s \to \infty]{} \sigma$ if, for every $k \in [0 : n+1]$,

$$t_k^{(s)} \xrightarrow[s \to \infty]{} t_k.$$

Lemma 2.1. If $\sigma_s \xrightarrow[s \to \infty]{} \sigma$, then

$$h(\sigma_s) \xrightarrow[s \to \infty]{} h(\sigma).$$

Proof. Recall that

$$h(\sigma_s) = \sum_{k=0}^{n+1} (-1)^k \alpha_k(\sigma_s) f(t_k^{(s)}).$$

The proof reduces to showing that

$$\alpha_k(\sigma_s) \to \alpha_k(\sigma), \quad k \in [0:n+1] \tag{2.1}$$

as $s \to \infty$.

Recalling the expressions for $\alpha_k(\sigma_s)$ and $\alpha_k(\sigma)$, we conclude that it will suffice to show that

$$\frac{1}{t_k^{(s)} - t_i^{(s)}} \to \frac{1}{t_k - t_i}, \quad i \neq k; \quad i, k \in [0:n+1]$$

as $s \to \infty$.

But this follows in an obvious manner from the convergence of the bases, proving the lemma.

Corollary. If $\sigma_s \xrightarrow[s\to\infty]{} \sigma$, *then*

$$\rho(\sigma_s) \xrightarrow[s\to\infty]{} \rho(\sigma).$$

This follows immediately from the fact that $\rho(\sigma_s) = |h(\sigma_s)|$ and $\rho(\sigma) = |h(\sigma)|$.

Lemma 2.2. Let $\{\sigma_s\}$ *be a sequence of bases,*

$$\sigma_s = \{t_0^{(s)} < t_1^{(s)} < \ldots < t_{n+1}^{(s)}\},$$

such that $\rho(\sigma_s) \geqslant \beta > 0$, $s = 1, 2, \ldots$, *and suppose there exist finite limits*

$$\lim_{s\to\infty} t_k^{(s)} = t_k, \quad k \in [0:n+1].$$

Then

$$t_0 < t_1 < \ldots < t_{n+1},$$

i.e., the sequence $\{\sigma_s\}$ *converges to the basis*

$$\sigma = \{t_0 < t_1 < \ldots < t_{n+1}\}.$$

Proof. It is obvious that

$$t_0 \leqslant t_1 \leqslant \ldots \leqslant t_{n+1}.$$

Suppose that one of these inequalities is in fact an equality. Then there are at most $n+1$ distinct points in the sequence $\{t_k\}$. Thus, we can construct an interpolating polynomial $L_n(t)$ such that

$$L_n(t_k) = f(t_k), \quad k \in [0:n+1].$$

The function $\Delta(t) = f(t) - L_n(t)$ is continuous on $[c, d]$. Consequently, setting $\varepsilon = \beta > 0$, we can find $\delta > 0$ such that $|t'_k - t_k| < \delta$ implies $|\Delta(t'_k)| < \beta$.

Take s_0 so large that

$$|t_k^{(s_0)} - t_k| < \delta, \quad k \in [0:n+1].$$

Then

$$\rho(\sigma_{s_0}) = \inf_{\{A\}} \max_{k \in [0:n+1]} |f(t_k^{(s_0)}) - P_n(A, t_k^{(s_0)})| \leqslant$$
$$\leqslant \max_{k \in [0:n+1]} |f(t_k^{(s_0)}) - L_n(t_k^{(s_0)})| = \max_{k \in [0:n+1]} |\Delta(t_k^{(s_0)})| < \beta,$$

contrary to assumption.

Lemma 2.3. There exists an extremal basis.

Proof. If $\bar{\rho} = 0$, any basis is extremal. We may therefore assume that $\bar{\rho} > 0$. By the definition of $\bar{\rho}$, there exists a sequence of bases $\{\sigma_s\}$, $s = 1, 2, \ldots$, such that

$$\rho(\sigma_s) \xrightarrow[s \to \infty]{} \bar{\rho}. \tag{2.2}$$

Take $\bar{s}$ so large that for $s > \bar{s}$

$$\rho(\sigma_s) \geqslant \frac{\bar{\rho}}{2} > 0.$$

Now for all $k \in [0 : n+1]$ and $s > \bar{s}$,

$$c \leqslant t_k^{(s)} \leqslant d,$$

and so, by Lemma 2.2, there is a subsequence $\{\sigma_{s_i}\}$ converging to some basis σ^*.

By the corollary to Lemma 2.1,

$$\rho(\sigma_{s_i}) \xrightarrow[i \to \infty]{} \rho(\sigma^*).$$

In view of (2.2), we obtain

$$\rho(\sigma^*) = \bar{\rho}.$$

Thus σ^* is an extremal basis.

3. We introduce the notation

$$\varphi(A) = \max_{c \leqslant t \leqslant d} |f(t) - P_n(A, t)|;$$
$$\Delta(A, t) = f(t) - P_n(A, t).$$

The following propositions may be proved in exactly the same way as in the general discrete case (see Chap. I, §3).

I. If $\varphi(A(\sigma)) = \rho(\sigma)$, then $P_n(A(\sigma), t)$ is a polynomial of best approximation for the continuous case.

II. If $\varphi(A(\sigma)) > \rho(\sigma)$, one can construct a new basis $\sigma_1 = S\sigma$ such that $\rho(\sigma_1) > \rho(\sigma)$.

The transformation S may be described in exactly the same way as for the discrete case; in the new basis σ_1 we replace one of the abscissas of the old basis σ by a point at which $|\Delta(A(\sigma), t)|$ assumes its maximum $\varphi(A(\sigma))$.

We can now state the fundamental theorem.

Theorem 2.1. (Chebyshev). There exists a unique polynomial of best approximation for $f(t)$ on the interval $[c, d]$. A polynomial $P_n(A^*, t)$ is a polynomial of best approximation if and only if it is a Chebyshev interpolating polynomial for some extremal basis.

Finally, $\bar{\rho} = \rho_n(f)$, *i.e.,*

$$\inf_{\{A\}} \max_{c \leqslant t \leqslant d} |\Delta(A, t)| = \sup_{\sigma \in \Xi} \rho(\sigma). \tag{2.3}$$

The proof duplicates that of Theorem 3.1 in the discrete case. Essentially, the only difference is that in the continuous case the set of bases is infinite. However, this obstacle has already been dealt with in Lemma 2.3, where we proved the existence of an extremal basis.

Corollary. A basis $\sigma \in \Xi$,

$$\sigma = \{t_0 < t_1 < \ldots < t_{n+1}\},$$

is extremal (and therefore $P_n(A(\sigma), t)$ is a polynomial of best approximation) if and only if

$$\varphi(A(\sigma)) = \rho(\sigma), \tag{2.4}$$

or, equivalently,

$$\max_{c \leqslant t \leqslant d} |\Delta(A(\sigma), t)| = \max_{k \in [0:n+1]} |\Delta(A(\sigma), t_k)|.$$

Figure 4 illustrates the characteristic shape of the function $\Delta(A(\sigma^*), t)$ when σ^* is an extremal basis.

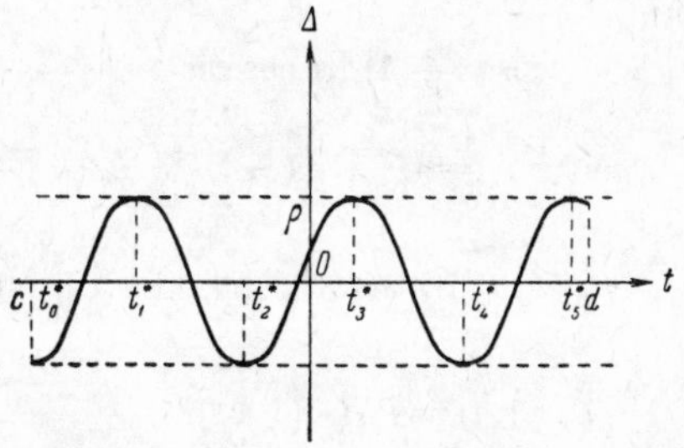

FIGURE 4.

4. Consider the function

$$f(t) = t^{n+1}$$

on the interval $[c, d]$. We wish to find a corresponding polynomial of best approximation of degree $\leqslant n$ and the value of the best fit $\rho_n(f)$. To this end, we set

$$P_n(A, t) = t^{n+1} - \frac{1}{2^n}\left(\frac{d-c}{2}\right)^{n+1} T_{n+1}\left(-1 + \frac{2}{d-c}(t-c)\right),$$

where $T_{n+1}(u)$ is the Chebyshev polynomial

$$T_{n+1}(u) = \cos(n+1)\arccos u, \qquad -1 \leqslant u \leqslant 1.$$

It is readily verified (Chap. I, Lemma 2.1) that $P_n(A, t)$ is indeed an algebraic polynomial of degree $\leqslant n$.

We have

$$\Delta(A, t) \overset{df}{=} t^{n+1} - P_n(A, t) = \\ = \frac{1}{2^n}\left(\frac{d-c}{2}\right)^{n+1} T_{n+1}\left(-1 + \frac{2}{d-c}(t-c)\right),$$

whence it follows that

$$\varphi(A) \overset{df}{=} \max_{c \leqslant t \leqslant d} |\Delta(A, t)| = \frac{1}{2^n}\left(\frac{d-c}{2}\right)^{n+1}.$$

We introduce a basis $\sigma^* = \{t_0 < t_1 < \ldots < t_{n+1}\}$, where

$$t_k = \frac{d+c}{2} + \frac{d-c}{2}\cos\frac{(n+1-k)\pi}{n+1}, \qquad k \in [0 : n+1].$$

The value of $\Delta(A, t)$ at the points of σ^* is found as follows:

$$\Delta(A, t_k) = \frac{1}{2^n}\left(\frac{d-c}{2}\right)^{n+1} T_{n+1}\left(\cos\frac{(n+1-k)\pi}{n+1}\right) = \\ = \frac{1}{2^n}\left(\frac{d-c}{2}\right)^{n+1} \cos(n+1)\arccos\cos\frac{(n+1-k)\pi}{n+1} = \\ = (-1)^k\left[(-1)^{n+1}\frac{1}{2^n}\left(\frac{d-c}{2}\right)^{n+1}\right].$$

We have thus shown that $P_n(A, t)$ is the Chebyshev interpolating polynomial for the basis σ^*:

$$P_n(A, t) \equiv P_n(A(\sigma^*), t),$$

and moreover

$$\rho(\sigma^*) = \left|(-1)^{n+1}\frac{1}{2^n}\left(\frac{d-c}{2}\right)^{n+1}\right| = \varphi(A(\sigma^*)).$$

Thus σ^* is an extremal basis and $P_n(A, t)$ is the required polynomial of best approximation. In particular,

$$\rho_n(t^{n+1}; c, d) = \frac{1}{2^n}\left(\frac{d-c}{2}\right)^{n+1}.$$

Another formulation of this result is as follows: for any algebraic polynomial $\tilde{P}_{n+1}(t)$ of degree $n+1$ with leading coefficient unity,

$$\max_{c \leqslant t \leqslant d} |\tilde{P}_{n+1}(t)| \geqslant \frac{1}{2^n}\left(\frac{d-c}{2}\right)^{n+1};$$

equality can hold here only for the polynomial

$$\tilde{P}^*_{n+1}(t) = \frac{1}{2^n}\left(\frac{d-c}{2}\right)^{n+1} T_{n+1}\left(-1 + \frac{2}{d-c}(t-c)\right).$$

$\tilde{P}^*_{n+1}(t)$ is known as the polynomial of least deviation from zero on $[c, d]$ (with respect to the set of polynomials of degree $n+1$ with leading coefficient unity).

§3. LIMIT THEOREMS

1. Let $A(\sigma)$ denote the coefficient vector of the polynomial of best approximation to a function f for the basis σ.

Lemma 3.1. If $\sigma_s, s = 1, 2, \ldots$, is a sequence of bases converging to a basis σ_0, then

$$A(\sigma_s) \xrightarrow[s\to\infty]{} A(\sigma_0)$$

(convergence of vectors is defined in coordinatewise fashion; see Appendix I, §2).

Proof. First, by Lemma 2.1,

$$h(\sigma_s) \xrightarrow[s\to\infty]{} h(\sigma_0). \tag{3.1}$$

We now observe that the polynomials $P_n(A(\sigma_0), t)$ and $P_n(A(\sigma_s), t)$ satisfy the equalities

$$\left.\begin{aligned} P_n(A(\sigma_0), t_k^{(0)}) &= y_k^{(0)}, \\ P_n(A(\sigma_s), t_k^{(s)}) &= y_k^{(s)}, \end{aligned}\right\} \tag{3.2}$$

where

$$\begin{aligned} y_k^{(0)} &= f(t_k^{(0)}) - (-1)^k h(\sigma_0), \\ y_k^{(s)} &= f(t_k^{(s)}) - (-1)^k h(\sigma_s), \\ k &\in [0:n], \quad s = 1, 2, \ldots \end{aligned}$$

(equalities (3.2) are also valid for $k = n+1$, but we do not need this here).

Thus, the polynomials $P_n(A(\sigma_0), t)$ and $P_n(A(\sigma_s), t)$, $s = 1, 2, \ldots$, are interpolating polynomials. Since by assumption $t_k^{(s)} \xrightarrow[s\to\infty]{} t_k^{(0)}$, $k \in [0:n]$, and moreover

$y_k^{(s)} \xrightarrow[s\to\infty]{} y_k^{(0)}$ by the continuity of $f(t)$ and (3.1), it follows by the continuity lemma (see Appendix I, §2) that

$$A(\sigma_s) \xrightarrow[s\to\infty]{} A(\sigma_0).$$

This completes the proof.

Now set

$$\varphi(A) = \max_{c\leqslant t\leqslant d} |f(t) - P_n(A, t)|.$$

Lemma 3.2. If A_s, $s = 1, 2, \ldots$, is a sequence of vectors converging to a vector A^*, then

$$\varphi(A_s) \xrightarrow[s\to\infty]{} \varphi(A^*).$$

Proof. For any two continuous functions $F_1(t)$ and $F_2(t)$ on $[c, d]$ (see Appendix III, §2):

$$\left| \max_{c\leqslant t\leqslant d} |F_1(t)| - \max_{c\leqslant t\leqslant d} |F_2(t)| \right| \leqslant \max_{c\leqslant t\leqslant d} |F_1(t) - F_2(t)|.$$

Applying this to the present case, we obtain

$$|\varphi(A_s) - \varphi(A^*)| \leqslant \max_{c\leqslant t\leqslant d} |P_n(A^*, t) - P_n(A_s, t)| =$$
$$= \max_{c\leqslant t\leqslant d} \left| \sum_{i=0}^{n} (a_i^* - a_i^{(s)}) t^i \right| \leqslant M \sum_{i=0}^{n} |a_i^* - a_i^{(s)}|.$$

It remains only to let $s \to \infty$, proving the lemma.

2. *Theorem 3.1. If $P_n(A_s, t)$, $s = 1, 2, \ldots$, is a sequence of algebraic polynomials such that*

$$\varphi(A_s) \to \rho_n(f) \tag{3.3}$$

as $s \to \infty$, then $A_s \to A^*$, where A^* is the coefficient vector of the polynomial of best approximation.

Proof. We must prove that for all $k \in [0:n]$

$$a_k^{(s)} \xrightarrow[s\to\infty]{} a_k^*. \tag{3.4}$$

Suppose that (3.4) fails to hold for some $k = k_0$. Then there exist $\varepsilon_0 > 0$ and a sequence $\{s_i\}$, $i = 1, 2, \ldots$, such that

$$|a_{k_0}^{(s_i)} - a_{k_0}^*| \geqslant \varepsilon_0. \tag{3.5}$$

Now the sequence of polynomials $\{P_n(A_{s_i}, t)\}$, $i = 1, 2, \ldots$, is bounded on $[c, d]$, unformly in i. Indeed, by (3.3),

$$\varphi(A_{s_i}) \leqslant M_1.$$

Hence

$$\max_{c\leqslant t\leqslant d} |P_n(A_{s_i}, t)| \leqslant$$
$$\leqslant \max_{c\leqslant t\leqslant d} |P_a(A_{s_i}, t) - f(t)| + \max_{c\leqslant t\leqslant d} |f(t)| =$$
$$= \varphi(A_{s_i}) + \max_{c\leqslant t\leqslant d} |f(t)| \leqslant M_2.$$

By the compactness lemma (see Appendix I, §2), there exist a subsequence of vectors $\{A_{s_{i_r}}\}$ and a vector A_0 such that

$$A_{s_{i_r}} \xrightarrow[r\to\infty]{} A_0.$$

By Lemma 3.2, as $r\to\infty$

$$\varphi(A_{s_{i_r}}) \to \varphi(A_0).$$

In view of (3.3), we see that $\varphi(A_0) = \rho_n(f)$, i.e., $P_n(A_0, t)$ is the polynomial of best approximation. Since the latter is unique, we have $A_0 = A^*$, so that

$$A_{s_{i_r}} \xrightarrow[r\to\infty]{} A^*.$$

This contradicts (3.5), completing the proof.

Theorem 3.2. Let $\{\sigma_s\}, s = 1, 2, \ldots,$ be a sequence of bases such that

$$\rho(\sigma_s) \to \rho_n(f)$$

as $s\to\infty$; then $A(\sigma_s)\to A^*$, where A^* is the coefficient vector of the polynomial of best approximation.

Proof. If $\rho_n(f) = 0$, the assertion is obvious. Let $\rho_n(f) > 0$. By Theorem 3.1, it will suffice to show that

$$\varphi(A(\sigma_s)) \xrightarrow[s\to\infty]{} \rho_n(f).$$

Suppose the contrary, so that for some $\varepsilon_0 > 0$ there exists a subsequence $\{\sigma_{s_i}\}$, $i = 1, 2, \ldots,$ such that

$$\varphi(A(\sigma_{s_i})) - \rho_n(f) \geqslant \varepsilon_0. \tag{3.6}$$

For sufficiently large s_i we have

$$\rho(\sigma_{s_i}) \geqslant \frac{1}{2}\rho_n(f) > 0,$$

and so, by Lemma 2.2, there exists a subsequence of bases $\{\sigma_{s_{i_r}}\}$, $r = 1, 2, \ldots,$ which converges to some basis σ_0:

$$\sigma_{s_{i_r}} \xrightarrow[r\to\infty]{} \sigma_0. \tag{3.7}$$

But then $\rho(\sigma_{s_{i_r}}) \to \rho(\sigma_0)$ (Corollary to Lemma 2.1) and so $\rho(\sigma_0) = \rho_n(f)$.

We have thus shown that σ_0 is an extremal basis, so that

$$\varphi(A(\sigma_0)) = \rho(\sigma_0) = \rho_n(f).$$

In addition, using Lemma 3.1 we infer from (3.7) that

$$A\left(\sigma_{s_{i_r}}\right) \xrightarrow[r\to\infty]{} A(\sigma_0).$$

It now follows from Lemma 3.2 that

$$\varphi\left(A\left(\sigma_{s_{i_r}}\right)\right) \to \varphi(A(\sigma_0)) = \rho_n(f),$$

contradicting (3.6).

§4. REMEZ' METHOD OF SUCCESSIVE CHEBYSHEV INTERPOLATIONS

1. Let $f(t)$ be a continuous function defined on an interval $[c, d]$. We shall now outline, in the most general setting, a method of successive Chebyshev interpolations for constructing the polynomial of best approximation to $f(t)$.

We first select an initial basis σ_1 arbitrarily, and then construct the corresponding Chebyshev interpolating polynomial $P_n(A(\sigma_1), t)$. Suppose that $\rho(\sigma_1) > 0$.

Suppose we have already constructed a basis σ_s and polynomial $P_n(A(\sigma_s), t)$. If $\varphi(A(\sigma_s)) = \rho(\sigma_s)$, then $P_n(A(\sigma_s), t)$ is the required polynomial of best approximation and the procedure terminates. Let $\varphi(A(\sigma_s)) > \rho(\sigma_s)$. We then construct a new basis σ_{s+1} so that the following conditions are satisfied:

I. $$m_s \overset{df}{=} \min_{k\in[0:n+1]} |\Delta_s(t_k^{(s+1)})| \geqslant \rho(\sigma_s) \overset{df}{=} \rho_s,$$

where $\Delta_s(t) = f(t) - P_n(A(\sigma_s), t)$.

II. $$\operatorname{sign}\Delta_s(t_k^{(s+1)}) = -\operatorname{sign}\Delta_s(t_{k+1}^{(s+1)}), \quad k \in [0:n].$$

III. $$l_s \overset{df}{=} \max_{k\in[0:n+1]} |\Delta_s(t_k^{(s+1)})| > \rho_s.$$

We now carry out Chebyshev interpolation on the basis σ_{s+1}, constructing the polynomial $P_n(A(\sigma_{s+1}), t)$.

Continuing in this way, we obtain a sequence

$$P_n(A(\sigma_1), t),\ P_n(A(\sigma_2), t),\ \ldots,\ P_n(A(\sigma_s), t),\ \ldots$$

If this sequence is finite, its last term is by construction the polynomial of best approximation for $f(t)$ on $[c, d]$.

We now consider the case that the sequence $\{P_n(A(\sigma_s), t)\}$ is infinite. Introduce the notation

$$\varphi_s = \max_{c\leqslant t\leqslant d} |\Delta_s(t)|.$$

Theorem 4.1. If

$$\lim_{s\to\infty} \frac{l_s - \rho_s}{\varphi_s - \rho_s} > 0, \tag{4.1}$$

then $A(\sigma_s) \to A^*$, *where* A^* *is the coefficient vector of the polynomial of best approximation.*

Moreover, for some θ, $0 < \theta < 1$,

$$\rho_n(f) - \rho_s \leqslant M\theta^s. \tag{4.2}$$

The constant M is independent of s.

2. We first prove a lemma.

Lemma 4.1. If $\{\sigma_s\}$, $s = 1, 2, \ldots$, *is a sequence of bases on* $[c, d]$ *such that*

$$\rho(\sigma_s) \geqslant \beta > 0,$$

then there exists $\eta > 0$ *such that for any* $k_1 \neq k_2$ $(k_1, k_2 \in [0 : n+1])$ *and* $s = 1, 2, \ldots$

$$\left| t_{k_1}^{(s)} - t_{k_2}^{(s)} \right| \geqslant \eta. \tag{4.3}$$

In addition,

$$\theta_1 \overset{df}{=} \inf_{\{s\}} \min_{k \in [0:n+1]} \alpha_k(\sigma_s) \geqslant \frac{1}{n+2} \left(\frac{\eta}{d-c} \right)^{n+1}. \tag{4.4}$$

Proof. Suppose that no such η exists. Then there are indices k_1^0 and k_2^0, $k_1^0 \neq k_2^0$, and a subsequence $\{s_r\}$, $r = 1, 2, \ldots$, such that

$$t_{k_1^0}^{(s_r)} - t_{k_2^0}^{(s_r)} \xrightarrow[r\to\infty]{} 0 \tag{4.5}$$

(remember that the indices k_1 and k_2, $k_1, k_2 \in [0: n+1]$, are fixed and finite). The subsequence $\{s_r\}$, in turn, contains a subsequence $\{s_{r_m}\}$, $m = 1, 2, \ldots$, for which there exist finite limits

$$\lim_{m\to\infty} t_k^{(s_{r_m})} = t_k^{(0)}, \quad k \in [0 : n+1].$$

Moreover, by (4.5) we have $t_{k_1^0}^{(0)} = t_{k_2^0}^{(0)}$. Since, as before,

$$\rho\left(\sigma_{s_{r_m}}\right) \geqslant \beta > 0,$$

we have a contradiction to Lemma 2.2.

We have thus proved the existence of η, and inequality (4.4) now follows readily from (4.3).

Indeed,

$$\alpha_k(\sigma_s) = \frac{\dfrac{(-1)^k}{\left(t_k^{(s)} - t_0^{(s)}\right)\dots(\;)\dots\left(t_k^{(s)} - t_{n+1}^{(s)}\right)}}{\sum\limits_{i=0}^{n} \dfrac{(-1)^i}{\left(t_i^{(s)} - t_0^{(s)}\right)\dots(\;)\dots\left(t_i^{(s)} - t_{n+1}^{(s)}\right)}} =$$

$$= \frac{\dfrac{1}{\left|t_k^{(s)} - t_0^{(s)}\right|\dots(\;)\dots\left|t_k^{(s)} - t_{n+1}^{(s)}\right|}}{\sum\limits_{i=0}^{n} \dfrac{1}{\left|t_i^{(s)} - t_0^{(s)}\right|\dots(\;)\dots\left|t_i^{(s)} - t_{n+1}^{(s)}\right|}} \geqslant \frac{1}{n+2}\left(\frac{\eta}{d-c}\right)^{n+1}.$$

This completes the proof.

3. Proof of Theorem 4.1. Condition (4.1) means that there exists ε_1, $0 < \varepsilon_1 < 1$, such that for all $s \geqslant \bar{s}$

$$\frac{l_s - \rho_s}{\varphi_s - \rho_s} \geqslant \varepsilon_1$$

or

$$l_s - \rho_s \geqslant \varepsilon_1(\varphi_s - \rho_s). \tag{4.6}$$

By III, we may assume that inequality (4.6) holds for all natural numbers s.

Set

$$\theta_1 = \inf_{\{s\}} \min_{k \in [0:n+1]} \alpha_k(\sigma_s).$$

By definition, $0 \leqslant \theta_1 < 1$. We claim that

$$\rho_{s+1} \geqslant \rho_s + \theta_1(l_s - \rho_s). \tag{4.7}$$

Indeed, let $l_s = \left|\Delta_s\left(t_{k_0}^{(s+1)}\right)\right|$. In view of I–III, we have

$$\rho_{s+1} = \left|\sum_{k=0}^{n+1}(-1)^k \alpha_k(\sigma_{s+1}) f\left(t_k^{(s+1)}\right)\right| =$$

$$= \sum_{k=0}^{n+1} \alpha_k(\sigma_{s+1}) \left|\Delta_s\left(t_k^{(s+1)}\right)\right| \geqslant$$

$$\geqslant \sum_{k=0}^{n+1} \alpha_k(\sigma_{s+1})\rho_s + \alpha_{k_0}(\sigma_{s+1})(l_s - \rho_s) \geqslant \rho_s + \theta_1(l_s - \rho_s).$$

It follows, in particular, from (4.7) that $\rho(\sigma_s) \geqslant \rho(\sigma_1)$ for all s. Since by assumption $\rho(\sigma_1) > 0$, we see by Lemma 4.1 that $0 < \theta_1 < 1$.

Now, replacing s by $s-1$ in (4.7) and noting (4.6), we get

$$\rho_s - \rho_{s-1} \geqslant \theta_1(l_{s-1} - \rho_{s-1}) \geqslant$$

$$\geqslant \theta_1\varepsilon_1(\varphi_{s-1} - \rho_{s-1}) \geqslant \theta_1\varepsilon_1(\rho_n(f) - \rho_{s-1}).$$

The inequality

$$\rho_s - \rho_{s-1} \geqslant \theta_1 \varepsilon_1 (\rho_n(f) - \rho_{s-1})$$

may be rewritten

$$\rho_n(f) - \rho_s \leqslant \theta (\rho_n(f) - \rho_{s-1}), \tag{4.8}$$

where $\theta = 1 - \theta_1 \varepsilon_1$, $0 < \theta < 1$.

It follows from (4.8) that

$$\rho_n(f) - \rho_s \leqslant \theta^{s-1} (\rho_n(f) - \rho_1) \overset{df}{=} M\theta^s.$$

This proves inequality (4.2). The relation $A(\sigma_s) \xrightarrow[s \to \infty]{} A^*$ now follows from (4.2) and Theorem 3.2. The proof is complete.

Remark. Condition (4.1) is surely true if $l_s = \varphi_s$ for all s. This corresponds to the situation in which, at each step of the procedure, the new basis σ_{s+1} includes a point at which $\max\limits_{c \leqslant t \leqslant d} |\Delta_s(t)|$ is achieved (compare the de la Vallée-Poussin algorithm and the R-algorithm).

In the general case, convergence is assured if one takes care that the quotient

$$\frac{l_s - \rho_s}{\varphi_s - \rho_s}$$

is bounded away from zero uniformly in s.

4. A few words concerning the choice of the initial basis σ_1. We are interested in a basis σ_1 for which $\rho(\sigma_1)$ is as large as possible. Recall that our function $f(t)$ is continuous on $[c, d]$ and

$$\rho(\sigma_1) = \left| \frac{f[t_0^{(1)}, t_1^{(1)}, \ldots, t_{n+1}^{(1)}]}{\chi[t_0^{(1)}, t_1^{(1)}, \ldots, t_{n+1}^{(1)}]} \right|.$$

The numerator depends on the function f, and no special choice of σ_1 can possibly increase it uniformly for all continuous functions. We therefore try to maximize

$$\frac{1}{|\chi[t_0^{(1)}, t_1^{(1)}, \ldots, t_{n+1}^{(1)}]|}.$$

If we set $\Phi(t) = t^{n+1}$, it follows from a property of divided differences (see Appendix I, §1) that

$$\frac{1}{|\chi[t_0^{(1)}, t_1^{(1)}, \ldots, t_{n+1}^{(1)}]|} = \left| \frac{\Phi[t_0^{(1)}, t_1^{(1)}, \ldots, t_{n+1}^{(1)}]}{\chi[t_0^{(1)}, t_1^{(1)}, \ldots, t_{n+1}^{(1)}]} \right| \overset{df}{=} \rho_\Phi(\sigma_1).$$

But

$$\rho_\Phi(\sigma_1) \leqslant \rho_n(\Phi; c, d),$$

and equality holds for an extremal basis σ^* in best approximation of the function $\Phi(t) = t^{n+1}$ on $[c, d]$ by polynomials of degree $\leqslant n$.

Using the results of §2.4, we get

$$\frac{1}{\left|\chi\left[t_0^{(1)}, t_1^{(1)}, \ldots, t_{n+1}^{(1)}\right]\right|} \leqslant \frac{1}{2^n}\left(\frac{d-c}{2}\right)^{n+1},$$

and equality will hold here if the components of the basis σ_1 are taken as

$$t_k^{(1)} = \frac{d+c}{2} + \frac{d-c}{2}\cos\frac{(n+1-k)\pi}{n+1}, \quad k \in [0 : n+1].$$

§5. METHOD OF GRIDS

1. As before, we consider a continuous function $f(t)$ on $[c, d]$ and wish to determine its polynomial of best approximation $P_n(A^*, t)$.

Set

$$z_k = c + k\frac{d-c}{N}, \quad w_k = f(z_k), \quad k \in [0 : N].$$

Denote the set of points $\{z_k\}$ by T_N.

Using the methods of Chapter I, we can construct an algebraic polynomial $P_n(A_N, t)$ solving the discrete problem

$$\max_{k \in [0:N]} |w_k - P_n(A_N, z_k)| = \rho_N, \tag{5.1}$$

where

$$\rho_N = \inf_{\{A\}} \max_{k \in [0:N]} |w_k - P_n(A, z_k)|.$$

Theorem 5.1. (Convergence of the grid process). As $N \to \infty$, we have

$$A_N \to A^*.$$

Proof. Let $\sigma^* = \{t_0^* < t_1^* < \ldots < t_{n+1}^*\}$ be an extremal basis for best approximation on the entire interval $[c, d]$ (its existence is assured by Lemma 2.3). For sufficiently large N, consider the bases $\sigma_N = \{z_0^{(N)} < < z_1^{(N)} < \ldots < z_{n+1}^{(N)}\}$, where $z_k^{(N)}$ is the point of the grid T_N nearest t_k^*, $k \in [0 : n+1]$.

Clearly, $\sigma_N \to \sigma^*$ as $N \to \infty$, so that

$$\rho(\sigma_N) \to \rho(\sigma^*). \tag{5.2}$$

Let σ_N^* be an extremal basis in T_N. Then (see (5.1))

$$\rho(\sigma_N^*) = \rho_N, \quad A(\sigma_N^*) = A_N.$$

Since σ_N is a basis in T_N, we have

$$\rho(\sigma_N) \leqslant \rho(\sigma_N^*). \tag{5.3}$$

On the other hand

$$\rho(\sigma_N^*) \leqslant \rho(\sigma^*), \tag{5.4}$$

since σ_N^* is a basis for $[c, d]$ as well.

Combining (5.3), (5.4) and (5.2), we get

$$\rho(\sigma_N^*) \to \rho(\sigma^*) = \rho_n(f),$$

whence it follows by Theorem 3.2 that

$$A_N \to A^*.$$

This completes the proof.

Note that inequality (5.4) is equivalent to

$$\rho_N \leqslant \rho_n(f). \tag{5.5}$$

2. In this subsection we develop an estimate for the rate of convergence of a grid process.

Theorem 5.2. Let $f(t)$ have a continuous second derivative on $[c, d]$. If A_N is the coefficient vector of the polynomial of best approximation on T_N, then

$$\varphi(A_N) - \rho_n(f) \leqslant M\left(\frac{d-c}{N}\right)^2, \tag{5.6}$$

where

$$\varphi(A_N) = \max_{c \leqslant t \leqslant d} |f(t) - P_n(A_N, t)|$$

and M is a constant independent of N.

Proof. Set

$$\Delta_N(t) = f(t) - P_n(A_N, t).$$

By (5.5),

$$\varphi(A_N) - \rho_n(f) \leqslant \varphi(A_N) - \rho_N = \max_{c \leqslant t \leqslant d} |\Delta_N(t)| - \max_{k \in [0:N]} |\Delta_N(z_k)|.$$

Let $t_N \in [c, d]$ be a point at which

$$|\Delta_N(t_N)| = \max_{c \leqslant t \leqslant d} |\Delta_N(t)|. \tag{5.7}$$

If $\Delta_N(t_N)=0$ or t_N is a point of the grid T_N, then

$$\varphi(A_N)-\rho_n(f)=0,$$

and inequality (5.6) is true for any $M\geqslant 0$. We may therefore assume that $\Delta_N(t_N)\neq 0$ and t_N is not a point of T_N. In particular, t_N is an interior point of $[c, d]$, and by (5.7),

$$\Delta'_N(t_N)=0. \tag{5.8}$$

Let z_s denote the point of T_N nearest t_N. By Taylor's formula,

$$\Delta_N(z_s)=\Delta_N(t_N)+ \\ +\frac{\Delta'_N(t_N)}{1!}(z_s-t_N)+\frac{\Delta''_N(t_N+\theta(z_s-t_N))}{2!}(z_s-t_N)^2,$$

where $0<\theta<1$. Hence, in view of (5.8), we obtain

$$|\Delta_N(t_N)|-|\Delta_N(z_s)|\leqslant\frac{M_{N2}}{8}\left(\frac{d-c}{N}\right)^2,$$

where $M_{N2}=\max\limits_{c\leqslant t\leqslant d}|\Delta''_N(t)|$. Moreover,

$$\max_{c\leqslant t\leqslant d}|\Delta_N(t)|-\max_{k\in[0:N]}|\Delta_N(z_k)|\leqslant\frac{M_{N2}}{8}\left(\frac{d-c}{N}\right)^2.$$

It remains to show that the sequence $\{M_{N2}\}$ is bounded uniformly in N. To this end, we use Theorem 5.1 and Lemma 2.3 of Appendix I. By Theorem 5.1, $A_N \xrightarrow[N\to\infty]{} A^*$, and so

$$\max_{c\leqslant t\leqslant d}|P_n(A_N, t)|\leqslant M_1.$$

Hence, by Lemma 2.3 in Appendix I,

$$\max_{c\leqslant t\leqslant d}|P''_n(A_N, t)|\leqslant M_1^{(2)}.$$

We now have

$$M_{N2}\leqslant\max_{c\leqslant t\leqslant d}|f''(t)|+\max_{c\leqslant t\leqslant d}|P''_n(A_N, t)|\leqslant \\ \leqslant\max_{c\leqslant t\leqslant d}|f''(t)|+M_1^{(2)}\overset{\text{df}}{=}M_2.$$

This completes the proof.

§6*. BEHAVIOR OF COEFFICIENTS OF POLYNOMIALS OF BEST APPROXIMATION

1. By a famous theorem of Weierstrass, any function $f(t)$ continuous on $[c, d]$ may be uniformly approximated to within arbitrary accuracy by

algebraic polynomials. This means that for any $\varepsilon>0$ there is a polynomial $P(t)$ such that

$$\max_{c\leqslant t\leqslant d} |f(t)-P(t)|<\varepsilon.$$

The following theorem complements this classical result.

Theorem 6.1. A function $f(t)$ continuous on $[-1, 1]$ may be uniformly approximated to within arbitrary accuracy by algebraic polynomials $Q_n(t)$,

$$Q_n(t)=\sum_{k=0}^{n} a_k^{(n)}t^k,$$

with uniformly bounded coefficients

$$|a_k^{(n)}|\leqslant M, \quad k\in[0:n],$$

if and only if, for any $t\in(-1, 1)$

$$f(t)=\sum_{k=0}^{\infty} a_k t^k; \quad |a_k|\leqslant M, \quad k=0, 1, 2, \ldots \tag{6.1}$$

Proof. Sufficiency. Let $f(t)$ be continuous on $[-1, 1]$, admitting a representation (6.1) for all $t\in(-1, 1)$.

Set

$$F_n(t)=\sum_{k=0}^{n} a_k t^k.$$

Given $\varepsilon>0$, we select r, $0<r<1$, so that for all $t\in[-1, 1]$

$$|f(t)-f(rt)|<\frac{\varepsilon}{2}.$$

Since the series (6.1) is uniformly convergent on $[-r, r]$, there exists a natural number n such that

$$|f(rt)-F_n(rt)|<\frac{\varepsilon}{2}$$

uniformly in $t\in[-1, 1]$. For $t\in[-1, 1]$,

$$|f(t)-F_n(rt)|<\varepsilon.$$

It remains only to observe that the coefficients of the algebraic polynomial $F_n(rt)$ are bounded by M in absolute value.

Necessity. Consider a sequence of polynomials

$$Q_n(t)=\sum_{k=0}^{n} a_k^{(n)}t^k, \quad |a_k^{(n)}|\leqslant M,$$

converging uniformly as $n\to\infty$ to $f(t)$ on $[-1, 1]$. We claim that for each k the sequence $\{a_k^{(n)}\}$ is convergent. Indeed,

$$a_0^{(n)} = Q_n(0) \xrightarrow[n\to\infty]{} f(0).$$

Supposing that the sequences $\{a_k^{(n)}\}$ are convergent for $k \in [0:\nu-1]$, we shall prove that the sequence also converges for $k=\nu$.

For $n \geqslant \nu$, set

$$q_n^{(\nu)}(t) \overset{df}{=} \frac{Q_n(t) - \sum_{k=0}^{\nu-1} a_k^{(n)} t^k}{t^\nu} = a_\nu^{(n)} + a_{\nu+1}^{(n)} t + \ldots + a_n^{(n)} t^{n-\nu}.$$

By the definition of $Q_n(t)$ and the induction hypothesis, the sequence $\{q_n^{(\nu)}(t)\}$ is convergent for any $t \in [-1, 1]$ other than zero. We claim that it is also convergent for $t=0$. To prove this, we observe that for $t \in (-1, 1)$

$$|q_n^{(\nu)}(t) - q_n^{(\nu)}(0)| = \left| t \sum_{k=\nu+1}^{n} a_k^{(n)} t^{k-\nu-1} \right| \leqslant \frac{M|t|}{1-|t|}.$$

And now

$$|q_n^{(\nu)}(0) - q_{n+m}^{(\nu)}(0)| \leqslant \frac{2M|t|}{1-|t|} + |q_n^{(\nu)}(t) - q_{n+m}^{(\nu)}(t)|. \tag{6.2}$$

Let $\varepsilon > 0$. Select some $t \neq 0$ for which the first term on the right of (6.2) is less than $\varepsilon/2$. Then, using the convergence of the sequence $\{q_n^{(\nu)}(t)\}$, select n such that for all natural numbers m the second term is less than $\varepsilon/2$. Then, for all m,

$$|q_n^{(\nu)}(0) - q_{n+m}^{(\nu)}(0)| < \varepsilon.$$

Hence it follows that $\{q_n^{(\nu)}(0)\}$ is convergent. Using the equality $a_\nu^{(n)} = q_n^{(\nu)}(0)$, we conclude that the sequence $\{a_\nu^{(n)}\}$ is also convergent.

Thus $a_k^{(n)} \xrightarrow[n\to\infty]{} a_k$ for all $k = 0, 1, 2, \ldots$, and moreover $|a_k| \leqslant M$. Consider the series $\sum_{k=0}^{\infty} a_k t^k$.

We shall show that for $-1 < t < 1$

$$f(t) = \sum_{k=0}^{\infty} a_k t^k. \tag{6.3}$$

Fix t, $|t| < 1$, and take an arbitrary $\varepsilon > 0$. Select N so that

$$\frac{M|t|^{N+1}}{1-|t|} < \frac{\varepsilon}{4}$$

and so that for $n > N$

$$|f(t) - Q_n(t)| < \frac{\varepsilon}{4}.$$

We now find $\bar{n} > N$ such that

$$\sum_{k=0}^{N} | a_k^{(\bar{n})} - a_k | < \frac{\varepsilon}{4}.$$

Then

$$\left| \sum_{k=0}^{\infty} a_k t^k - f(t) \right| \leqslant \left| \sum_{k=0}^{\infty} a_k t^k - Q_{\bar{n}}(t) \right| + | Q_{\bar{n}}(t) - f(t) | \leqslant$$
$$\leqslant \sum_{k=0}^{N} | a_k^{(\bar{n})} - a_k | + \frac{2M |t|^{N+1}}{1 - |t|} + \frac{\varepsilon}{4} < \varepsilon.$$

Since ε was arbitrary, this implies (6.3), completing the proof.

2. Let $P_n(A_n, t)$, $A_n = (a_0^{(n)}, \ldots, a_n^{(n)})$, denote the polynomial of best approximation of degree $\leqslant n$ for a continuous function $f(t)$ on $[-1, 1]$. By Weierstrass' theorem, the sequence $\{P_n(A_n, t)\}$ converges uniformly to $f(t)$ on $[-1, 1]$ as $n \to \infty$.

Set

$$|A_n| = \max_{k \in [0:n]} |a_k^{(n)}|.$$

By Theorem 6.1, a necessary condition for the sequence $\{|A_n|\}$ to be bounded is that the function $f(t)$ admit a representation (6.1) for $-1 < t < 1$. Otherwise, $|A_n| \xrightarrow[n \to \infty]{} \infty$.

Set

$$\delta_n = \frac{|A_n|}{\max\limits_{-1 \leqslant t \leqslant 1} |P_n(A_n, t)|}.$$

As $n \to \infty$, we have

$$\max_{-1 \leqslant t \leqslant 1} |P_n(A_n, t)| \to \max_{-1 \leqslant t \leqslant 1} |f(t)|.$$

Therefore, if $|A_n| \to \infty$, then also $\delta_n \to \infty$.

When δ_n is large, we say that the polynomial $P_n(A_n, t)$ is ill-conditioned.

An example of an ill-conditioned polynomial is the polynomial of best approximation of degree 14 for the function $|t|$ on $[-1, 1]$:

$$P_{14}(t) = 198.46t^{14} - 714.90t^{12} + 1022.98t^{10} - 739.91t^8 +$$
$$+ 285.33t^6 - 57.607t^4 + 6.6443t^2 + 0.019949.$$

In this case $\delta_{14} \approx 10^3$.

In practice, however, one rarely encounters polynomials of best approximation of degree higher than 10, and ill-conditioned polynomials appear only for high degrees n.

Chapter III

THE DISCRETE MINIMAX PROBLEM

§1. STATEMENT OF THE PROBLEM

Let $f_i(X)$, $i \in [0:N]$, $X = (x_1, x_2, \ldots, x_n)$ be functions defined on euclidean n-space E_n. Set

$$\mu = \inf_{\{X\}} \max_{i \in [0:N]} f_i(X).$$

Our problem is to find a point X^* at which

$$\max_{i \in [0:N]} f_i(X^*) = \mu.$$

We introduce the notation

$$\varphi(X) = \max_{i \in [0:N]} f_i(X).$$

Since

$$\mu = \inf_{X \in E_n} \varphi(X),$$

the original minimax problem is equivalent to minimization of the function $\varphi(X)$ on E_n.

In view of this interpretation, we shall devote considerable space to a detailed investigation of the properties of the maximum function $\varphi(X)$. In particular, we shall show that under natural assumptions $\varphi(X)$ is differentiable in all directions. We shall establish necessary conditions for $\varphi(X)$ to have a minimum and interpret these conditions from the geometric viewpoint.

Much attention will be centered on successive approximation techniques for determining stationary points of $\varphi(X)$, i. e., points satisfying the necessary conditions for a minimum. Among other things, we shall give an example showing that the method of steepest descent may converge to a nonstationary point.

One section will be devoted to sufficient conditions for $\varphi(X)$ to have a local minimum.

All the requisite definitions, notation and properties relating to euclidean n-space and functions of n variables may be found in Appendices II and III.

§2. PROPERTIES OF THE MAXIMUM FUNCTION

1. We consider functions $f_i(X)$, $i \in [0:N]$, $X \in E_n$, and set

$$\varphi(X) = \max_{i \in [0:N]} f_i(X).$$

We first list the simplest properties of $\varphi(X)$.

I. *If all the functions* $f_i(X)$, $i \in [0:N]$ *are continuous at a point* X_0, *then* $\varphi(X)$ *is also continuous there.*

This follows from the inequalities (see Appendix III, §2)

$$|\varphi(X) - \varphi(X_0)| \overset{df}{=} |\max_{i \in [0:N]} f_i(X) - \max_{i \in [0:N]} f_i(X_0)| \leqslant \leqslant \max_{i \in [0:N]} |f_i(X) - f_i(X_0)|.$$

II. *If all the functions* $f_i(X)$, $i \in [0:N]$ *are continuous on* E_n *and for some* $X_0 \in E_n$ *the set*

$$M(X_0) = \{X \in E_n \mid \varphi(X) \leqslant \varphi(X_0)\}$$

is bounded, there exists a point X^* *at which* $\varphi(X)$ *assumes its minimum value.*

Proof. By Property I, $\varphi(X)$ is continuous on E_n. Thus the set $M(X_0)$ is closed. Since it is by assumption bounded, there is a point X^* at which

$$\varphi(X^*) = \min_{X \in M(X_0)} \varphi(X).$$

It remains to observe that

$$\min_{X \in M(X_0)} \varphi(X) = \inf_{X \in E_n} \varphi(X),$$

so that

$$\varphi(X^*) = \inf_{X \in E_n} \varphi(X).$$

III. *Let* $\Omega \subset E_n$ *be a convex set. If all the functions* $f_i(X)$, $i \in [0:N]$ *are convex on* Ω, *then* $\varphi(X)$ *is also convex on* Ω.

Proof. Let $X_1, X_2 \in \Omega$ and $\alpha \in [0, 1]$. Then

$$f_i(\alpha X_1 + (1-\alpha) X_2) \leqslant \alpha f_i(X_1) + (1-\alpha) f_i(X_2) \leqslant \leqslant \alpha\varphi(X_1) + (1-\alpha)\varphi(X_2). \quad (2.1)$$

Since these inequalities are valid for all $i \in [0:N]$, it follows that

$$\varphi(\alpha X_1 + (1-\alpha) X_2) \leqslant \alpha\varphi(X_1) + (1-\alpha)\varphi(X_2), \quad (2.2)$$

as required.

2. Fixing X, let us consider the set of indices $R(X)$ defined by

$$R(X)=\{i\in[0:N]\mid f_i(X)=\varphi(X)\}.$$

In other words, $R(X)$ contains every index i for which $f_i(X)$ equals the maximum function $\varphi(X)$.

Lemma 2.1. If the functions $f_i(X)$, $i\in[0:N]$ are continuous at a point X_0, then there exists a number $\alpha_0>0$ such that for all $\alpha\in[0,\ \alpha_0]$ and $g\in E_n$, $\|g\|=1$,

$$\varphi(X_0+\alpha g)\overset{df}{=}\max_{i\in[0:N]} f_i(X_0+\alpha g)=\max_{i\in R(X_0)} f_i(X_0+\alpha g). \tag{2.3}$$

Proof. It will suffice to consider the case that $R(X_0)\neq[0:N]$. Set

$$\beta=\varphi(X_0)-\max_{i\notin R(X_0)} f_i(X_0).$$

It is clear that $\beta>0$. By the continuity of $f_i(X)$ at X_0, there exists $\alpha_0>0$ such that for all $g\in E_n$, $\|g\|=1$, and $\alpha\in[0,\ \alpha_0]$

$$f_i(X_0+\alpha g)\geqslant\varphi(X_0)-\frac{1}{3}\beta \quad \text{if} \quad i\in R(X_0),$$

$$f_i(X_0+\alpha g)\leqslant\varphi(X_0)-\frac{2}{3}\beta \quad \text{if} \quad i\notin R(X_0).$$

This clearly implies (2.3), proving the lemma.

3. Definition. A function $\varphi(X)$, $X\in E_n$, is said to be differentiable at X_0 in the direction $g\in E_n$, $\|g\|=1$, if there exists a finite limit

$$\lim_{\alpha\to+0}\frac{\varphi(X_0+\alpha g)-\varphi(X_0)}{\alpha}.$$

This limit is known as the derivative of $\varphi(X)$ at X_0 in the direction g, denoted by $\frac{\partial\varphi(X_0)}{\partial g}$.

Theorem 2.1. Let the functions $f_i(X)$, $i\in[0:N]$, be continuously differentiable in a neighborhood $S_\delta(X_0)$ of X_0,

$$S_\delta(X_0)=\{X\mid\|X-X_0\|<\delta\},\qquad \delta>0.$$

Then the maximum function $\varphi(X)$ is differentiable at X_0 in any direction g, $\|g\|=1$, and

$$\frac{\partial\varphi(X_0)}{\partial g}=\max_{i\in R(X_0)}\left(\frac{\partial f_i(X_0)}{\partial X},\ g\right) \tag{2.4}$$

(the outer parentheses denote the scalar product in E_n).

Proof. For any $i\in[0:N]$, $\alpha\in(0,\delta)$ and g, $\|g\|=1$, we have (see Appendix III, §3)

$$f_i(X_0+\alpha g)=f_i(X_0)+\alpha\left(\frac{\partial f_i(X_0)}{\partial X},\ g\right)+o_i(g;\ \alpha), \tag{2.5}$$

where $o_i(g;\alpha)$ denotes a quantity such that

$$\frac{o_i(g;\ \alpha)}{\alpha}\xrightarrow[\alpha\to 0]{}0$$

uniformly in g, $\|g\|=1$.

By Lemma 2.1, there exists $\alpha_0>0$, $\alpha_0<\delta$ such that for $\alpha\in(0,\ \alpha_0)$

$$\varphi(X_0+\alpha g)=\max_{i\in R(X_0)}\left\{f_i(X_0)+\alpha\left(\frac{\partial f_i(X_0)}{\partial X},\ g\right)+o_i(g;\ \alpha)\right\}\leqslant$$
$$\leqslant\varphi(X_0)+\alpha\max_{i\in R(X_0)}\left(\frac{\partial f_i(X_0)}{\partial X},\ g\right)+\max_{i\in[0:N]}o_i(g;\ \alpha).$$

We now use the inequality (see Appendix III, §2)

$$\max_{i\in[0:N]}(\beta_i+\gamma_i)\geqslant\max_{i\in[0:N]}\beta_i+\max_{i\in R}\gamma_i, \tag{2.6}$$

where $R=\{i\,|\,\beta_i=\max_{k\in[0:N]}\beta_k\}$.

In view of (2.5), we obtain

$$\varphi(X_0+\alpha g)\geqslant\varphi(X_0)+\max_{i\in R(X_0)}\left\{\alpha\left(\frac{\partial f_i(X_0)}{\partial X},\ g\right)+o_i(g;\ \alpha)\right\}\geqslant$$
$$\geqslant\varphi(X_0)+\alpha\max_{i\in R(X_0)}\left(\frac{\partial f_i(X_0)}{\partial X},\ g\right)+\min_{i\in[0:N]}o_i(g;\ \alpha).$$

Thus, for $\alpha\in(0,\ \alpha_0)$ and any g, $\|g\|=1$, we have

$$\min_{i\in[0:N]}o_i(g;\ \alpha)\leqslant$$
$$\leqslant\varphi(X_0+\alpha g)-\varphi(X_0)-\alpha\max_{i\in R(X_0)}\left(\frac{\partial f_i(X_0)}{\partial X},\ g\right)\leqslant$$
$$\leqslant\max_{i\in[0:N]}o_i(g;\ \alpha).$$

Dividing by $\alpha>0$ and letting $\alpha\to+0$, we obtain the required assertion. In fact, we have proved even more:

$$\varphi(X_0+\alpha g)=\varphi(X_0)+\alpha\frac{\partial\varphi(X_0)}{\partial g}+o(g;\ \alpha), \tag{2.7}$$

where

$$\frac{o(g;\ \alpha)}{\alpha}\xrightarrow[\alpha\to+0]{}0$$

uniformly in g, $\|g\|=1$.

4. Example. Let

$$f_0(x) = \sin x, \quad f_1(x) = \cos x,$$

$$\varphi(x) = \max\{\sin x, \ \cos x\}$$

(see Figure 5). Since the function $\varphi(x)$ is 2π-periodic, we may confine attention to $x \in [0, 2\pi]$. We observe first that

$$R(x) = \begin{cases} \{1\}, & \text{if} \quad x \in \left[0, \frac{\pi}{4}\right) \quad \text{or} \quad x \in \left(\frac{5\pi}{4}, 2\pi\right]; \\ \{0, 1\}, & \text{if} \quad x = \frac{\pi}{4} \quad \text{or} \quad x = \frac{5\pi}{4}; \\ \{0\}, & \text{if} \quad x \in \left(\frac{\pi}{4}, \frac{5\pi}{4}\right). \end{cases}$$

$\varphi(x)$ is continuous but not continuously differentiable. Its first derivative is discontinuous at $x = \frac{\pi}{4}$ and $x = \frac{5\pi}{4}$. Even at these points, however, $\varphi(x)$ has derivatives in the two possible directions $g = (+1)$ and $g = (-1)$.

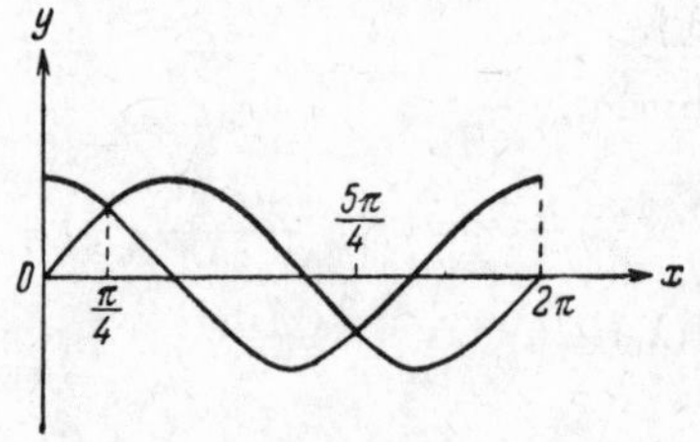

FIGURE 5.

For example, consider $x = \frac{\pi}{4}$. Setting $g = (-1)$ in (2.4), we get

$$\frac{\partial \varphi\left(\frac{\pi}{4}\right)}{\partial g} = \max_{i \in R\left(\frac{\pi}{4}\right)} \left(\frac{df_i\left(\frac{\pi}{4}\right)}{dx}, \ g \right) =$$

$$= \max\left\{ -\cos\frac{\pi}{4}, \ \sin\frac{\pi}{4} \right\} = \frac{\sqrt{2}}{2}.$$

Similarly, for $g = (+1)$,

$$\frac{\partial \varphi\left(\frac{\pi}{4}\right)}{\partial g} = \frac{\sqrt{2}}{2}.$$

Note that the function $\varphi(x)$ increases at $x = \frac{\pi}{4}$ in both admissible directions.

It is readily seen that in the general case, for $g=(+1)$ and $g=(-1)$,

$$\frac{\partial\varphi(x)}{\partial g}=\begin{cases}-g\sin x, & \text{if } x\in\left[0, \frac{\pi}{4}\right) \text{ or } x\in\left(\frac{5\pi}{4}, 2\pi\right];\\ \max\{g\cos x, -g\sin x\}, & \text{if } x=\frac{\pi}{4} \text{ or } x=\frac{5\pi}{4};\\ g\cos x, & \text{if } x\in\left(\frac{\pi}{4}, \frac{5\pi}{4}\right).\end{cases}$$

5*. To conclude this section, we generalize formula (2.7).

Let the functions $f_i(X)$, $i\in[0:N]$, $X=(x_1, x_2, \ldots, x_n)$ be continuous and have continuous derivatives of order up to and including l in some neighborhood $S_\delta(X_0)$, $\delta>0$, of a point X_0. In this case, for any $\alpha\in(0, \delta)$ and $g=(g_1, g_2, \ldots, g_n)$, $\|g\|=1$, we have the equality (see Appendix III, §3)

$$f_i(X_0+\alpha g)=f_i(X_0)+\sum_{k=1}^{l}\frac{\alpha^k}{k!}\frac{\partial^k f_i(X_0)}{\partial g^k}+o_l(g; \alpha^l), \qquad (2.8)$$

where

$$\frac{\partial^k f_i(X_0)}{\partial g^k}=\sum_{j_1, \ldots, j_k=1}^{n}\frac{\partial^k f_i(X_0)}{\partial x_{j_1}\ldots\partial x_{j_k}}g_{j_1}\ldots g_{j_k}$$

and

$$\frac{o_l(g; \alpha^l)}{\alpha^l}\xrightarrow[\alpha\to 0]{}0$$

uniformly in g, $\|g\|=1$.

We introduce the notation

$$\frac{\partial^0 f_i(X)}{\partial g^0}=f_i(X), \quad i\in[0:N];$$

$$R_0(X, g)=[0:N];$$

$$R_k(X, g)=\Big\{i\mid i\in R_{k-1}(X, g), \frac{\partial^{k-1}f_i(X)}{\partial g^{k-1}}=\max_{j\in R_{k-1}(X, g)}\frac{\partial^{k-1}f_j(X)}{\partial g^{k-1}}\Big\}, \qquad k\in[1:l].$$

Since $R_0(X, g)$ is independent of X and g, and $R_1(X, g)$ is independent of g, we shall write

$$R_0(X, g)=R_0, \ R_1(X, g)=R(X).$$

Clearly,

$$R_0\supset R(X)\supset R_2(X, g)\supset\ldots$$

Theorem 2.2. Let $f_i(X)$, $i \in [0:N]$ be continuous functions with continuous derivatives of order up to and including l in some neighborhood $S_\delta(X_0)$ of X_0. Then the maximum function $\varphi(X)$ admits the following representation near X_0 in the direction g, $\|g\|=1$:

$$\varphi(X_0+\alpha g)=\varphi(X_0)+\sum_{k=1}^{l}\frac{\alpha^k}{k!}\frac{\partial^k\varphi(X_0)}{\partial g^k}+o(g;\ \alpha^l),$$

where

$$\frac{\partial^k\varphi(X_0)}{\partial g^k}=\max_{i\in R_k(X_0,\ g)}\frac{\partial^k f_i(X_0)}{\partial g^k}$$

and

$$\frac{o(g;\ \alpha^l)}{\alpha^l}\xrightarrow[\alpha\to+0]{}0$$

uniformly in g, $\|g\|=1$.

Proof. As in the proof of Theorem 2.1, we find a number $\alpha_0>0$, $\alpha_0<\delta$, such that for any $\alpha\in(0,\ \alpha_0)$, by (2.8),

$$\varphi(X_0+\alpha g)\leqslant$$
$$\leqslant\varphi(X_0)+\max_{i\in R(X_0)}\left\{\sum_{k=1}^{l}\frac{\alpha^k}{k!}\frac{\partial^k f_i(X_0)}{\partial g^k}\right\}+\max_{i\in R_0}o_i(g;\ \alpha^l). \tag{2.9}$$

It is readily shown that there exists $\alpha_1>0$, $\alpha_1\leqslant\alpha_0$, so small that for $\alpha\in(0,\ \alpha_1)$

$$\max_{i\in R(X_0)}\left\{\sum_{k=1}^{l}\frac{\alpha^k}{k!}\frac{\partial^k f_i(X_0)}{\partial g^k}\right\}=\max_{i\in R(X_0)}\left\{\frac{\alpha}{1!}\frac{\partial f_i(X_0)}{\partial g}+\sum_{k=2}^{l}\frac{\alpha^k}{k!}\frac{\partial^k f_i(X_0)}{\partial g^k}\right\}=$$

$$=\max_{i\in R_1(X_0,\ g)}\left\{\frac{\alpha}{1!}\frac{\partial f_i(X_0)}{\partial g}+\sum_{k=2}^{l}\frac{\alpha^k}{k!}\frac{\partial^k f_i(X_0)}{\partial g^k}\right\}=$$
$$=\frac{\alpha}{1!}\max_{i\in R(X_0)}\frac{\partial f_i(X_0)}{\partial g}+\max_{i\in R_1(X_0,\ g)}\left\{\sum_{k=2}^{l}\frac{\alpha^k}{k!}\frac{\partial^k f_i(X_0)}{\partial g^k}\right\}.$$

Continuing in this way, we find a number $\alpha_{l-1}>0$, $\alpha_{l-1}\leqslant\alpha_{l-2}$, such that for $\alpha\in(0,\ \alpha_{l-1})$

$$\max_{i\in R(X_0)}\left\{\sum_{k=1}^{l}\frac{\alpha^k}{k!}\frac{\partial^k f_i(X_0)}{\partial g^k}\right\}=\sum_{k=1}^{l}\frac{\alpha^k}{k!}\max_{i\in R_k(X_0,\ g)}\frac{\partial^k f_i(X_0)}{\partial g^k}.$$

Hence, using (2.9), we obtain

$$\varphi(X_0+\alpha g)\leqslant\varphi(X_0)+\sum_{k=1}^{l}\frac{\alpha^k}{k!}\max_{i\in R_k(X_0,\,g)}\frac{\partial^k f_i(X_0)}{\partial g^k}+\max_{i\in R_0}o_i(g;\,\alpha^l).$$

On the other hand, successive application of inequality (2.6) yields

$$\varphi(X_0+\alpha g)\geqslant\varphi(X_0)+\max_{i\in R(X_0)}\left\{\sum_{k=1}^{l}\frac{\alpha^k}{k!}\frac{\partial^k f_i(X_0)}{\partial g^k}\right\}+\min_{i\in R_0}o_i(g;\,\alpha^l)\geqslant\varphi(X_0)+\sum_{k=1}^{l}\frac{\alpha^k}{k!}\max_{i\in R_k(X_0,\,g)}\frac{\partial^k f_i(X_0)}{\partial g^k}+\min_{i\in R_0}o_i(g;\,\alpha^l).$$

Thus, for $\alpha\in(0,\,\alpha_{l-1})$ and any g, $\|g\|=1$,

$$\min_{i\in R_0}o_i(g,\,\alpha^l)\leqslant\varphi(X_0+\alpha g)-\varphi(X_0)-\sum_{k=1}^{l}\frac{\alpha^k}{k!}\frac{\partial^k\varphi(X_0)}{\partial g^k}\leqslant\max_{i\in R_0}o_i(g;\,\alpha^l).$$

This clearly implies the required assertion, completing the proof.

§3. NECESSARY CONDITIONS FOR A MINIMAX

1. In the sequel we shall assume that the functions $f_i(X)$, $i\in[0:N]$, are continuously differentiable on E_n and that for some X_0 the set

$$M(X_0)=\{X\in E_n\,|\,\varphi(X)\leqslant\varphi(X_0)\}$$

is bounded. As before, we set

$$\varphi(X)=\max_{i\in[0:N]}f_i(X).$$

Theorem 3.1. In order that $\varphi(X)$ have a minimum on E_n at a point X^*, it is necessary, and if $\varphi(X)$ is convex, also sufficient that

$$\inf_{\|g\|=1}\max_{i\in R(X^*)}\left(\frac{\partial f_i(X^*)}{\partial X},\,g\right)\geqslant 0, \tag{3.1}$$

or, equivalently (see (2.4)),

$$\inf_{\|g\|=1}\frac{\partial\varphi(X^*)}{\partial g}\geqslant 0. \tag{3.2}$$

Proof. Necessity. Suppose that $\varphi(X)$ has a minimum at X^* but inequality (3.2) fails to hold. Then there exists a vector $g_1\in E_n$, $\|g_1\|=1$, such that

$$\frac{\partial\varphi(X^*)}{\partial g_1} = -a < 0. \tag{3.3}$$

By (2.7),

$$\varphi(X^* + \alpha g_1) = \varphi(X^*) + \alpha\frac{\partial\varphi(X^*)}{\partial g_1} + o(g_1;\ \alpha), \tag{3.4}$$

where

$$\frac{o(g_1;\ \alpha)}{\alpha} \xrightarrow[\alpha\to+0]{} 0.$$

Fix some $\alpha_1 > 0$, so small that

$$|o(g_1;\ \alpha_1)| \leqslant \frac{a}{2}\alpha_1.$$

Then, in view of (3.4) and (3.3), we obtain

$$\varphi(X^* + \alpha_1 g_1) \leqslant \varphi(X^*) - \frac{a}{2}\alpha_1,$$

contradicting the assumption that X^* is a minimum point of $\varphi(X)$.

Sufficiency. Let $\varphi(X)$ be a convex function and suppose condition (3.2) satisfied at X^*. Suppose that $\varphi(X)$ does not assume its minimum on E_n at X^*. Then there is a point X_0 at which $\varphi(X_0) < \varphi(X^*)$. By the convexity of $\varphi(X)$, for any $\alpha \in [0,\ 1]$

$$\begin{aligned}\varphi(X^* + \alpha(X_0 - X^*)) = \varphi((1-\alpha)X^* + \alpha X_0) \leqslant \\ \leqslant (1-\alpha)\varphi(X^*) + \alpha\varphi(X_0).\end{aligned} \tag{3.5}$$

Set $g_1 = \frac{X_0 - X^*}{\|X_0 - X^*\|}$, $\|g_1\| = 1$, and evaluate $\frac{\partial\varphi(X^*)}{\partial g_1}$:

$$\frac{\partial\varphi(X^*)}{\partial g_1} = \lim_{\alpha\to+0}\frac{1}{\alpha}\left[\varphi\left(X^* + \frac{\alpha}{\|X_0 - X^*\|}(X_0 - X^*)\right) - \varphi(X^*)\right].$$

In view of (3.5), it follows that for $0 < \alpha < \|X_0 - X^*\|$

$$\begin{aligned}&\varphi\left(X^* + \frac{\alpha}{\|X_0 - X^*\|}(X_0 - X^*)\right) - \varphi(X^*) \leqslant \\ &\leqslant \left(1 - \frac{\alpha}{\|X_0 - X^*\|}\right)\varphi(X^*) + \frac{\alpha}{\|X_0 - X^*\|}\varphi(X_0) - \varphi(X^*) = \\ &= \frac{\alpha}{\|X_0 - X^*\|}(\varphi(X_0) - \varphi(X^*)).\end{aligned}$$

Hence

$$\frac{\partial\varphi(X^*)}{\partial g_1} \leqslant \frac{\varphi(X_0) - \varphi(X^*)}{\|X_0 - X^*\|} < 0,$$

contrary to (3.2). This completes the proof.

A point X^* at which inequalities (3.1) and (3.2) hold is called a *stationary point* of the maximum function $\varphi(X)$ on E_n.

2. We now present a geometric interpretation of the necessary condition (3.1) for a minimax. To this end, for any fixed $X \in E_n$ we consider the set

$$H(X) = \left\{ Z \in E_n \mid Z = \frac{\partial f_i(X)}{\partial X},\ i \in R(X) \right\}.$$

Let $L(X)$ denote the convex hull of $H(X)$ (see Appendix II, §1):

$$L(X) = \left\{ Z = \sum_{i \in R(X)} \alpha_i \frac{\partial f_i(X)}{\partial X} \,\middle|\, \alpha_i \geqslant 0,\ \sum_{i \in R(X)} \alpha_i = 1 \right\}.$$

Note that $L(X)$ is a bounded, closed, convex set.

Theorem 3.2. Inequality (3.1) is equivalent to the inclusion

$$\mathbf{0} \in L(X^*). \tag{3.6}$$

Proof. We first show that (3.1) implies (3.6). Suppose on the contrary that $\mathbf{0} \notin L(X^*)$. Then, by the separation theorem (see Appendix II, §1), there exist a vector g_0, $\|g_0\| = 1$, and a number $a > 0$ such that for all $Z \in L(X^*)$

$$(Z, g_0) \leqslant -a < 0.$$

Since $H(X^*) \subset L(X^*)$, this inequality implies

$$\max_{i \in R(X^*)} \left(\frac{\partial f_i(X^*)}{\partial X},\ g_0 \right) \leqslant -a < 0,$$

contradicting (3.1).

We now show that (3.6) implies (3.1). Suppose the contrary. Then there exist a vector g_0, $\|g_0\| = 1$, and a number $a > 0$ such that for all $i \in R(X^*)$

$$\left(\frac{\partial f_i(X^*)}{\partial X},\ g_0 \right) \leqslant -a < 0. \tag{3.7}$$

Take an arbitrary vector $Z \in L(X^*)$. By (3.7) and the definition of $L(X^*)$, we have

$$(Z, g_0) = \left(\sum_{i \in R(X^*)} \alpha_i \frac{\partial f_i(X^*)}{\partial X},\ g_0 \right) = \sum_{i \in R(X^*)} \alpha_i \left(\frac{\partial f_i(X^*)}{\partial X},\ g_0 \right) \leqslant -a \sum_{i \in R(X^*)} \alpha_i = -a < 0.$$

But then $Z = \mathbf{0}$ cannot be in $L(X^*)$, which contradicts (3.6), proving the theorem.

Corollary. A necessary condition for $\varphi(X)$ to have a minimum at X^* is that

$$0 \in L(X^*).$$

If $\varphi(X)$ is convex, this condition is also sufficient.

Remark. Condition (3.6) generalizes the classical necessary condition for a minimum in the case of a continuously differentiable function. The latter is obtained from (3.6) by setting $N=0$. Indeed, in that case

$$\varphi(X)=f_0(X).$$

The sets $H(X)$ and $L(X)$ then reduce to the singleton $\left\{\frac{\partial f_0(X)}{\partial X}\right\}$. Instead of the inclusion relation (3.6) we have the equality

$$\frac{\partial f_0(X^*)}{\partial X}=0. \tag{3.8}$$

Thus, a necessary condition for a continuously differentiable function $f_0(X)$ to have a minimum on E_n at X^* is that (3.8) be satisfied, and if $f_0(X)$ is convex this condition is also sufficient.

3. For fixed $X \in E_n$, we introduce the function

$$\chi(g)=\max_{i \in R(X)}\left(\frac{\partial f_i(X)}{\partial X},\ g\right).$$

In view of the general properties of the maximum function (see §2.1), $\chi(g)$ is a continuous and convex function on E_n. In addition, it is obvious that for any $\lambda \geqslant 0$

$$\chi(\lambda g)=\lambda\chi(g). \tag{3.9}$$

We are interested in the value of $\chi(g)$ when $g \in S^0$, where

$$S^0=\{g \in E_n \mid \|g\|=1\}.$$

Since S^0 is a bounded closed set, $\chi(g)$ assumes its minimum on S^0. Hence it follows, in particular, that the infimum on the left of inequality (3.1) is attained, and we may therefore rewrite (3.1) as

$$\min_{\|g\|=1}\ \max_{i \in R(X^*)}\left(\frac{\partial f_i(X^*)}{\partial X},\ g\right)\geqslant 0. \tag{3.1'}$$

Lemma 3.1.

$$\chi(g)=\max_{Z \in L(X)}(Z,\ g),$$

where $L(X)$ is the set defined in the preceding subsection.

Proof. Clearly,

$$\chi(g) = \max_{Z \in H(X)} (Z, g). \tag{3.10}$$

Since $H(X) \subset L(X)$, we have

$$\chi(g) \leqslant \max_{Z \in L(X)} (Z, g). \tag{3.11}$$

On the other hand, any point $Z' \in L(X)$ may be expressed as

$$Z' = \sum_{i \in R(X)} \alpha_i' Z_i; \; Z_i \in H(X); \; \alpha_i' \geqslant 0; \; \sum_{i \in R(X)} \alpha_i' = 1.$$

Therefore

$$(Z', g) = \sum_{i \in R(X)} \alpha_i' (Z_i, g) \leqslant \max_{Z \in H(X)} (Z, g) \sum_{i \in R(X)} \alpha_i' = \max_{Z \in H(X)} (Z, g).$$

Since this is true for any $Z' \in L(X)$, we have

$$\max_{Z \in L(X)} (Z, g) \leqslant \max_{Z \in H(X)} (Z, g).$$

In view of (3.10), we obtain

$$\max_{Z \in L(X)} (Z, g) \leqslant \chi(g). \tag{3.12}$$

The lemma now follows from (3.11) and (3.12)

4. Consider the function

$$\psi(X) = \min_{\|g\|=1} \max_{i \in R(X)} \left(\frac{\partial f_i(X)}{\partial X}, g \right).$$

In terms of $\psi(X)$, the necessary minimax condition (3.1) becomes

$$\psi(X^*) \geqslant 0. \tag{3.1''}$$

Let $\chi(X, g) = \chi(g)$. By Lemma 3.1,

$$\psi(X) = \min_{\|g\|=1} \chi(X, g) = \min_{\|g\|=1} \max_{Z \in L(X)} (Z, g). \tag{3.13}$$

Lemma 3.2. If $\psi(X^*) \geqslant 0$, i.e., X^* is a stationary point of the maximum function $\varphi(X)$, then

$$\psi(X^*) = r(X^*),$$

where $r(X^*)$ is the radius of the largest ball centered at the origin that can be inscribed in $L(X^*)$ (Figure 6).

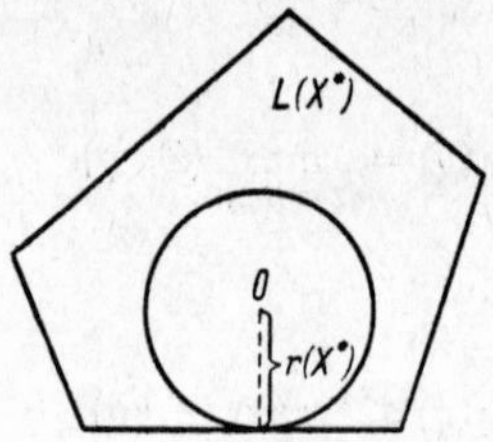

FIGURE 6.

Proof. Let $\bar{S}_\delta$ denote the ball of radius $\delta \geqslant 0$ with center at the origin:

$$\bar{S}_\delta = \{Z \in E_n \mid \|Z\| \leqslant \delta\}.$$

If $\bar{S}_\delta \subset L(X^*)$, then for any g, $\|g\| = 1$,

$$\delta = \max_{Z \in \bar{S}_\delta} (Z, g) \leqslant \max_{Z \in L(X^*)} (Z, g).$$

Hence, via (3.13), it follows that $\delta \leqslant \psi(X^*)$. In particular,

$$r(X^*) \leqslant \psi(X^*). \tag{3.14}$$

We now claim that

$$\bar{S}_{\psi(X^*)} \subset L(X^*). \tag{3.15}$$

Suppose on the contrary that there is a point $Z_0 \in \bar{S}_{\psi(X^*)}$

$$\|Z_0\| \leqslant \psi(X^*), \tag{3.16}$$

which is not in $L(X^*)$. By the separation theorem (Appendix II, §1), there exist a vector g_0, $\|g_0\| = 1$, and a number $a > 0$ such that, for any $Z \in L(X^*)$,

$$(Z - Z_0, g_0) \leqslant -a.$$

Hence in view of (3.16), we get

$$\max_{Z \in L(X^*)} (Z, g_0) \leqslant (Z_0, g_0) - a \leqslant \psi(X^*) - a,$$

contradicting (3.13).

This proves (3.15), and hence also

$$\psi(X^*) \leqslant r(X^*). \tag{3.17}$$

The lemma now follows from (3.14) and (3.17).

5. We now consider the points $X \in E_n$ for which $\psi(X) < 0$. These points are not stationary, and by Theorem 3.2 $0 \notin L(X)$.

Lemma 3.3. If $\psi(X) < 0$, then

$$\psi(X) = \max_{Z \in L(X)} \left(Z, -\frac{Z^*}{\|Z^*\|}\right) = -\|Z^*\|,$$

where Z^* is the point of $L(X)$ nearest the origin.

Proof. For any $Z \in L(X)$, we have the inequality (see Appendix II, Lemma 1.3)

$$(Z, Z^*) \geqslant (Z^*, Z^*).$$

Set $\bar{g} = -\frac{Z^*}{\|Z^*\|}$, $\|\bar{g}\| = 1$.

For any $Z \in L(X)$, we have

$$(Z, \bar{g}) = -\frac{1}{\|Z^*\|}(Z, Z^*) \leqslant -\frac{1}{\|Z^*\|}(Z^*, Z^*) = -\|Z^*\|.$$

Hence it follows that

$$\max_{Z \in L(X)} (Z, \bar{g}) = -\|Z^*\|. \tag{3.18}$$

Next, for any g, $\|g\| = 1$,

$$-\|Z^*\| \leqslant (Z^*, g) \leqslant \max_{Z \in L(X)} (Z, g).$$

Thus, for any g, $\|g\| = 1$, it follows from (3.18) that

$$\max_{Z \in L(X)} (Z, \bar{g}) \leqslant \max_{Z \in L(X)} (Z, g),$$

that is,

$$\max_{Z \in L(X)} (Z, \bar{g}) = \min_{\|g\|=1} \max_{Z \in L(X)} (Z, g) = \psi(X). \tag{3.19}$$

Combining (3.18) and (3.19) and using the definition of $\bar{g}$, we obtain the desired assertion.

Equality (3.19) may be given the following meaning (see (3.13)): for fixed $X \in E_n$, $\psi(X) < 0$, the function $\chi(X, g)$ assumes a minimum on the unit sphere S^0 at $g = \bar{g}$. This statement will now be made rigorous.

Lemma 3.4. If $\psi(X) < 0$, the function $\chi(X, g)$ assumes a minimum with respect to $g \in S^0$ at a unique point.

Proof. Since X is fixed, we shall write $\chi(X, g) = \chi(g)$. Suppose there are two vectors $g_1 \neq g_2$, $\|g_1\| = \|g_2\| = 1$, such that

$$\chi(g_1) = \chi(g_2) = \min_{\|g\|=1} \chi(g) = \psi(X).$$

Note that then

$$(g_1, g_2) < 1. \tag{3.20}$$

Since $\chi(g)$ is convex throughout E_n (see subsection 3), it follows that for any $\alpha \in (0, 1)$

$$\chi(\alpha g_1 + (1-\alpha) g_2) \leqslant \alpha\chi(g_1) + (1-\alpha)\chi(g_2) = \psi(X). \tag{3.21}$$

Fix $\alpha_0 \in (0, 1)$ so that

$$\beta \overset{df}{=} \| \alpha_0 g_1 + (1-\alpha_0) g_2 \| > 0.$$

By (3.20), we have

$$\beta^2 = \|\alpha_0 g_1 + (1-\alpha_0) g_2\|^2 = \alpha_0^2 + 2\alpha_0(1-\alpha_0)(g_1, g_2) + \\ + (1-\alpha_0)^2 < \alpha_0^2 + 2\alpha_0(1-\alpha_0) + (1-\alpha_0)^2 = 1.$$

Thus, $0 < \beta < 1$. We now set

$$g_0 = \frac{1}{\beta}(\alpha_0 g_1 + (1-\alpha_0) g_2), \quad \| g_0 \| = 1.$$

Since $\frac{1}{\beta} > 1$ and $\psi(X) < 0$, we obtain by (3.9) and (3.21)

$$\chi(g_0) = \frac{1}{\beta}\chi(\alpha_0 g_1 + (1-\alpha_0) g_2) \leqslant \frac{1}{\beta}\psi(X) < \psi(X),$$

which is impossible. This completes the proof.

6. Definition. A vector $g(\bar{X})$, $\| g(\bar{X}) \| = 1$ will be called a direction of steepest descent of the maximum function $\varphi(X)$ at $\bar{X}$ if

$$\frac{\partial \varphi(\bar{X})}{\partial g(\bar{X})} = \min_{\|g\|=1} \frac{\partial \varphi(\bar{X})}{\partial g}.$$

A vector $g_1(\bar{X})$, $\| g_1(\bar{X}) \| = 1$ will be called a direction of steepest ascent of the maximum function $\varphi(X)$ at $\bar{X}$ if

$$\frac{\partial \varphi(\bar{X})}{\partial g_1(\bar{X})} = \max_{\|g\|=1} \frac{\partial \varphi(\bar{X})}{\partial g}.$$

Theorem 3.3. If $\psi(\bar{X}) < 0$, the maximum function $\varphi(X)$ has a unique direction of steepest descent $g(\bar{X})$, $\| g(\bar{X}) \| = 1$ at $\bar{X}$. Moreover,

$$g(\bar{X}) = -\frac{Z^*}{\| Z^* \|}, \quad \frac{\partial \varphi(\bar{X})}{\partial g(\bar{X})} = -\| Z^* \|,$$

where Z^* is the point of $L(\bar{X})$ nearest the origin.

In essence, we have already proved this theorem. It suffices to observe that

$$\frac{\partial \varphi(\bar{X})}{\partial g} = \chi(\bar{X}, g),$$

and to use the results of Lemmas 3.3 and 3.4.

Remark. Let $N=0$, i. e.,

$$\varphi(X) = f_0(X).$$

Then, for all $X \in E_n$,

$$L(X) = \left\{ \frac{\partial f_0(X)}{\partial X} \right\}.$$

If $\psi(\bar{X}) < 0$, i. e., $\frac{\partial f(\bar{X})}{\partial X} \neq 0$, it follows from Theorem 3.3 that the direction of steepest descent of $f_0(X)$ at $\bar{X}$ is the vector

$$g(\bar{X}) = -\frac{\partial f_0(\bar{X})}{\partial X} \Big/ \left\| \frac{\partial f_0(\bar{X})}{\partial X} \right\|.$$

The direction of steepest ascent of $f_0(X)$ at $\bar{X}$ will clearly be

$$g_1(\bar{X}) = \frac{\partial f_0(\bar{X})}{\partial X} \Big/ \left\| \frac{\partial f_0(\bar{X})}{\partial X} \right\|.$$

Thus, if $f_0(X)$ is continuously differentiable, we have

$$g_1(\bar{X}) = -g(\bar{X}). \tag{3.22}$$

at any point $\bar{X}$ for which $\frac{\partial f_0(\bar{X})}{\partial X} \neq 0$.

Note that in the general case ($N > 0$) equality (3.22) need not hold. Moreover, in the general case the direction of steepest ascent need not be unique.

In the situation illustrated in Figure 7, $g(\bar{X})$ is the direction of steepest descent, $g_1^{(1)}(\bar{X})$ and $g_1^{(2)}(\bar{X})$ are directions of steepest ascent.

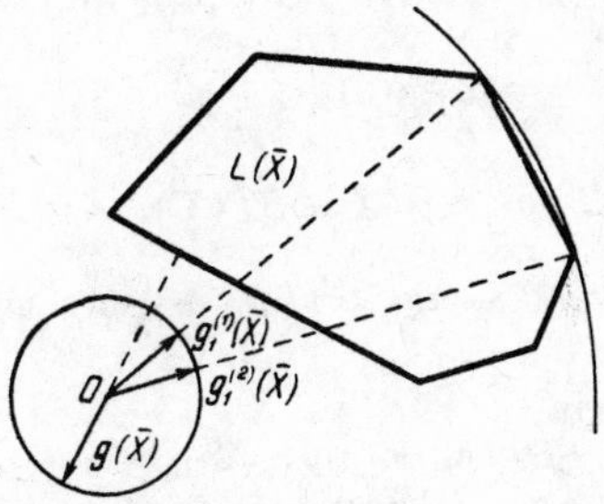

FIGURE 7.

§4. SUFFICIENT CONDITIONS FOR A LOCAL MINIMAX. SOME ESTIMATES

1. In the last section we showed that if the maximum function $\varphi(X)$ is convex, any stationary point X^* is at the same time a minimum point for $\varphi(X)$ on E_n. We now wish to find conditions for a stationary point X^* to be a local minimum point of $\varphi(X)$.

As before, we assume that the functions $f_i(X), i \in [0:N]$ are continuously differentiable on E_n. Let

$$\varphi(X) = \max_{i \in [0:N]} f_i(X),$$

$$\psi(X) = \min_{\|g\|=1} \max_{i \in R(X)} \left(\frac{\partial f_i(X)}{\partial X}, g \right).$$

2. Definition. A point $X^* \in E_n$ is called a strict local minimum point for $\varphi(X)$ if there is a neighborhood $S_\delta(X^*)$, $\delta > 0$, of X^*,

$$S_\delta(X^*) = \{X \in E_n \mid \| X - X^* \| < \delta\},$$

such that for all $X \in S_\delta(X^*)$, $X \neq X^*$,

$$\varphi(X) > \varphi(X^*).$$

Theorem 4.1. If $\psi(X^*) = r > 0$, then X^* is a strict local minimum point for $\varphi(X)$.

Proof. We know (see (2.7)) that

$$\varphi(X^* + \alpha g) = \varphi(X^*) + \alpha \frac{\partial \varphi(X^*)}{\partial g} + o(g; \alpha), \tag{4.1}$$

where $\frac{o(g;\alpha)}{\alpha} \xrightarrow[\alpha \to +0]{} 0$, uniformly in g, $\| g \| = 1$.

By the definition of $\psi(X)$, we have for all g, $\| g \| = 1$,

$$\frac{\partial \varphi(X^*)}{\partial g} \geqslant r.$$

Select $\alpha_1 > 0$ so small that, for all $\alpha \in (0, \alpha_1)$ and all g, $\| g \| = 1$,

$$| o(g; \alpha) | \leqslant \frac{r}{2} \alpha.$$

By (4.1), this shows that for any $X \in S_{\alpha_1}(X^*)$, $X \neq X^*$,

$$\varphi(X) = \varphi(X^* + \alpha g) \geqslant \varphi(X^*) + \frac{r}{2} \alpha > \varphi(X^*).$$

This proves the theorem.

Remark. If $\psi(X^*) = r > 0$, there exists $\alpha_0 > 0$ such that, for all $X \in S_{\alpha_0}(X^*)$, $X \neq X^*$, the index set $R(X)$ is a proper subset of $R(X^*)$:

$$R(X) \subset R(X^*), \qquad R(X) \neq R(X^*).$$

Indeed, by Lemma 2.1 there exists $\alpha_1 > 0$ such that for all $X \in S_{\alpha_1}(X^*)$ we have $R(X) \subset R(X^*)$.

We note in addition that for any g, $\|g\| = 1$, there is an index $i_1(g) \in R(X^*)$ such that

$$\left(\frac{\partial f_{i_1(g)}(X^*)}{\partial X}, g\right) \geqslant r. \tag{4.2}$$

Set $i_2(g) = i_1(-g) \in R(X^*)$. Then

$$\left(\frac{\partial f_{i_2(g)}(X^*)}{\partial X}, g\right) \leqslant -r. \tag{4.3}$$

We now write

$$f_i(X^* + \alpha g) = f_i(X^*) + \alpha \left(\frac{\partial f_i(X^*)}{\partial X}, g\right) + o_i(g; \alpha). \tag{4.4}$$

We choose the required number α_0, $0 < \alpha_0 \leqslant \alpha_1$, so that for any $\alpha \in (0, \alpha_0)$

$$|o_i(g; \alpha)| \leqslant \frac{r}{2}\alpha \tag{4.5}$$

uniformly in $i \in [0:N]$ and g, $\|g\| = 1$.

In this case, for any $X \in S_{\alpha_0}(X^*)$, $X = X^* + \alpha g$, $X \neq X^*$, it follows from (4.4), (4.2), (4.3) and (4.5) that

$$f_{i_1(g)}(X) \geqslant f_{i_1(g)}(X^*) + \frac{1}{2} r\alpha = \varphi(X^*) + \frac{1}{2} r\alpha;$$
$$f_{i_2(g)}(X) \leqslant f_{i_2(g)}(X^*) - \frac{1}{2} r\alpha = \varphi(X^*) - \frac{1}{2} r\alpha.$$

Thus, $f_{i_1(g)}(X) > f_{i_2(g)}(X)$. Thus $i_2(g) \notin R(X)$, but $i_2(g) \in R(X^*)$.

3*. *Theorem 4.2. Let X^* be a stationary point of $\varphi(X)$ on E_n, with $\psi(X^*) = 0$. Let the functions $f_i(X)$, $i \in [0:N]$ be twice continuously differentiable in some neighborhood $S_\delta(X^*)$, $\delta > 0$ of X^*.*
Set

$$\frac{\partial^2 f_i(X)}{\partial X^2} = \begin{Vmatrix} \frac{\partial^2 f_i(X)}{\partial x_1 \partial x_1} & \frac{\partial^2 f_i(X)}{\partial x_1 \partial x_2} & \cdots & \frac{\partial^2 f_i(X)}{\partial x_1 \partial x_n} \\ \frac{\partial^2 f_i(X)}{\partial x_2 \partial x_1} & \frac{\partial^2 f_i(X)}{\partial x_2 \partial x_2} & \cdots & \frac{\partial^2 f_i(X)}{\partial x_2 \partial x_n} \\ \cdots & \cdots & \cdots & \cdots \\ \frac{\partial^2 f_i(X)}{\partial x_n \partial x_1} & \frac{\partial^2 f_i(X)}{\partial x_n \partial x_2} & \cdots & \frac{\partial^2 f_i(X)}{\partial x_n \partial x_n} \end{Vmatrix}.$$

Suppose that for some $\gamma > 0$

$$\min_{g \in G_\gamma} \max_{i \in R_2(X^*, g)} \left(\frac{\partial^2 f_i(X^*)}{\partial X^2} g, g\right) \stackrel{df}{=} a(X^*) > 0, \tag{4.6}$$

where

$$G_\gamma = \left\{ g \mid \| g \| = 1,\ 0 \leqslant \frac{\partial \varphi(X^*)}{\partial g} \leqslant \gamma \right\},$$

$$R_2(X^*, g) = \left\{ i \in R(X^*) \mid \left(\frac{\partial f_i(X^*)}{\partial X}, g \right) = \frac{\partial \varphi(X^*)}{\partial g} \right\},$$

$$\left(\frac{\partial^2 f_i(X^*)}{\partial X^2} g, g \right) = \sum_{j_1, j_2 = 1}^{n} \frac{\partial^2 f_i(X^*)}{\partial x_{j_1} \partial x_{j_2}} g_{j_1} g_{j_2}.$$

Then X^* is a strict local minimum point for $\varphi(X)$.

Proof. We note first that for all $i \in [0:N]$, any g, $\|g\| = 1$, and $\alpha \in (0, \delta)$, we have (see Appendix III, §3)

$$f_i(X^* + \alpha g) = f_i(X^*) + \alpha \left(\frac{\partial f_i(X^*)}{\partial X}, g \right) + \\ + \frac{1}{2} \alpha^2 \left(\frac{\partial^2 f_i(X^* + \theta_i g)}{\partial X^2} g, g \right), \tag{4.7}$$

where $\theta_i \in (0, \alpha)$.

Next, it follows from (4.6) that for any $g \in G_\gamma$ (the set G_γ is not empty, for by assumption $\psi(X^*) = 0$) there is an index $i(g) \in R_2(X^*, g)$ such that

$$\left(\frac{\partial^2 f_{i(g)}(X^*)}{\partial X^2} g, g \right) \geqslant a(X^*).$$

Since the function $\left(\frac{\partial^2 f_i(X)}{\partial X^2} g, g \right)$ is continuous with respect to X at X^*, uniformly in g, $\|g\| = 1$, and $i \in [0:N]$, it follows that there exists α_0, $0 < \alpha_0 < \delta$, such that for all $\alpha \in (0, \alpha_0)$ and all $g \in G_\gamma$

$$\left(\frac{\partial^2 f_{i(g)}(X^* + \alpha g)}{\partial X^2} g, g \right) \geqslant \frac{1}{2} a(X^*).$$

In addition, the definition of $R_2(X^*, g)$ and the equality $\psi(X^*) = 0$ entail

$$\left(\frac{\partial f_{i(g)}(X^*)}{\partial X}, g \right) \geqslant 0.$$

It now follows from (4.7) that, for any $X = X^* + \alpha g$, $\alpha \in (0, \alpha_0)$, $g \in G_\gamma$,

$$f_{i(g)}(X) \geqslant f_{i(g)}(X^*) + \frac{1}{4} \alpha^2 a(X^*) = \varphi(X^*) + \frac{1}{4} \alpha^2 a(X^*).$$

A fortiori, $\varphi(X) > \varphi(X^*)$.

We introduce the notation

$$\tilde{G}_\gamma = \left\{ g \mid \| g \| = 1,\ \frac{\partial \varphi(X^*)}{\partial g} \geqslant \gamma \right\}.$$

Since $\psi(X^*) = 0$, i. e.,

$$\min_{\|g\|=1} \frac{\partial \varphi(X^*)}{\partial g} = 0,$$

we see that $G_\gamma \cup \tilde{G}_\gamma = S^0 \overset{df}{=} \{g \mid \|g\| = 1\}$.

It is readily seen from (4.7) that there exists α_1, $0 < \alpha_1 < \alpha_0$, so small that for all $X = X^* + \alpha g$, $\alpha \in (0, \alpha_1)$, $g \in \tilde{G}_\gamma$, and any $i \in R_2(X^*, g)$,

$$f_i(X) \geqslant \varphi(X^*) + \frac{1}{2}\alpha\gamma.$$

A fortiori, $\varphi(X) > \varphi(X^*)$.

We have thus shown that for any $X \in S_{\alpha_1}(X^*)$, $X \neq X^*$, we have $\varphi(X) > \varphi(X^*)$, completing the proof.

Remark 1. Inequality (4.6) holds, for example, if the matrices $\frac{\partial^2 f_i(X^*)}{\partial X^2}$ are positive definite for all $i \in R(X^*)$, in other words, for all $i \in R(X^*)$ and $V \in E_n$,

$$\left(\frac{\partial^2 f_i(X^*)}{\partial X^2} V, V\right) \geqslant m_i \|V\|^2,$$

where $m_i > 0$ are independent of V.

Remark 2. Theorem 4.2 clearly remains in force if we replace (4.6) by the stronger inequality

$$\min_{\|g\|=1} \max_{i \in R_2(X^*, g)} \left(\frac{\partial^2 f_i(X^*)}{\partial X^2} g, g\right) > 0.$$

Remark 3. The following example will show that the assumption $\gamma > 0$ in Theorem 4.2 is essential, i. e., the inequality

$$\min_{g \in G_0} \max_{i \in R_2(X^*, g)} \left(\frac{\partial^2 f_i(X^*)}{\partial X^2} g, g\right) > 0, \tag{4.8}$$

where $G_0 = \left\{g \mid \|g\| = 1, \frac{\partial \varphi(X^*)}{\partial g} = 0\right\}$, may not suffice to assure that X^* is a strict local minimum point for $\varphi(X)$.

Example. Let

$$X = (x_1, x_2),$$
$$f_0(X) = -x_1 - x_2^2, \; f_1(X) = 3x_1 + x_1^2 + x_2^2.$$

Consider the point $X^* = (0, 0)$. Clearly,

$$\varphi(X^*) = \max_{i \in [0:1]} f_i(X^*) = 0,$$
$$R(X^*) = \{0, 1\}, \; H(X^*) = \{(-1, 0), (3, 0)\},$$
$$L(X^*) = \{Z_\alpha \mid Z_\alpha = (3 - 4\alpha, 0), \; 0 \leqslant \alpha \leqslant 1\}.$$

Since $0 = Z_{3/4} \in L(X^*)$, it follows that X^* is a stationary point of $\varphi(X)$, and moreover

$$\psi(X^*) = \min_{\|g\|=1} \max_{Z \in L(X^*)} (Z, g) = 0.$$

It is readily seen that the set

$$G_0 \stackrel{df}{=} \left\{ g \mid \| g \| = 1, \frac{\partial \varphi(X^*)}{\partial g} = 0 \right\}$$

contains exactly two vectors: $g_1 = (0, 1)$ and $g_2 = (0, -1)$, and

$$R_2(X^*, g_1) = R_2(X^*, g_2) = \{0, 1\}.$$

Further,

$$\frac{\partial^2 f_0(X^*)}{\partial X^2} = \begin{pmatrix} 0 & 0 \\ 0 & -2 \end{pmatrix}, \quad \frac{\partial^2 f_1(X^*)}{\partial X^2} = \begin{pmatrix} 2 & 0 \\ 0 & 2 \end{pmatrix}.$$

Since

$$\left(\frac{\partial^2 f_0(X^*)}{\partial X^2} g_1, g_1 \right) = \left(\frac{\partial^2 f_0(X^*)}{\partial X^2} g_2, g_2 \right) = -2,$$
$$\left(\frac{\partial^2 f_1(X^*)}{\partial X^2} g_1, g_1 \right) = \left(\frac{\partial^2 f_1(X^2)}{\partial X^2} g_2, g_2 \right) = 2,$$

we obtain

$$\min_{g \in G_0} \max_{i \in R_2(X^*, g)} \left(\frac{\partial^2 f_i(X^*)}{\partial X^2} g, g \right) = 2 > 0.$$

Thus inequality (4.8) is fulfilled. We shall now see that despite this the point X^* is not a local minimum point for $\varphi(X)$.

Let $V_\varepsilon = (-\varepsilon, 1)$, $\varepsilon > 0$. For $\alpha \geqslant 0$,

$$f_0(X^* + \alpha V_\varepsilon) = \alpha\varepsilon - \alpha^2,$$
$$f_1(X^* + \alpha V_\varepsilon) = -3\alpha\varepsilon + \alpha^2\varepsilon^2 + \alpha^2.$$

Let us examine the behavior of the maximum function

$$\varphi(X^* + \alpha V_\varepsilon) = \max \{f_0(X^* + \alpha V_\varepsilon), f_1(X^* + \alpha V_\varepsilon)\}.$$

It is readily verified that (Figure 8)

$$\varphi(X^* + \alpha V_\varepsilon) = \begin{cases} f_0(X^* + \alpha V_\varepsilon) & \text{if} \quad \alpha \in \left[0, \frac{4\varepsilon}{2 + \varepsilon^2}\right], \\ f_1(X^* + \alpha V_\varepsilon) & \text{if} \quad \alpha \geqslant \frac{4\varepsilon}{2 + \varepsilon^2}. \end{cases}$$

If we denote $\alpha_\varepsilon = \frac{4\varepsilon}{2+\varepsilon^2}$, then for any $0<\varepsilon<1$

$$\varphi(X^* + \alpha_\varepsilon V_\varepsilon) = \frac{4\varepsilon^2(-2+\varepsilon^2)}{(2+\varepsilon^2)^2} < 0 = \varphi(X^*).$$

It remains only to observe that

$$\alpha_\varepsilon V_\varepsilon \xrightarrow[\varepsilon \to +0]{} \mathbf{0}.$$

4. We shall now establish certain estimates for

$$\mu = \min_{X \in E_n} \max_{i \in [0:N]} f_i(X) = \min_{X \in E_n} \varphi(X).$$

As always, we assume that the functions $f_i(X)$, $i \in [0:N]$ are continuously differentiable throughout E_n.

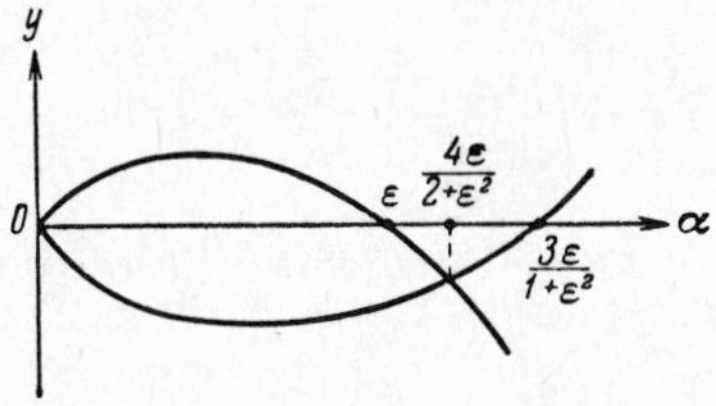

FIGURE 8.

Theorem 4.3. Assume that for some point $X_0 \in E_n$ there is a set of indices $Q \subset [0:N]$ such that

$$\min_{\|g\|=1} \max_{i \in Q} \left(\frac{\partial f_i(X_0)}{\partial X}, g \right) \geqslant 0. \tag{4.9}$$

If the functions $f_i(X)$ are convex for all $i \in Q$, then

$$\min_{i \in Q} f_i(X_0) \leqslant \mu \leqslant \varphi(X_0).$$

Proof. The inequality $\mu \leqslant \varphi(X_0)$ is obvious by the definition of μ. We prove the second inequality. To this end, we set

$$\tilde{f}_i(X) = f_i(X) - f_i(X_0), \quad i \in Q,$$
$$\tilde{\varphi}(X) = \max_{i \in Q} \tilde{f}_i(X).$$

Clearly,

$$\frac{\partial \tilde{f}_i(X)}{\partial X} = \frac{\partial f_i(X)}{\partial X}, \quad \tilde{\varphi}(X_0) = 0$$

and

$$\tilde{R}(X_0) \stackrel{df}{=} \{i \in Q \mid \tilde{f}_i(X_0) = \tilde{\varphi}(X_0)\} = Q. \tag{4.10}$$

Since $\tilde{f}_i(X)$ is convex for each $i \in Q$, we conclude by virtue of (4.9), (4.10) and Theorem 3.1 that X_0 is a minimum point for $\tilde{\varphi}(X)$ on E_n.

We now have that for all $X \in E_n$

$$0 = \tilde{\varphi}(X_0) \leqslant \tilde{\varphi}(X) = \max_{i \in Q} (f_i(X) - f_i(X_0)) \leqslant \leqslant \max_{i \in [0:N]} f_i(X) - \min_{i \in Q} f_i(X_0), \tag{4.11}$$

whence we obtain the required inequality

$$\min_{i \in Q} f_i(X_0) \leqslant \mu.$$

5. *Theorem 4.4. Assume that for some point $X_0 \in E_n$ there exists a set $Q \subset [0:N]$ such that*

$$\min_{\|g\|=1} \max_{i \in Q} \left(\frac{\partial f_i(X_0)}{\partial X}, g \right) \stackrel{df}{=} -\varepsilon < 0. \tag{4.12}$$

Let the functions $f_i(X)$ be twice continuously differentiable on E_n for all $i \in Q$, and suppose that for all $i \in Q$, $X \in E_n$ and $V \in E_n$

$$\left(\frac{\partial^2 f_i(X)}{\partial X^2} V, V \right) \geqslant m \| V \|^2, \tag{4.13}$$

where $m > 0$ is independent of i, X and V. Then

$$\min_{i \in Q} f_i(X_0) - \frac{\varepsilon^2}{2m} \leqslant \mu \leqslant \varphi(X_0).$$

Proof. For any $i \in [0:N]$, $\alpha \in (-\infty, \infty)$ and g, $\|g\| = 1$,

$$f_i(X_0 + \alpha g) = f_i(X_0) + \alpha \left(\frac{\partial f_i(X_0)}{\partial X}, g \right) + + \frac{1}{2} \alpha^2 \left(\frac{\partial^2 f_i(X_0 + \theta_i g)}{\partial X^2} g, g \right), \tag{4.14}$$

where $\theta_i \in (0, \alpha)$.

By (4.12), for any g, $\|g\| = 1$, there is an index $i(g) \in Q$ such that

$$\left(\frac{\partial f_{i(g)}(X_0)}{\partial X}, g \right) \geqslant -\varepsilon. \tag{4.15}$$

Using (4.14), (4.15) and (4.13), we get

$$f_{i(g)}(X_0 + \alpha g) \geqslant f_{i(g)}(X_0) - \alpha\varepsilon + \frac{1}{2} \alpha^2 m \geqslant \geqslant \min_{i \in Q} f_i(X_0) - \alpha\varepsilon + \frac{1}{2} \alpha^2 m.$$

Now for any $\alpha \in (-\infty, \infty)$ we have

$$\frac{1}{2}\alpha^2 m - \alpha\varepsilon \geqslant \min_{-\infty<\beta<\infty}\left[\frac{1}{2}\beta^2 m - \beta\varepsilon\right] = -\frac{\varepsilon^2}{2m},$$

and so

$$\varphi(X_0 + \alpha g) \geqslant f_{i(g)}(X_0 + \alpha g) \geqslant \min_{i \in Q} f_i(X_0) - \frac{\varepsilon^2}{2m}.$$

Finally, we obtain

$$\mu \geqslant \min_{i \in Q} f_i(X_0) - \frac{\varepsilon^2}{2m}.$$

The inequality $\varphi(X_0) \geqslant \mu$ is obvious by the definition of μ.

Corollary. Under the assumptions of the theorem, if $Q \subset R(X_0)$, then

$$0 \leqslant \varphi(X_0) - \mu \leqslant \frac{\varepsilon^2}{2m}.$$

§5. METHOD OF COORDINATEWISE DESCENT. METHOD OF STEEPEST DESCENT. COUNTEREXAMPLES

1. In this section we examine some methods for minimizing the maximum function $\varphi(X)$. We begin with the simplest method — coordinatewise descent. Set

$$e_1 = (1, 0, \ldots, 0), \quad e_2 = (0, 1, 0, \ldots, 0), \ldots$$
$$\ldots, e_n = (0, \ldots, 0, 1).$$

Select an initial approximation $X_0 \in E_n$ arbitrarily. Assume that the set

$$M(X_0) = \{X \mid \varphi(X) \leqslant \varphi(X_0)\}$$

is bounded. Since $\varphi(X)$ is continuous, $M(X_0)$ is also closed.

Suppose that the k-th approximation $X_k \in M(X_0)$ has already been determined. To construct X_{k+1}, we consider the straight line

$$X = X_{k1}(\alpha) \stackrel{df}{=} X_k + \alpha e_1, \qquad -\infty < \alpha < \infty,$$

and determine $\alpha_{k1} \in (-\infty, \infty)$ such that

$$\varphi(X_{k1}(\alpha_{k1})) = \min_{\alpha \in (-\infty, \infty)} \varphi(X_{k1}(\alpha)).$$

Since $M(X_0)$ is a bounded closed set, the minimum is achieved and moreover $X_{k1}(\alpha_{k1}) \in M(X_0)$.

Now consider the straight line

$$X = X_{k2}(\alpha) \stackrel{df}{=} X_{k1}(\alpha_{k1}) + \alpha e_2, \qquad -\infty < \alpha < \infty,$$

and determine $\alpha_{k2} \in (-\infty, \infty)$ such that

$$\varphi(X_{k2}(\alpha_{k2})) = \min_{\alpha \in (-\infty, \infty)} \varphi(X_{k2}(\alpha)).$$

It is obvious that $X_{k2}(\alpha_{k2}) \in M(X_0)$. Continuing in this way, we obtain a straight line

$$X = X_{kn}(\alpha) \stackrel{df}{=} X_{k,n-1}(\alpha_{k,n-1}) + \alpha e_n, \qquad -\infty < \alpha < \infty.$$

Determine $\alpha_{kn} \in (-\infty, \infty)$ for which

$$\varphi(X_{kn}(\alpha_{kn})) = \min_{\alpha \in (-\infty, \infty)} \varphi(X_{kn}(\alpha)).$$

Set $X_{k+1} = X_{kn}(\alpha_{kn})$. It is clear that $X_{k+1} \in M(X_0)$ and $\varphi(X_{k+1}) \leqslant \varphi(X_k)$.

Continuing the procedure, we obtain a sequence of points $X_k \in M(X_0)$, $k = 0, 1, 2, \ldots$, with

$$\varphi(X_0) \geqslant \varphi(X_1) \geqslant \ldots \geqslant \varphi(X_k) \geqslant \ldots$$

One might expect the limit point of the sequence $\{X_k\}$ to be a local minimum point for $\varphi(X)$. In general, however, this need not be the case. In fact, the limit point of the sequence $\{X_k\}$ may not even be a stationary point of $\varphi(X)$.

Example 1. Let

$$X = (x_1, x_2); \quad e_1 = (1, 0), \quad e_2 = (0, 1).$$
$$f_1(X) = 2x_1 + x_2; \quad f_2(X) = -x_1 - 2x_2;$$
$$f_3(X) = -x_1 + x_2 - 3.$$
$$\varphi(X) = \max_{i \in [1:3]} f_i(X).$$

As initial approximation we take $X_0 = (0, 0)$. Note that

$$\varphi(X_0 + \alpha e_1) = \max\{2\alpha, -\alpha, -\alpha - 3\} = \begin{cases} 2\alpha, & \text{if} \quad \alpha \geqslant 0, \\ -\alpha, & \text{if} \quad \alpha \leqslant 0. \end{cases}$$

Hence $\alpha_{01} = 0$, $X_{01}(\alpha_{01}) = X_0$.

Now,

$$\varphi(X_{01}(\alpha_{01}) + \alpha e_2) = \max\{\alpha, -2\alpha, \alpha - 3\} = \begin{cases} \alpha, & \text{if} \quad \alpha \geqslant 0, \\ -2\alpha, & \text{if} \quad \alpha \leqslant 0. \end{cases}$$

Hence $\alpha_{02} = 0$, $X_1 \stackrel{df}{=} X_{02}(\alpha_{02}) = X_0$.

Continuing, we obtain $X_k = X_0$, $k = 1, 2, \ldots$, so that the only limit point of the sequence $\{X_k\}$ is X_0.

Let us check whether X_0 is a stationary point. We have

$$R(X_0)=\{1,\ 2\},\qquad H(X_0)=\{(2,\ 1),\ (-1,\ -2)\},$$
$$L(X_0)=\{Z_\alpha \mid Z_\alpha=(3\alpha-1,\ 3\alpha-2),\ 0\leqslant\alpha\leqslant 1\}.$$

Clearly, $\mathbf{0}\notin L(X_0)$. Thus X_0 is not a stationary point of $\varphi(X)$.

Remark. It is easily shown that the point of $L(X_0)$ nearest the origin is $Z_{1/2}=\left(\frac{1}{2},\ -\frac{1}{2}\right)$ (Figure 9). Using Theorem 3.3, we conclude that $\varphi(X)$, which is nondecreasing at X_0 as a function of either coordinate separately, has the following direction of steepest descent:

$$g(X_0)=\left(-\frac{\sqrt{2}}{2},\ \frac{\sqrt{2}}{2}\right).$$

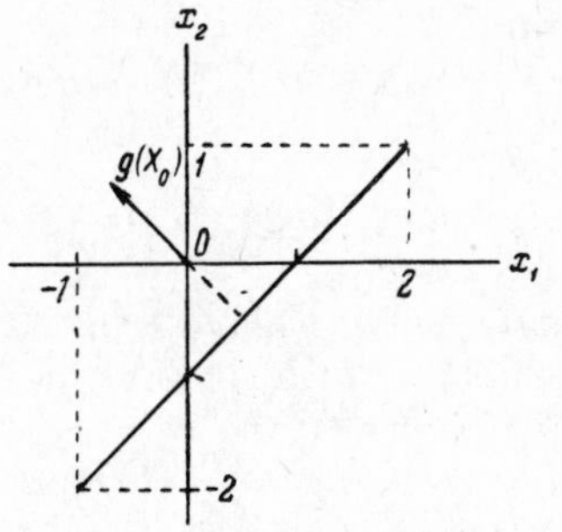

FIGURE 9.

2. We next describe the method of steepest descent. This method is based on the fact that if $\bar{X}$ is not a stationary point for $\varphi(X)$, i. e., $\psi(\bar{X})<0$, then the function $\varphi(X)$ has a unique direction of steepest descent $g(\bar{X})$ at $\bar{X}$ (see Theorem 3.3).

Let $X_0\in E_n$ be the initial approximation. Suppose that the set $M(X_0)=\{X\mid\varphi(X)\leqslant\varphi(X_0)\}$ is bounded. By the continuity of $\varphi(X)$, $M(X_0)$ is also closed.

Suppose we have already found the k-th approximation $X_k\in M(X_0)$. If $\psi(X_k)\geqslant 0$, then X_k is a stationary point and the procedure terminates. Otherwise, we construct the vector $g_k=g(X_k)$. Consider the ray

$$X=X_k(\alpha)\overset{df}{=}X_k+\alpha g_k,\qquad \alpha\geqslant 0,$$

and find $\alpha_k\in[0,\ \infty)$ for which

$$\varphi(X_k(\alpha_k))=\min_{\alpha\in[0,\ \infty)}\varphi(X_k(\alpha))$$

(the minimum is achieved, because $M(X_0)$ is bounded and closed).

Set $X_{k+1}=X_k(\alpha_k)$. It is clear that $X_{k+1}\in M(X_0)$ and $\varphi(X_{k+1})<\varphi(X_k)$. Proceeding in this way, we obtain a sequence $\{X_k\}$ such that $X_k\in M(X_0)$, $k=0, 1,\ 2,\ \ldots,$

$$\varphi(X_0)>\varphi(X_1)>\ \ldots\ >\varphi(X_k)>\ \ldots,$$

and if the sequence $\{X_k\}$ is finite its last element is by construction a stationary point of $\varphi(X)$.

Let us use steepest descent to minimize the function $\varphi(X)$ considered in Example 1. The initial approximation will again be $X_0=(0,0)$. As we have already seen, the direction of steepest descent is

$$g_0 = g(X_0) = \left(-\frac{\sqrt{2}}{2},\ \frac{\sqrt{2}}{2}\right).$$

We have

$$\varphi(X_0+\alpha g_0) = \max\left\{-\frac{\sqrt{2}}{2}\alpha,\ -\frac{\sqrt{2}}{2}\alpha,\ \sqrt{2}\alpha-3\right\} = \\ = \frac{\sqrt{2}}{2}\max\left\{-\alpha,\ 2\alpha-3\sqrt{2}\right\}.$$

Clearly, $\alpha_0=\sqrt{2}$ and $X_1 \overset{df}{=} X_0+\alpha_0 g_0 = (-1,\ 1)$. Continuing, we get $f_1(X_1)=f_2(X_1)=f_3(X_1)=-1$,

$$R(X_1)=\{1,\ 2,\ 3\},\qquad H(X_1)=\{(2,\ 1),\ (-1,\ -2),\ (-1,\ 1)\}.$$

The origin lies in the triangle $L(X_1)$ (Figure 10). Thus X_1 is a stationary point. In fact, X_1 is a minimum point for $\varphi(X)$, since in this case $\varphi(X)$ is convex. Finally, we note that $\varphi(X_1)=-1$.

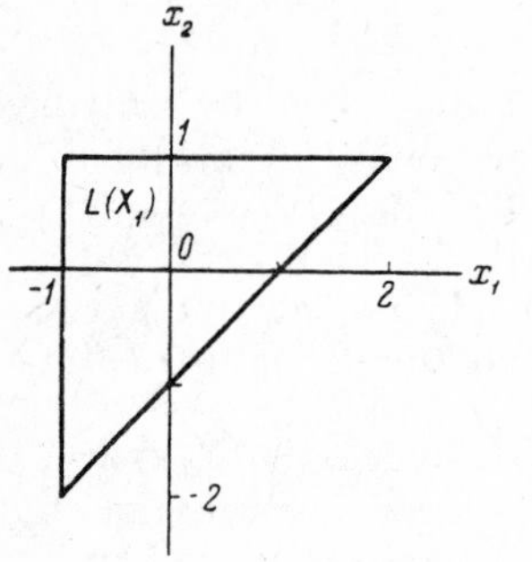

FIGURE 10.

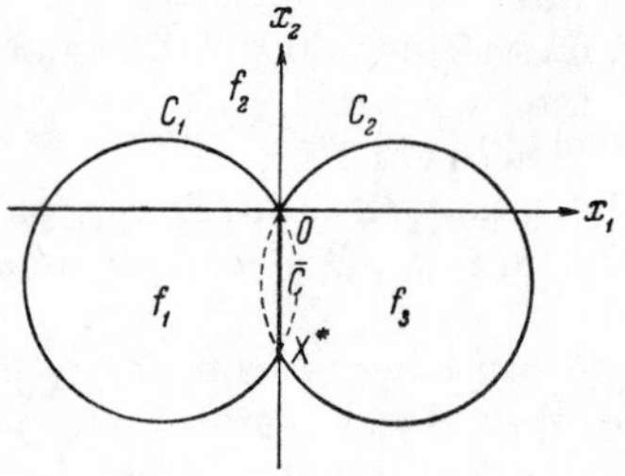

FIGURE 11.

The method of steepest descent is more flexible than the method of coordinatewise descent. Nevertheless, it too does not necessarily exhibit a fundamental property — convergence — in the general case. In other words, the limit point of the sequence $\{X_k\}$ constructed by steepest descent need not be a stationary point of $\varphi(X)$. The reason is that the maximum function $\varphi(X)$ is not smooth (continuously differentiable).

The next example will demonstrate the "jamming" effect that may occur in this method.

3*. Example 2. Let

$$X = (x_1,\ x_2);$$
$$f_1(X) = -5x_1 + x_2; \quad f_2(X) = 4x_2 + x_1^2 + x_2^2;$$
$$f_3(X) = 5x_1 + x_2;$$
$$\varphi(X) = \max_{i \in [1:3]} f_i(X).$$

Since the functions $f_i(X)$, $i \in [1:3]$ are convex, so is the function $\varphi(X)$. We denote

$$C_1 \overset{df}{=} \{X \mid f_1(X) = f_2(X)\} = \left\{X \,\Big|\, \left(x_1 + \frac{5}{2}\right)^2 + \left(x_2 + \frac{3}{2}\right)^2 = \frac{34}{4}\right\},$$
$$C_2 \overset{df}{=} \{X \mid f_2(X) = f_3(X)\} = \left\{X \,\Big|\, \left(x_1 - \frac{5}{2}\right)^2 + \left(x_2 + \frac{3}{2}\right)^2 = \frac{34}{4}\right\},$$
$$\overline{C} \overset{df}{=} \{X \mid f_1(X) = f_3(X)\} = \{X \mid x_1 = 0\}.$$

It is readily verified that (Figure 11)

$$\varphi(X) = \begin{cases} f_1(X), & \text{if } X \text{ is inside } C_1 \text{ to the left of } \overline{C}, \\ f_3(X), & \text{if } X \text{ is inside } C_2 \text{ to the right of } \overline{C}, \\ f_2(X), & \text{if } X \text{ is outside both } C_1 \text{ and } C_2. \end{cases}$$

At the points $\overline{X} = (0,\ 0)$ and $X^* = (0,\ -3)$, we have

$$f_1(\overline{X}) = f_2(\overline{X}) = f_3(\overline{X}) = 0,$$
$$f_1(X^*) = f_2(X^*) = f_3(X^*) = -3.$$

We claim that X^* is a stationary point of $\varphi(X)$, but $\overline{X}$ is not. Indeed,

$$\frac{\partial f_1(X)}{\partial X} = (-5, 1), \quad \frac{\partial f_2(X)}{\partial X} = (2x_1,\ 4 + 2x_2),$$
$$\frac{\partial f_3(X)}{\partial X} = (5, 1).$$

Next,

$$R(X^*) = \{1, 2, 3\}, \quad H(X^*) = \{(-5,\ 1),\ (0,\ -2),\ (5,\ 1)\}.$$

The origin lies in the triangle $L(X^*)$ (Figure 12). Thus $X^* = (0, -3)$ is a stationary point. Since $\varphi(X)$ is a convex function, X^* is a minimum point of $\varphi(X)$. Note that $\varphi(X^*) = -3$.

We now consider the point $\overline{X}$. We have

$$R(\overline{X}) = \{1, 2, 3\}, \quad H(\overline{X}) = \{(-5,\ 1),\ (0,\ 4),\ (5,\ 1)\}.$$

The origin is not an interior point of the triangle $L(\overline{X})$ (Figure 13), and so $\overline{X} = (0, 0)$ is not a stationary point of $\varphi(X)$.

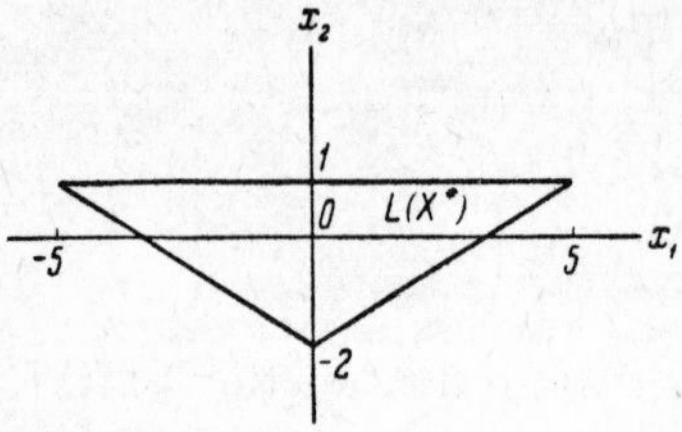

FIGURE 12.

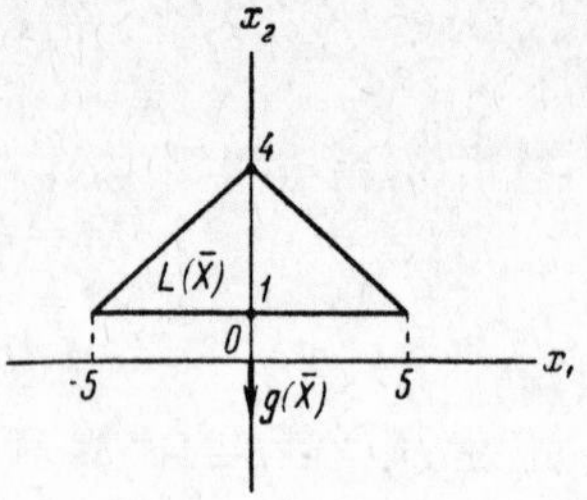

FIGURE 13.

We shall now use the method of steepest descent to construct a sequence $\{X_k\}$ converging to $\bar{X}$. The principal difficulty in doing this is to choose the initial approximation X_0.

We introduce the notation (Figure 14):

$G_1 = \left(-\frac{5}{2}, -\frac{3}{2}\right)$ is the center of the circle C_1;

$G = (0, -2)$ is the minimum point of the function $f_2(X)$;

$Z_1 = (-5, 1)$ is the gradient of the function $f_1(X)$;

$Z(X) \overset{df}{=} (2x_1, \ 4 + 2x_2) = 2(X - G)$ is the gradient of the function $f_2(X)$ (a point is also viewed as a vector whose components are the coordinates of the point);

$$C_1' = \left\{X = (x_1, x_2) \mid X \in C_1, \ -\frac{5}{2} < x_1 < 0, \ x_2 > 0\right\};$$

A is a point on C_1' satisfying the condition $(A - G_1, Z_1) = 0$ (the segment $\overline{AG_1}$ is perpendicular to $\overline{OZ_1}$);

B is a point on C_1', satisfying the following condition: if P is the first point at which the tangent to C_1 at B intersects C_2, then $(P - B, P - G) = 0$;

$\tilde{C}_1$ is the set of points of C_1 lying to the right of A and B and to the left of O.

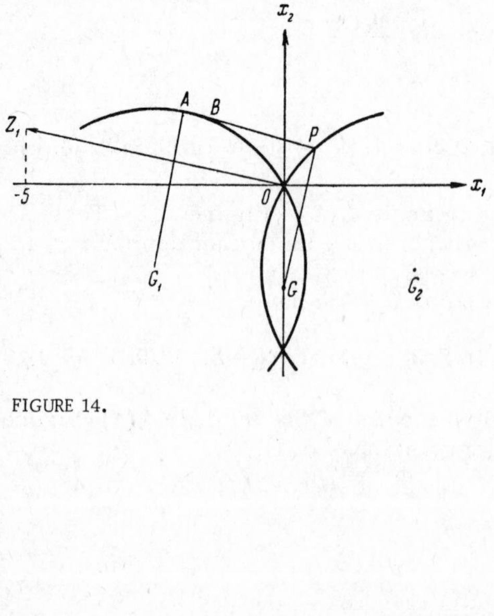

FIGURE 14.

We claim that if $X_0 \in \tilde{C}_1$ and $\{X_k\}$ is the sequence constructed by steepest descent, the sequence will converge to $\overline{X} = (0, 0)$.

The proof falls into three steps. We shall first show that the direction of steepest descent of $\varphi(X)$ at X_0 lies along the tangent to the circle C_1 at X_0, pointing toward C_2. At the second step we shall show that the minimum of $\varphi(X)$ in the direction of steepest descent is attained at the first point of intersection of the tangent just mentioned with the circle C_2. Thus, X_1 is the first point of intersection of the tangent to C_1 at X_0 with the circle C_2. The construction of the entire sequence $\{X_k\}$ should now be clear. At the third step we show that $X_k \xrightarrow[k \to \infty]{} \overline{X}$.

We now consider the proof in detail.

Step I. Let $X_0 = \left(x_1^{(0)}, x_2^{(0)}\right) \in \tilde{C}_1$. Then

$$R(X_0) = \{1, 2\}, \qquad H(X_0) = \{Z_1, Z(X_0)\}.$$

The set $L(X_0)$ is the segment joining Z_1 and $Z(X_0)$.

It is easy to see that the vector

$$V_0 = \left(-1, \frac{2x_1^{(0)} + 5}{2x_2^{(0)} + 3}\right) \tag{5.1}$$

is perpendicular to the straight line through Z_1 and $Z(X_0)$; in other words, $(Z_1 - Z(X_0), V_0) = 0$. On the other hand, V_0 is also perpendicular to the radius-vector through X_0 from G_1: $(X_0 - G_1, V_0) = 0$. Thus V_0, being perpendicular to the straight line through the points Z_1 and $Z(X_0)$, is tangent to the circle C_1 at X_0. Noting that X_0 lies on C_1 to the right of A, we infer that the vector V_0 points toward the point of $L(X_0)$ nearest the origin (see Figure 15, where the vectors Z_1, $Z(X_0) \overset{df}{=} 2(X_0 - G)$ and V_0 are drawn from the point X_0). Thus the direct of steepest descent of $\varphi(X)$ at X_0 is the vector

$$g(X_0) = -\frac{V_0}{\|V_0\|}. \tag{5.2}$$

This is the direction of the tangent to C_1 at X_0 toward the circle C_2.

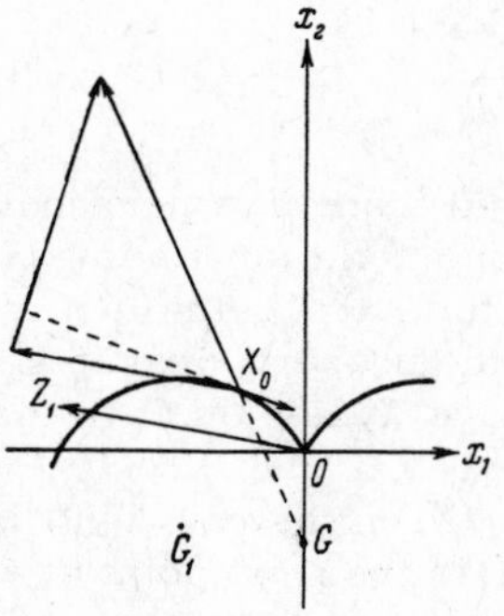

FIGURE 15.

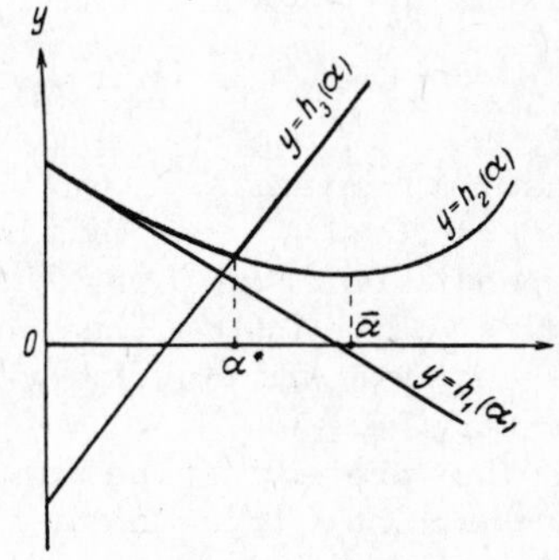

FIGURE 16.

Step II. Set $g_0 = g(X_0)$ and consider the function $\varphi(X)$ on the ray

$$X = X_0(\alpha) \stackrel{df}{=} X_0 + \alpha g_0, \quad \alpha \geqslant 0.$$

Denote

$$h_i(\alpha) = f_i(X_0 + \alpha g_0), \quad i \in [1:3].$$

Clearly,

$$\varphi(X_0 + \alpha g_0) = \max_{i \in [1:3]} h_i(\alpha).$$

Simple manipulations, utilizing (5.1) and (5.2), give (Figure 16)

$$h_1(\alpha) = f_1(X_0) + 2\alpha(X_0 - G, g_0),$$
$$h_2(\alpha) = f_2(X_0) + 2\alpha(X_0 - G, g_0) + \alpha^2(g_0, g_0),$$
$$h_3(\alpha) = f_3(X_0) - 2\alpha(X_0 - G_3, g_0),$$

where $G_3 = (0, -1)$.

Note that the straight line $y = h_1(\alpha)$ is tangent to the parabola $y = h_2(\alpha)$ at $\alpha = 0$, since $f_1(X_0) = f_2(X_0)$ and $h_1'(0) = h_2'(0)$. Now $h_2(\alpha)$ has a minimum at

$$\bar{\alpha} = -\frac{(X_0 - G, g_0)}{(g_0, g_0)}.$$

Let α^* be the first point at which the straight line $y = h_3(\alpha)$ cuts the parabola $y = h_2(\alpha)$ (the corresponding point $X_0(\alpha^*) \stackrel{df}{=} X_0 + \alpha^* g_0$ is the intersection of the tangent to C_1 at X_0 and the circle C_2).

If we can prove that

$$\alpha^* < \bar{\alpha}, \tag{5.3}$$

this will show that $\varphi(X)$ attains a minimum on the ray $X = X_0(\alpha)$ at the point $X_0(\alpha^*)$.

Inequality (5.3) follows from the fact that X_0 lies on C_1 to the right of B. Indeed, it is readily shown that

$$(X_0 + \bar{\alpha} g_0 - G, g_0) = 0.$$

This means that the radius-vector through $X_0(\bar{\alpha})$ from G is perpendicular to the tangent to C_1 at X_0. We thus have a simple rule for finding $X_0(\bar{\alpha})$: drop a perpendicular from G onto the tangent to C_1 at X_0; the foot of this perpendicular will be $X_0(\bar{\alpha})$. The point B has the property that if $X_0 = B$ the corresponding $\bar{\alpha}$ will coincide with α^*. Since X_0 lies on C_1 to the right of B, we have $\alpha^* < \bar{\alpha}$.

We have thus proved that the minimum of $\varphi(X)$ on the ray $X_0(\alpha)$ is attained at the point $X_0(\alpha^*)$. In other words, X_1 is the first point at which the tangent to C_1 at X_0 cuts C_2.

Proceeding in a similar fashion for X_1, we determine X_2 as the first point at which the tangent to C_2 at X_1 cuts the circle C_1, and so on (Figure 17).

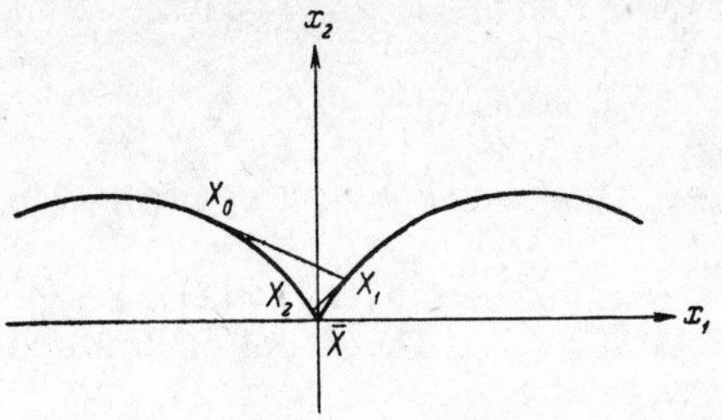

FIGURE 17.

We have thus demonstrated construction of the entire sequence $\{X_k\}$. It remains to show that

$$X_k \xrightarrow[k\to\infty]{} \bar{X}.$$

Step III. Let the points P_0, P_1, P_2, P_3 be as illustrated in Figure 18. We denote $\angle X_1X_0P_0=\alpha_0$, $\angle X_0OP_3=\beta$. It is obvious that $\|X_1\|\leqslant\|P_1\|$ and

$$\frac{\|X_1\|}{\|X_0\|}\leqslant\frac{\|P_1\|}{\|P_0\|}=\frac{\|P_2\|}{\|P_3\|}=1-\operatorname{tg}\beta\operatorname{tg}\alpha_0\leqslant 1-\operatorname{tg}\alpha_0\operatorname{tg}\beta_0,$$

where β_0 is the acute angle between the tangent to C_1 at O and the axis $x_1=0$.

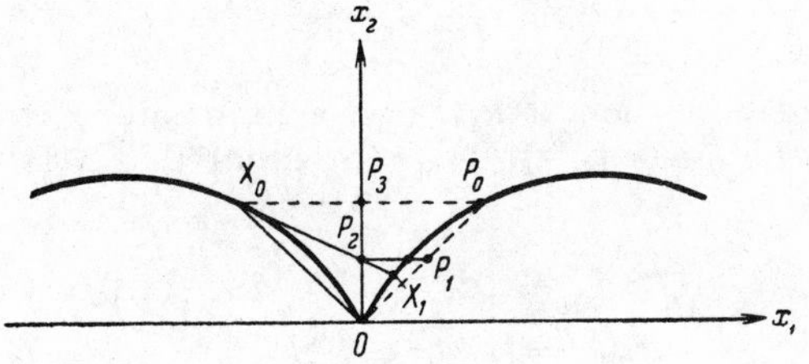

FIGURE 18.

Setting $1-\operatorname{tg}\alpha_0\operatorname{tg}\beta_0=q$, $0<q<1$, we obtain

$$\frac{\|X_1\|}{\|X_0\|}\leqslant q.$$

One shows similarly that for any natural number k

$$\frac{\|X_k\|}{\|X_{k-1}\|}\leqslant q.$$

Hence $\|X_k\| \leqslant q^k \|X_0\|$, $0<q<1$. Thus $X_k \xrightarrow[k\to\infty]{} 0$ or, what is the same,

$$X_k \xrightarrow[k\to\infty]{} \bar{X}.$$

Recalling that $\bar{X}$ is not a stationary point of $\varphi(X)$, we have the promised counterexample.

Remark. In this example the function $\psi(X)$ has a discontinuity at all points $X=\bar{X}+\alpha(X^*-\bar{X}), 0\leqslant\alpha\leqslant 1$. The proof of this statement (based on Lemma 3.3) is left to the reader.

§6. FIRST METHOD OF SUCCESSIVE APPROXIMATIONS

1. For the sequel we shall need some new notions.

As before, we assume that the functions $f_i(X)$, $i\in[0:N]$, are continuously differentiable throughout E_n, and denote

$$\varphi(X)=\max_{i\in[0:N]} f_i(X).$$

For fixed $X\in E_n$, we set

$$R_\varepsilon(X)=\{i \mid \varphi(X)-f_i(X)\leqslant\varepsilon\}, \quad \varepsilon\geqslant 0.$$

Clearly, $R_0(X)=R(X)$ and if $\varepsilon'\geqslant\varepsilon$ then

$$R_{\varepsilon'}(X)\supset R_\varepsilon(X). \tag{6.1}$$

Let us study the dependence of $R_\varepsilon(X)$ on ε in greater detail. To this end, we re-index the numbers $f_i \overset{df}{=} f_i(X)$ (remember that X is fixed!) in decreasing order:

$$f_{i_{01}}=\ldots=f_{i_{0l_0}}>f_{i_{11}}=\ldots$$
$$\ldots=f_{i_{1l_1}}>\ldots>f_{i_{m1}}=\ldots=f_{i_{ml_m}}.$$

It is clear that $0\leqslant m\leqslant N$, and $m\overset{df}{=}m(X)$ depends on X. Denote

$$J_k(X)=\{i_{k1},\ldots,i_{kl_k}\}, \quad k\in[0:m];$$
$$a_k(X)=\varphi(X)-f_i(X), \quad i\in J_k(X).$$

Obviously, $a_0(X)=0$ and (Figure 19)

$$0=a_0(X)<a_1(X)<\ldots<a_m(X).$$

We set $a_{m+1}(X)=\infty$ by definition.

Now introduce the index sets

$$R^s(X)=\bigcup_{k=0}^{s} J_k(X), \quad s\in[0:m].$$

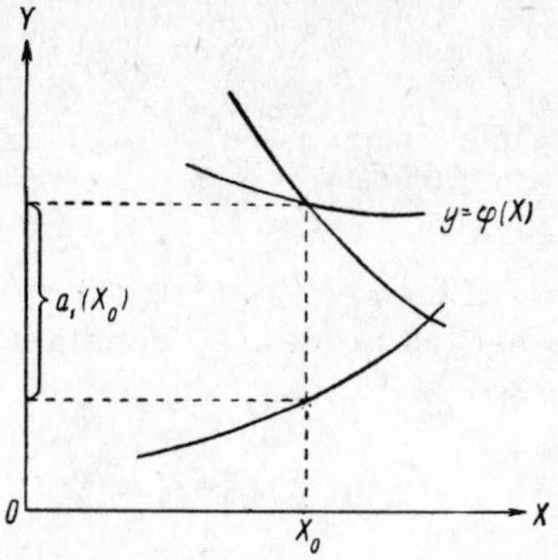

FIGURE 19.

We have $R^0(X)=R(X)$, $R^m(X)=[0:N]$. We can now present the fundamental description of $R_\varepsilon(X)$:

$$R_\varepsilon(X)=R^s(X), \quad \text{if} \quad a_s(X)\leqslant\varepsilon<a_{s+1}(X); \; s\in[0:m]. \qquad (6.2)$$

In particular,

$$\begin{aligned} R_\varepsilon(X)&=R(X), && \text{if} \quad 0\leqslant\varepsilon<a_1(X); \\ R_\varepsilon(X)&=[0:N], && \text{if} \quad a_m(X)\leqslant\varepsilon<\infty. \end{aligned}$$

Lemma 6.1. For any $X_0\in E_n$ there exist numbers $\gamma_0>0$ and $\delta_0>0$ such that, for any $\varepsilon\in[0,\gamma_0]$ and $X'\in E_n$, $\|X'-X_0\|<\delta_0$,

$$R_\varepsilon(X')\subset R(X_0).$$

Proof. If $R(X_0)=[0:N]$, any positive numbers will suffice for γ_0 and δ_0.

Let $R(X_0)\neq[0:N]$. Then we set $\gamma_0=\frac{1}{2}a_1(X_0)$ and choose $\delta_0>0$ in such a way that if $\|X'-X_0\|<\delta_0$ then

$$|f_i(X')-f_i(X_0)|<\frac{1}{4}a_1(X_0)$$

uniformly in $i\in[0:N]$. Note that then it is also true that

$$|\varphi(X')-\varphi(X_0)|<\frac{1}{4}a_1(X_0).$$

Now let $\varepsilon\in[0,\gamma_0]$ and $\|X'-X_0\|<\delta_0$. For $i\in R_\varepsilon(X')$,

$$\begin{aligned} \varphi(X_0)-f_i(X_0)&=[\varphi(X_0)-\varphi(X')]+[\varphi(X')-f_i(X')]+ \\ &+[f_i(X')-f_i(X_0)]<\frac{1}{4}a_1(X_0)+\varepsilon+\frac{1}{4}a_1(X_0)\leqslant a_1(X_0). \end{aligned}$$

Thus, $\varphi(X_0)-f_i(X_0)<a_1(X_0)$, which is possible only if $i\in R(X_0)$.

2. We now define

$$\psi_\varepsilon(X) = \min_{\|g\|=1} \max_{i \in R_\varepsilon(X)} \left(\frac{\partial f_i(X)}{\partial X}, g \right), \quad \varepsilon \geq 0.$$

Clearly, $\psi_0(X) = \psi(X)$.

By (6.1) and (6.2), if we fix X and regard $\psi_\varepsilon(X)$ as a function of ε, $0 \leq \varepsilon < \infty$, it is nondecreasing and piecewise constant (i. e., a step function), and moreover (see Figure 20)

$$\psi_\varepsilon(X) = \psi_{a_s(X)}(X), \quad \text{if} \quad a_s(X) \leq \varepsilon < a_{s+1}(X), \ s \in [0:m]. \tag{6.3}$$

In particular,

$$\psi_\varepsilon(X) = \psi(X), \qquad \text{if} \qquad 0 \leq \varepsilon < a_1(X). \tag{6.4}$$

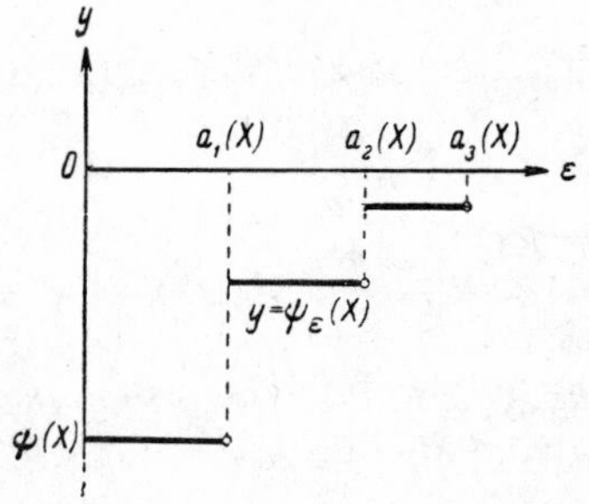

FIGURE 20.

Definition 1. A point $X^* \in E_n$ for which

$$\psi_\varepsilon(X^*) \geq 0,$$

will be called an ε-stationary point of $\varphi(X)$.

Definition 2. A vector $g_\varepsilon(\bar{X})$, $\|g_\varepsilon(\bar{X})\| = 1$, will be called a direction of ε-steepest descent of $\varphi(X)$ at the point $\bar{X}$ if

$$\max_{i \in R_\varepsilon(\bar{X})} \left(\frac{\partial f_i(\bar{X})}{\partial X}, g_\varepsilon(\bar{X}) \right) = \psi_\varepsilon(\bar{X}).$$

We set

$$H_\varepsilon(X) = \left\{ Z = \frac{\partial f_i(X)}{\partial X} \,\middle|\, i \in R_\varepsilon(X) \right\}.$$

Let $L_\varepsilon(X)$ denote the convex hull of the set $H_\varepsilon(X)$.

We have the following two propositions.

I. *A point X^* is ε-stationary for $\varphi(X)$ if and only if the origin is a point of $L_\varepsilon(X^*)$:*

$$0 \in L_\varepsilon(X^*).$$

II. *If $\bar{X}$ is not an ε-stationary point of $\varphi(X)$, i.e., $\psi_\varepsilon(\bar{X}) < 0$, then $\varphi(X)$ has a unique direction of ε-steepest descent $g_\varepsilon(\bar{X})$ at $\bar{X}$:*

$$g_\varepsilon(\bar{X}) = -\frac{Z_\varepsilon^*}{\|Z_\varepsilon^*\|},$$

where Z_ε^* is the point of $L_\varepsilon(\bar{X})$ nearest the origin, and we have $\psi_\varepsilon(\bar{X}) = -\|Z_\varepsilon^*\|$.

The proofs are analogous to those of Theorems 3.2 and 3.3.

3. We can now describe the first method of successive approximations for finding stationary points of $\varphi(X)$.

We first fix two parameters $\varepsilon_0 > 0$ and $\rho_0 > 0$, and take an initial approximation $X_0 \in E_n$. Assume that the set $M(X_0) = \{X \mid \varphi(X) \leqslant \varphi(X_0)\}$ is bounded. Since $\varphi(X)$ is continuous, $M(X_0)$ must also be closed.

Suppose we have already found the k-th approximation $X_k \in M(X_0)$. If $\psi(X_k) \geqslant 0$, then X_k is a stationary point and we are done.

Suppose, then, that $\psi(X_k) < 0$. Checking the number sequence $\varepsilon_\nu = \varepsilon_0/2^\nu$, $\nu = 0, 1, 2, \ldots$ we find the first ν for which (Figure 21)

$$\psi_{\varepsilon_\nu}(X_k) \leqslant -\frac{\rho_0}{\varepsilon_0}\varepsilon_\nu \tag{6.5}$$

(this may be the case for $\nu = 0$), and denote it by ν_k. That ν_k is finite follows from the fact that the right-hand side of inequality (6.5) tends to zero as $\nu \to \infty$, while by (6.4) the left-hand side tends to $\psi(X_k) < 0$.

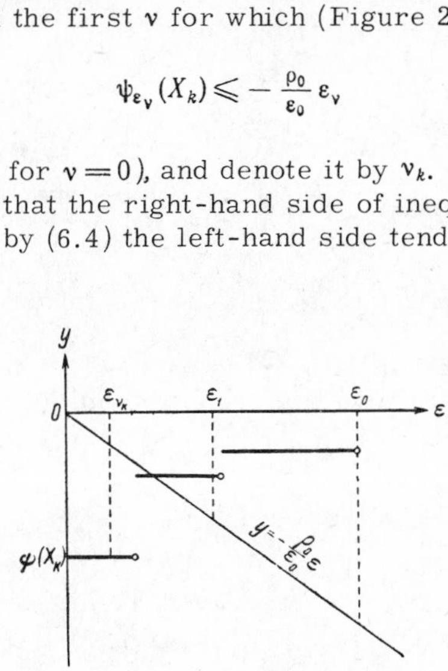

FIGURE 21.

Thus,

$$\psi_{\varepsilon_{\nu_k}}(X_k) \leqslant -\frac{\rho_0}{\varepsilon_0}\varepsilon_{\nu_k},$$

and if $\nu_k > 0$ then

$$\psi_{\varepsilon_{\nu_k - 1}}(X_k) > -\frac{\rho_0}{\varepsilon_0}\varepsilon_{\nu_k - 1}.$$

To simplify the notation, we put

$$\hat{\varepsilon}_k = \varepsilon_{\nu_k}.$$

Set $g_k = g_{\hat{\varepsilon}_k}(X_k)$, where $g_{\hat{\varepsilon}_k}(X_k)$ is the direction of $\hat{\varepsilon}_k$-steepest descent of $\varphi(X)$ at X_k, and consider the ray

$$X = X_k(\alpha) \stackrel{df}{=} X_k + \alpha g_k, \quad \alpha \geqslant 0.$$

Determine $\alpha_k \in [0, \infty)$ such that

$$\varphi(X_k(\alpha_k)) = \min_{\alpha \in [0, \infty)} \varphi(X_k(\alpha))$$

(the infimum is attained since the set $M(X_0)$ is bounded and closed).

We define the $(k+1)$-th approximation X_{k+1} to be $X_k(\alpha_k)$:

$$X_{k+1} = X_k(\alpha_k).$$

Clearly, $X_{k+1} \in M(X_0)$. It is readily shown that

$$\varphi(X_{k+1}) < \varphi(X_k). \tag{6.6}$$

Indeed, for sufficiently small α,

$$\begin{aligned}
&\varphi(X_k + \alpha g_k) = \\
&= \max_{i \in R(X_k)} \left\{ f_i(X_k) + \alpha \left(\frac{\partial f_i(X_k)}{\partial X}, g_k \right) + o_i(g_k; \alpha) \right\} \leqslant \\
&\leqslant \varphi(X_k) + \alpha \max_{i \in R_{\hat{\varepsilon}_k}(X_k)} \left(\frac{\partial f_i(X_k)}{\partial X}, g_{\hat{\varepsilon}_k}(X_k) \right) + \max_{i \in [0:N]} o_i(g_k; \alpha) = \\
&= \varphi(X_k) + \alpha \psi_{\hat{\varepsilon}_k}(X_k) + \max_{i \in [0:N]} o_i(g_k; \alpha) \leqslant \varphi(X_k) - \frac{1}{2} \alpha \frac{\rho_0}{\varepsilon_0} \hat{\varepsilon}_k.
\end{aligned}$$

Inequality (6.6 now follows in an obvious manner.

We may thus assume that we have constructed the sequence $\{X_k\}$ and together with it sequences

$$\{\hat{\varepsilon}_k\}, \ \{g_k\}, \ \{\alpha_k\},$$

in such a way that $X_k \in M(X_0)$, $k = 0, 1, 2, \ldots$,

$$\varphi(X_0) > \varphi(X_1) > \ldots > \varphi(X_k) > \ldots, \tag{6.7}$$

and if the sequence $\{X_k\}$ is finite our construction shows that the last term is a stationary point of $\varphi(X)$.

We now consider the case that the sequence $\{X_k\}$ is infinite.

Lemma 6.2. $\hat{\varepsilon}_k \to 0$ *as* $k \to \infty$.

Proof. If this is false, there exist a number $\varepsilon^* > 0$ and a subsequence $\{\hat{\varepsilon}_{k_j}\}$ such that

$$\hat{\varepsilon}_{k_j} \geqslant \varepsilon^*.$$

Using the inclusion $R_{\varepsilon^*}(X_{k_j}) \subset R_{\varepsilon_{k_j}}(X_{k_j})$, we see that if $i \in R_{\varepsilon^*}(X_{k_j})$,

$$f_i(X_{k_j} + \alpha g_{k_j}) \leqslant f_i(X_{k_j}) + \alpha\psi_{\varepsilon_{k_j}}(X_{k_j}) + o_{ik_j}(\alpha) \leqslant$$
$$\leqslant f_i(X_{k_j}) - \alpha\rho^* + o_{ik_j}(\alpha),$$

where $\rho^* = \frac{\rho_0}{\varepsilon_0}\varepsilon^*$. Since $X_{k_j} \in M(X_0)$ and $M(X_0)$ is a bounded closed set, it follows from Remark 2 to Lemma 3.1 in Appendix III that there exists $\alpha_0 > 0$, independent of i and k_j, such that for $i \in R_{\varepsilon^*}(X_{k_j})$ and $\alpha \in [0, \alpha_0]$

$$f_i(X_{k_j} + \alpha g_{k_j}) \leqslant \varphi(X_{k_j}) - \frac{1}{2}\alpha\rho^*. \tag{6.8}$$

We now set

$$C_1 = \max_{X \in M(X_0)} \max_{i \in [0:N]} \left\| \frac{\partial f_i(X)}{\partial X} \right\|.$$

If $i \notin R_{\varepsilon^*}(X_{k_j})$,

$$f_i(X_{k_j} + \alpha g_{k_j}) \leqslant f_i(X_{k_j}) + \alpha C_1 + o_{ik_j}(\alpha) \leqslant$$
$$\leqslant \varphi(X_{k_j}) - \varepsilon^* + \alpha C_1 + o_{ik_j}(\alpha).$$

There exists α_1, $0 < \alpha_1 \leqslant \alpha_0$, independent of i and k_j, so small that for $i \notin R_{\varepsilon^*}(X_{k_j})$ and $\alpha \in [0, \alpha_1]$

$$f_i(X_{k_j} + \alpha g_{k_j}) \leqslant \varphi(X_{k_j}) - \frac{1}{2}\varepsilon^*. \tag{6.9}$$

Now fix some α', $0 < \alpha' \leqslant \alpha_1$; it follows from (6.8) and (6.9) that

$$\varphi(X_{k_j} + \alpha' g_{k_j}) \leqslant \varphi(X_{k_j}) - \beta, \tag{6.10}$$

where $\beta = \min\left\{\frac{1}{2}\alpha'\rho^*, \frac{1}{2}\varepsilon^*\right\}$, uniformly in k_j.

This inequality readily implies a contradiction: by virtue of (6.7), the whole sequence $\{\varphi(X_k)\}$, which is bounded below by the number $\bar{\varphi} = \min_{X \in M(X_0)} \varphi(X)$, converges to a limit

$$\varphi(X_k) \xrightarrow[k \to \infty]{} \varphi^*,$$

such that for all $k = 0, 1, 2, \ldots$

$$\varphi(X_k) > \varphi^*. \tag{6.11}$$

We choose $\bar{k}_j$ so large that

$$\varphi(X_{\bar{k}_j}) < \varphi^* + \frac{1}{2}\beta.$$

It follows from (6.10) that

$$\varphi(X_{k_{j+1}}) \leqslant \varphi(X_{k_j} + \alpha' g_{k_j}) \leqslant \varphi^* - \frac{1}{2}\beta,$$

contradicting (6.11) and proving the lemma.

4. If the sequence $\{X_k\}$ constructed in the preceding subsection is infinite, it has at least one limit point, since all its terms lie in the bounded closed set $M(X_0)$.

Theorem 6.1. *Any limit point of the sequence $\{X_k\}$ is a stationary point of $\varphi(X)$.*

Proof. Let

$$X_{k_j} \xrightarrow[k_j \to \infty]{} X^*.$$

It is clear that $X^* \in M(X_0)$. We must show that X^* is a stationary point of $\varphi(X)$, i. e., that

$$\psi(X^*) \geqslant 0.$$

Suppose the contrary:

$$\psi(X^*) \overset{df}{=} \min_{\|g\|=1} \max_{i \in R(X^*)} \left(\frac{\partial f_i(X^*)}{\partial X}, g \right) = -b < 0.$$

By Lemma 6.1, there exists $\gamma^* > 0$ such that for sufficiently large $k_j > K_1$ and $\varepsilon \in [0, \gamma^*]$

$$R_\varepsilon(X_{k_j}) \subset R(X^*).$$

Hence it follows that for $\varepsilon \in [0, \gamma^*]$ and $k_j > K_1$

$$\psi_\varepsilon(X_{k_j}) \leqslant \min_{\|g\|=1} \max_{i \in R(X^*)} \left(\frac{\partial f_i(X_{k_j})}{\partial X}, g \right).$$

Choose a natural number $K \geqslant K_1$ such that for any $k_j > K$

$$\left| \min_{\|g\|=1} \max_{i \in R(X^*)} \left(\frac{\partial f_i(X_{k_j})}{\partial X}, g \right) - \min_{\|g\|=1} \max_{i \in R(X^*)} \left(\frac{\partial f_i(X^*)}{\partial X}, g \right) \right| < \frac{b}{2}.$$

Thus, for $k_j > K$ and $\varepsilon \in [0, \gamma^*]$

$$\psi_\varepsilon(X_{k_j}) \leqslant -\frac{b}{2}.$$

Set

$$\varepsilon' = \min\left\{ \gamma^*, \frac{\varepsilon_0}{\rho_0} \cdot \frac{b}{2} \right\}.$$

Let ε^* denote the largest term of the sequence $\varepsilon_\nu=\varepsilon_0/2^\nu$, $\nu=0, 1, 2, \ldots$, smaller than ε'. Then, for all $k_j>K$,

$$\psi_{\varepsilon^*}(X_{k_j})\leqslant -\frac{b}{2}\leqslant -\frac{\rho_0}{\varepsilon_0}\varepsilon'\leqslant -\frac{\rho_0}{\varepsilon_0}\varepsilon^*.$$

Recalling that $\hat{\varepsilon}_{k_j}$ is the first term of the sequence $\{\varepsilon_\nu\}$ satisfying an analogous inequality, we see that for all $k_j>K$

$$\hat{\varepsilon}_{k_j}\geqslant\varepsilon^*,$$

contrary to Lemma 6.2.

Remark. Each step of this successive approximation procedure involves the choice of $\hat{\varepsilon}_k$, g_k and α_k. The most laborious task is the choice of $\hat{\varepsilon}_k=\varepsilon_{\nu_k}$, the first term of the sequence $\varepsilon_\nu=\varepsilon_0/2^\nu, \nu=0, 1, 2, \ldots$, for which inequality (6.5) is satisfied. By Lemma 6.2, $\hat{\varepsilon}_k\xrightarrow[k\to\infty]{}0$. Hence it follows that as k increases the number of trials (comparisons of $\psi_{\varepsilon_\nu}(X_k)$ and $-\frac{\rho_0}{\varepsilon_0}\varepsilon_\nu$, $\nu=0, 1, 2, \ldots$) necessary to choose $\hat{\varepsilon}_k$ also increases.

A convenient strategy for the choice of $\hat{\varepsilon}_k$, $k=1,2, \ldots$, is to begin each trial not from ε_0 but from the result $\hat{\varepsilon}_{k-1}$ of the previous step.

If

$$\psi_{\hat{\varepsilon}_{k-1}}(X_k)>-\frac{\rho_0}{\varepsilon_0}\hat{\varepsilon}_{k-1},$$

examine the sequence $\hat{\varepsilon}_{k-1}/2^\nu$, $\nu=1, 2, \ldots$, until the desired $\hat{\varepsilon}_k$ is found. If on the other hand

$$\psi_{\hat{\varepsilon}_{k-1}}(X_k)\leqslant -\frac{\rho_0}{\varepsilon_0}\hat{\varepsilon}_{k-1},$$

find the largest $\nu\in\left[0:\log_2\frac{\varepsilon_0}{\hat{\varepsilon}_{k-1}}\right]$ for which

$$\psi_{\hat{\varepsilon}_{k-1}2^\nu}(X_k)\leqslant -\frac{\rho_0}{\varepsilon_0}\hat{\varepsilon}_{k-1}2^\nu.$$

The corresponding $\hat{\varepsilon}_{k-1}2^\nu$ is $\hat{\varepsilon}_k$.

5*. In connection with the above remark, we now describe a modified version of the first method of successive approximations.

Suppose we have already found the k-th approximation X'_k, $k=0, 1, 2, \ldots$, and $\psi(X'_k)<0$. To construct X'_{k+1}, we propose to examine the sequence $\bar{\varepsilon}_{k-1}/2^\nu, \nu=0, 1, 2, \ldots; \bar{\varepsilon}_{-1}=\varepsilon_0$. Let $\bar{\varepsilon}_k$ be the first term of this sequence such that

$$\psi_{\bar{\varepsilon}_k}(X'_k)\leqslant -\frac{\rho_0}{\varepsilon_0}\bar{\varepsilon}_k.$$

The rest of the procedure is the same as before: we set $g'_k=g_{\bar{\varepsilon}_k}(X'_k)$ and let X'_{k+1} be the minimum point of the function $\varphi(X)$ on the ray $X=X'_k+\alpha g'_k$, $\alpha\geqslant 0$. Clearly, $\bar{\varepsilon}_k\leqslant\bar{\varepsilon}_{k-1}$ and $\varphi(X'_{k+1})<\varphi(X'_k)$.

If the sequence $\{X'_k\}$ is infinite, then

$$\bar{\varepsilon}_k \to 0 \quad \text{as} \quad k \to \infty. \tag{6.12}$$

The proof of this statement coincides word for word with that of Lemma 6.2.

Theorem 6.2. Let $\{X'_{k_p}\}$ be a subsequence of $\{X'_k\}$ such that

$$\bar{\varepsilon}_{k_p} < \bar{\varepsilon}_{k_p-1}. \tag{6.13}$$

If $M(X'_0)$ is a bounded set, then any limit point of $\{X'_{k_p}\}$ is a stationary point of $\varphi(X)$.

Proof. We first observe that inequality (6.13) is equivalent to

$$\psi_{\bar{\varepsilon}_{k_p-1}}\left(X'_{k_p}\right) > -\frac{\rho_0}{\varepsilon_0}\bar{\varepsilon}_{k_p-1}. \tag{6.14}$$

Next, by (6.12), the subsequence $\{X'_{k_p}\}$ is infinite. We may assume without loss of generality that the entire sequence $\{X'_{k_p}\}$ is convergent:

$$X'_{k_p} \xrightarrow[k_p \to \infty]{} X^*.$$

We must prove that X^* is a stationary point of $\varphi(X)$. Suppose the contrary, and let

$$\psi(X^*) = -b < 0.$$

Repeating the arguments of the proof of Theorem 6.1, we see that there exist $K > 0$ and $\gamma^* > 0$ such that for $k_p > K$ and $\varepsilon \in [0, \gamma^*]$,

$$\psi_\varepsilon\left(X'_{k_p}\right) \leqslant -\frac{b}{2}. \tag{6.15}$$

It follows from (6.12) and (6.15) that for sufficiently large k_p

$$\psi_{\bar{\varepsilon}_{k_p-1}}\left(X'_{k_p}\right) \leqslant -\frac{\rho_0}{\varepsilon_0}\bar{\varepsilon}_{k_p-1},$$

which contradicts (6.14), proving the theorem.

Corollary. If the maximum function $\varphi(X)$ is convex and the set $M(X'_0)$ bounded, then any limit point of the sequence $\{X'_k\}$ constructed by the modified method is a minimum point of $\varphi(X)$.

Indeed, the sequence $\{\varphi(X'_k)\}$ is convergent:

$$\varphi(X'_k) \xrightarrow[k \to \infty]{} \varphi^*.$$

By Theorems 6.2 and 3.1, some subsequence of $\{X'_k\}$ converges to a minimum point of $\varphi(X)$. Thus, since $\varphi(X)$ is continuous, we get

$$\varphi^* = \min_{X \in E_n} \varphi(X) \stackrel{df}{=} \mu.$$

Thus $\varphi(X'_k) \xrightarrow[k\to\infty]{} \mu$.
For any limit point X^* of the sequence $\{X'_k\}$, we have

$$\varphi(X^*) = \mu.$$

Thus X^* is a minimum point of $\varphi(X)$.

§7. ε-STATIONARY POINTS. SECOND METHOD OF SUCCESSIVE APPROXIMATIONS

1. Recall that $X^* \in E_n$ is an ε-stationary point of $\varphi(X)$ if

$$\psi_\varepsilon(X^*) \geqslant 0,$$

or, equivalently,

$$\min_{\|g\|=1} \max_{i \in R_\varepsilon(X^*)} \left(\frac{\partial f_i(X^*)}{\partial X}, g\right) \geqslant 0.$$

Let us examine this definition more closely. Let X^* be an ε-stationary point and let $f_i(X)$ be convex functions on E_n for $i \in R_\varepsilon(X^*)$. Then

$$0 \leqslant \varphi(X^*) - \mu \leqslant \varepsilon,$$

where $\mu = \inf_{X \in E_n} \varphi(X)$. Indeed, by Theorem 4.3,

$$\min_{i \in R_\varepsilon(X^*)} f_i(X^*) \leqslant \mu \leqslant \varphi(X^*).$$

However, for $i \in R_\varepsilon(X^*)$,

$$\varphi(X^*) - f_i(X^*) \leqslant \varepsilon.$$

We now have

$$0 \leqslant \varphi(X^*) - \mu \leqslant \varphi(X^*) - \min_{i \in R_\varepsilon(X^*)} f_i(X^*) \leqslant \varepsilon,$$

as required.

Thus, if X^* is an ε-stationary point and $f_i(X)$, $i \in R_\varepsilon(X^*)$ are convex functions, then $\varphi(X^*)$ is an approximation to μ, the absolute error being at most ε.

Fix some $\varepsilon > 0$. We now proceed to describe a successive approximation procedure for finding ε-stationary points.

The initial approximation will be any point $X_0 \in E_n$. Assume that the set $M(X_0)$ is bounded. Suppose we have already determined the k-th approximation $X_k \in M(X_0)$. If $\psi_\varepsilon(X_k) \geqslant 0$, then X_k is an ε-stationary point and the procedure terminates. Suppose, then, that $\psi_\varepsilon(X_k) < 0$. We set $g_k = g_\varepsilon(X_k)$, where $g_\varepsilon(X_k)$ is the direction of ε-steepest descent of $\varphi(X)$ at X_k.

On the ray

$$X = X_k(\alpha) \overset{df}{=} X_k + \alpha g_k, \quad \alpha \geqslant 0,$$

we find a point $X_k(\alpha_k)$ for which

$$\varphi(X_k(\alpha_k)) = \min_{\alpha \geqslant 0} \varphi(X_k(\alpha)).$$

We now define the $(k+1)$-th approximation X_{k+1} as the point $X_k(\alpha_k)$:

$$X_{k+1} = X_k(\alpha_k).$$

It is clear that $X_{k+1} \in M(X_0)$ and $\varphi(X_{k+1}) < \varphi(X_k)$.

The sequel is analogous. If the sequence $\{X_k\}$ resulting from the procedure is finite, its term is an ε-stationary point of $\varphi(X)$ by construction. Otherwise, we have

Theorem 7.1. Any limit point of the sequence $\{X_k\}$ is an ε-stationary point of $\varphi(X)$.

We shall need two lemmas.

Lemma 7.1. For any $X \in E_n$ and $\varepsilon > 0$, there exists $\delta > 0$ such that whenever $\| X' - X \| < \delta$,

$$R_\varepsilon(X') \subset R_\varepsilon(X).$$

Proof. By (6.2), there exists $\varepsilon' > \varepsilon$ such that

$$R_{\varepsilon'}(X) = R_\varepsilon(X). \tag{7.1}$$

Let $\delta > 0$ be such that whenever $\| X' - X \| < \delta$

$$| f_i(X') - f_i(X) | < \frac{1}{2}(\varepsilon' - \varepsilon)$$

uniformly in $i \in [0:N]$. For any such point X', we have

$$| \varphi(X') - \varphi(X) | < \frac{1}{2}(\varepsilon' - \varepsilon).$$

Now let $i \in R_\varepsilon(X')$. If $\| X' - X \| < \delta$, we obtain

$$\varphi(X) - f_i(X) = [\varphi(X) - \varphi(X')] + [\varphi(X') - f_i(X')] + \\ + [f_i(X') - f_i(X)] \leqslant \frac{1}{2}(\varepsilon' - \varepsilon) + \varepsilon + \frac{1}{2}(\varepsilon' - \varepsilon) = \varepsilon'.$$

Consequently, $i \in R_{\varepsilon'}(X)$, and it follows from (7.1) that $i \in R_\varepsilon(X)$.

Lemma 7.2. The sequence $\{X_k\}$ constructed above satisfies the relation

$$\varliminf_{k \to \infty} \psi_\varepsilon(X_k) \geqslant 0. \tag{7.2}$$

Proof. Suppose the contrary. Then there exist a subsequence $\{X_{k_j}\}$ and a number $b>0$ such that

$$\psi_\varepsilon(X_{k_j}) \leqslant -b.$$

We can find $\alpha_1>0$ so small that for $i \in R_\varepsilon(X_{k_j})$, $\alpha \in [0, \alpha_1]$ and all k_j

$$f_i(X_{k_j}+\alpha g_{k_j}) \leqslant \varphi(X_{k_j}) - \frac{1}{2}\alpha b,$$

and for $i \notin R_\varepsilon(X_{k_j})$, $\alpha \in [0, \alpha_1]$ and all k_j

$$f_i(X_{k_j}+\alpha g_{k_j}) \leqslant \varphi(X_{k_j}) - \frac{1}{2}\varepsilon$$

(see the proof of Lemma (6.2).

Hence it follows that

$$\varphi(X_{k_j}+\alpha_1 g_{k_j}) \leqslant \varphi(X_{k_j}) - \beta, \tag{7.3}$$

where $\beta = \min\left\{\frac{1}{2}\alpha_1 b, \frac{1}{2}\varepsilon\right\}$, uniformly in k_j.

Inequality (7.3) now implies a contradiction, as in Lemma 6.2.

Proof of Theorem 7.1. Let

$$X_{k_j} \xrightarrow[k_j\to\infty]{} X^*.$$

We have to show that $\psi_\varepsilon(X^*) \geqslant 0$.

Suppose the contrary:

$$\psi_\varepsilon(X^*) = -b < 0.$$

By Lemma 7.1, for sufficiently large $k_j > K_1$, we have

$$R_\varepsilon(X_{k_j}) \subset R_\varepsilon(X^*).$$

We now show, as in the proof of Theorem 6.1, that there exists a natural number $K \geqslant K_1$ such that for all $k_j > K$

$$\psi_\varepsilon(X_{k_j}) \leqslant -\frac{b}{2}.$$

But this inequality contradicts (7.2), completing the proof.

2. Example. Consider the function

$$F(X, t) = (\cos t + \cos(2t + x_1) + \cos(3t + x_2))^2,$$

where $X = (x_1, x_2)$, $t \in [0, 2\pi]$.

Set

$$t_i = \frac{2\pi i}{N}, \quad i \in [0:N]; \quad f_i(X) = F(X, t_i).$$

Let $N = 100$ and $\varepsilon = 0.001$. We shall determine an ε-stationary point of the function

$$\varphi(X) = \max_{i \in [0:N]} f_i(X)$$

by the method described in subsection 1. Let the initial approximation be the point $X_0 = (0.785; 0.785)$. The results of the computations are presented in Table 2. Note that r_k is the distance from the origin to the polyhedron $L_\varepsilon(X_k)$; s_k is the number of indices i contained in $R_\varepsilon(X_k)$.

TABLE 2.

k	X_k		$\varphi(X_k)$	r_k	s_k	α_k
0	0.7850	0.7850	8.5762	1.6419	1	1.3530
1	2.1297	0.9347	4.6316	4.8784	1	0.0020
2	2.1281	0.9359	4.6286	4.5216	1	0.0002
3	2.1283	0.9358	4.6228	0.2381	2	0.0578
4	2.1606	0.9837	4.6045	0.0989	3	1.4549
5	2.8870	2.2443	3.9573	3.9563	1	0.0005
6	2.8874	2.2441	3.9558	0.3282	2	0.0766
7	2.8634	2.1713	3.9215	3.8887	1	0.0005
8	2.8639	2.1712	3.9204	0.3377	2	0.0221
9	2.8595	2.1495	3.9120	0.4220	2	0.0005
10	2.8595	2.1491	3.9118	0	3	—

At the tenth step, we obtain an ε-stationary point $X^* = (2.8595; 2.1491)$, with $\varphi(X^*) = 3.9118$.

Figure 22 illustrates the functions $F(X_k, t)$ for $k = 0, 4, 10$.

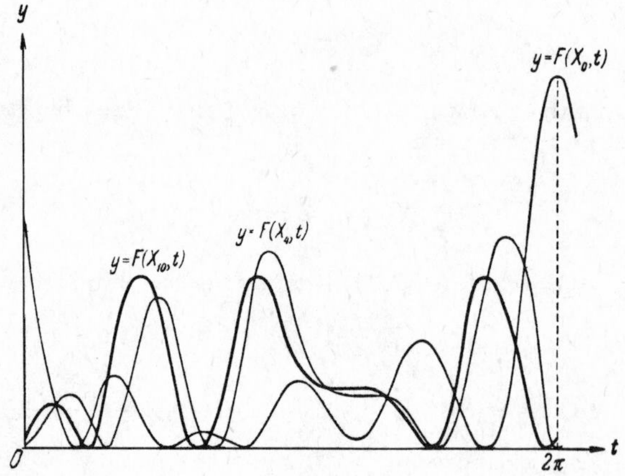

FIGURE 22.

3*. We now describe a modification of the above method of successive approximations. We shall assume here that the functions $f_i(X)$, $i \in [0:N]$, are twice continuously differentiable throughout E_n, so that for any $X \in E_n$ and $Y \in E_n$

$$f_i(X+Y) = f_i(X) + \left(\frac{\partial f_i(X)}{\partial X}, Y\right) + \\ + \frac{1}{2}\left(\frac{\partial^2 f_i(X+\theta_i Y)}{\partial X^2} Y, Y\right), \tag{7.4}$$

where $\theta_i \in (0, 1)$ and $i \in [0:N]$.

Fix some $\varepsilon > 0$, and take an initial approximation $X_0' \in E_n$. We shall assume again that the set $M(X_0')$ is bounded.

Let $\alpha > 0$ be some constant,

$$C_1 = \max_{X \in M(X_0')} \max_{i \in [0:N]} \left\| \frac{\partial f_i(X)}{\partial X} \right\|,$$

$$C_2 = \max_{X \in M(X_0'),\ \|\bar{Y}\| \leqslant \alpha C_1} \max_{i \in [0:N]} \left\| \frac{\partial^2 f_i(X+\bar{Y})}{\partial X^2} \right\|,$$

where

$$\left\| \frac{\partial^2 f_i(X+\bar{Y})}{\partial X^2} \right\| = \sqrt{\sum_{j_1, j_2=1}^{n} \left(\frac{\partial^2 f_i(X+\bar{Y})}{\partial x_{j_1} \partial x_{j_2}} \right)^2}.$$

It is easy to see, using the Cauchy inequality, that for $X \in M(X_0')$ and $\|\bar{Y}\| \leqslant \alpha C_1$

$$\left| \left(\frac{\partial^2 f_i(X+\bar{Y})}{\partial X^2} Y, Y \right) \right| \leqslant C_2 \|Y\|^2.$$

Hence, by (7.4), it follows that for all $i \in [0:N]$, $X \in M(X_0')$, $\|Y\| \leqslant \alpha C_1$,

$$f_i(X+Y) \leqslant f_i(X) + \left(\frac{\partial f_i(X)}{\partial X}, Y\right) + \frac{1}{2} C_2 \|Y\|^2. \tag{7.5}$$

We set

$$\bar{\alpha} = \min\left\{ \alpha, \frac{2}{C_2}, \frac{-C_1 + \sqrt{C_1^2 + C_2 \varepsilon}}{C_1 C_2} \right\}$$

(if $C_2 = 0$ we set $\bar{\alpha} = \min\left\{ \alpha, \frac{\varepsilon}{2C_1^2} \right\}$).

Fix some $\alpha_0 \in (0, \bar{\alpha})$.

Suppose we have already found the k-th approximation $X_k' \in M(X_0')$. If $\psi_\varepsilon(X_k') \geqslant 0$, then X_k' is an ε-stationary point and we are done.

If $\psi_\varepsilon(X_k') < 0$, we set $g_k' = g_\varepsilon(X_k')$ and

$$\delta_k = -\psi_\varepsilon(X_k') > 0.$$

We now define the $(k+1)$-th approximation to be

$$X'_{k+1} = X'_k + \alpha_0 \delta_k g'_k.$$

To complete our description of the method, it remains to prove that $X'_{k+1} \in M(X'_0)$. We shall prove even more:

$$\varphi(X'_{k+1}) < \varphi(X'_k). \tag{7.6}$$

Since $\delta_k \leqslant C_1$ and $\alpha_0 \leqslant \alpha$, it follows from (7.5) that, for all $i \in [0:N]$,

$$f_i(X'_{k+1}) \leqslant f_i(X'_k) + \alpha_0 \delta_k \left(\frac{\partial f_i(X'_k)}{\partial X}, \; g'_k \right) + \frac{1}{2} \alpha_0^2 \delta_k^2 C_2.$$

Let $i \in R_\varepsilon(X'_k)$. Then

$$f_i(X'_{k+1}) \leqslant f_i(X'_k) - \alpha_0 \delta_k^2 + \frac{1}{2} \alpha_0^2 \delta_k^2 C_2 \leqslant$$
$$\leqslant \varphi(X'_k) - \frac{1}{2} \alpha_0 \delta_k^2 C_2 \left(\frac{2}{C_2} - \alpha_0 \right). \tag{7.7}$$

Noting that $\alpha_0 < \frac{2}{C_2}$, we conclude that if $i \in R_\varepsilon(X'_k)$

$$f_i(X'_{k+1}) < \varphi(X'_k).$$

If $i \notin R_\varepsilon(X'_k)$, then

$$f_i(X'_{k+1}) \leqslant \varphi(X'_k) - \varepsilon + \alpha_0 C_1^2 + \frac{1}{2} \alpha_0^2 C_1^2 C_2.$$

Since $\alpha_0 < \dfrac{-C_1 + \sqrt{C_1^2 + C_2 \varepsilon}}{C_1 C_2}$, we have

$$\frac{1}{2} \alpha_0^2 C_1^2 C_2 + \alpha_0 C_1^2 \leqslant \frac{1}{2} \varepsilon.$$

Thus, if $i \notin R_\varepsilon(X'_k)$,

$$f_i(X'_{k+1}) \leqslant \varphi(X'_k) - \frac{1}{2} \varepsilon. \tag{7.8}$$

Combining (7.7) and (7.8), we get (7.6), proving the assertion.

Let $\{X'_k\}$ be the sequence constructed by the modified method. If it is finite, its last term is by construction an ε-stationary point of $\varphi(X)$. Otherwise:

Theorem 7.2. Any limit point of the sequence $\{X'_k\}$ is an ε-stationary point of $\varphi(X)$.

Proof. The proof is similar to that of Theorem 7.1. Let

$$X'_{k_p} \xrightarrow[k_p \to \infty]{} X^*,$$

and suppose that X^* is not an ε-stationary point of $\varphi(X)$. Then

$$\psi_\varepsilon(X^*) = -b < 0.$$

For sufficiently large $k_p > K$,

$$-\delta_k \stackrel{df}{=} \psi_\varepsilon(X'_{k_p}) \leqslant -\frac{b}{2}.$$

In view of (7.7) and (7.8), we see that for $k_p > K$

$$\varphi(X'_{k_p+1}) \leqslant \varphi(X'_{k_p}) - \beta,$$

where $\beta = \min\left\{\frac{1}{8}\alpha_0 b^2\left(\frac{2}{C_2} - \alpha_0\right), \frac{1}{2}\varepsilon\right\}$.

This inequality again leads to a contradiction, proving the theorem.

4. We now show how the ε-algorithm described in subsection 1 may be utilized in determining stationary points of $\varphi(X)$.

Fix some $\varepsilon_0 > 0$ and $\rho_0 > 0$. Choose an initial approximation $X_{00} \in E_n$. Let $M(X_{00})$ be bounded.

Applying the ε-algorithm with $\varepsilon = \varepsilon_0$, we obtain (after finitely many steps, by Lemma 7.2) a point $X_{0l_0} \in M(X_{00})$ such that

$$\psi_{\varepsilon_0}(X_{0l_0}) \geqslant -\rho_0.$$

Set $\varepsilon_1 = \frac{1}{2}\varepsilon_0$, $\rho_1 = \frac{1}{2}\rho_0$, $X_{10} = X_{0l_0}$.

Taking X_{10} as a new initial approximation, we apply the ε-algorithm with $\varepsilon = \varepsilon_1$. Finitely many steps lead to a point $X_{1l_1} \in M(X_{10}) \subset M(X_{00})$ such that

$$\psi_{\varepsilon_1}(X_{1l_1}) \geqslant -\rho_1.$$

Continuing in this way, we obtain a sequence of points $\{X_{kl_k}\}$ whose terms satisfy the inequality

$$\psi_{\varepsilon_k}(X_{kl_k}) \geqslant -\rho_k, \tag{7.9}$$

where $\varepsilon_k = \frac{\varepsilon_0}{2^k}$, $\rho_k = \frac{\rho_0}{2^k}$, $k = 0, 1, 2, \ldots$.

Theorem 7.3. Any limit point of the sequence $\{X_{kl_k}\}$ is a stationary point of $\varphi(X)$.

Proof. We may assume for simplicity's sake that the entire sequence $\{X_{kl_k}\}$ is convergent:

$$X_{kl_k} \xrightarrow[k\to\infty]{} X^*.$$

It remains to prove that $\psi(X^*) \geqslant 0$. Suppose the contrary: $\psi(X^*) = -b < 0$. By (6.4), there exists $\varepsilon' > 0$ such that

$$\psi_{\varepsilon'}(X^*) = -b < 0.$$

Hence it follows that for sufficiently large k

$$\psi_{\varepsilon_k}(X_{kl_k}) = \psi_{\varepsilon'}(X_{kl_k}) \leqslant -\frac{b}{2},$$

contradicting (7.9) and proving the theorem.

§8. THE D-FUNCTION. THIRD METHOD OF SUCCESSIVE APPROXIMATIONS

1. Fix some $\bar{\varepsilon}>0$ and set

$$D(X)=\inf_{\varepsilon\in[0,\,\bar{\varepsilon}]}\varepsilon\psi_{\varepsilon}(X) \tag{8.1}$$

for all $X\in E_n$. Recall (see §6.2) that

$$\psi_{\varepsilon}(X)=\psi_{a_s(X)}(X), \qquad \text{if} \qquad a_s(X)\leqslant\varepsilon<a_{s+1}(X),$$

and moreover

$$0=a_0(X)<a_1(X)<\ldots<a_m(X)<a_{m+1}(X)=\infty.$$

Let $\bar{m}(X)$ denote the largest s, $s\in[0:m]$, such that

$$a_s(X)<\bar{\varepsilon}.$$

If $\psi(X)<0$, we obtain the following representation for $D(X)$ (Figure 23):

$$D(X)=\min\{a_1\psi_{a_0}(X),\ \ldots,\ a_{\bar{m}}\psi_{a_{\bar{m}-1}}(X),\ \bar{\varepsilon}\psi_{a_{\bar{m}}}(X)\}, \tag{8.2}$$

where $a_s=a_s(X)$, $s\in[0:\bar{m}]$; $\bar{m}=\bar{m}(X)$.

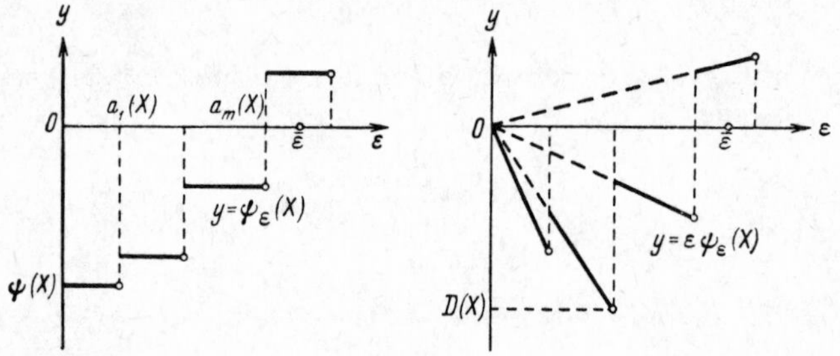

FIGURE 23.

It also follows from (8.1) that for any $X\in E_n$

$$D(X)\leqslant 0. \tag{8.3}$$

Lemma 8.1. The inequality $\psi(X)\geqslant 0$ is equivalent to

$$D(X)=0.$$

Proof. If $\psi(X)\geqslant 0$, it is clear that the infimum in (8.1) is attained for $\varepsilon=0$. Consequently, $D(X)=0$.

Now let $D(X)=0$. Then for $0<\varepsilon\leqslant\bar{\varepsilon}$

$$\psi_\varepsilon(X)\geqslant 0.$$

Since

$$\psi(X)=\psi_\varepsilon(X),$$

for any $0\leqslant\varepsilon<a_1(X)$, we see that $\psi(X)\geqslant 0$.

Theorem 8.1. A necessary condition for the maximum function $\varphi(X)$ to have a minimum at the point X^* is that

$$D(X^*)=0. \tag{8.4}$$

If $\varphi(X)$ is convex, this condition is also sufficient.

This theorem is an obvious corollary of Theorem 3.1 and Lemma 8.1.

In view of (8.3) and Lemma 8.1, we conclude that our original problem — determination of stationary points of $\varphi(X)$— reduces to maximization of the function $D(X)$.

2. Our next aim is to prove that the function $D(X)$ is continuous throughout E_n.

Let γ be any number such that

$$0<\gamma<\frac{1}{4}\min_{s\in[0:m]}(a_{s+1}(X)-a_s(X)), \tag{8.5}$$

and set

$$\Gamma'_\gamma(X)=[0,\ \infty)\setminus\bigcup_{k=0}^{m}(a_k(X)-\gamma,\ a_k(X)+\gamma).$$

Lemma 8.2. For any $X\in E_n$ and any γ satisfying (8.5), there exists $\delta>0$ such that for any $X'\in E_n$ with $\|X'-X\|<\delta$ and any $\varepsilon\in\Gamma'_\gamma(X)$

$$R_\varepsilon(X')=R_\varepsilon(X).$$

Proof. To fix ideas, suppose that

$$a_{\bar{k}}(X)+\gamma\leqslant\varepsilon\leqslant a_{\bar{k}+1}(X)-\gamma.$$

Choose $\delta>0$ so that whenever $\|X'-X\|<\delta$

$$|f_i(X')-f_i(X)|<\frac{1}{4}\gamma$$

uniformly in $i\in[0:N]$. Then it is also true that

$$|\varphi(X')-\varphi(X)|<\frac{1}{4}\gamma.$$

Let $i\in R_\varepsilon(X')$, where $\|X'-X\|<\delta$. Then

$$\varphi(X) - f_i(X) =$$
$$= [\varphi(X) - \varphi(X')] + [\varphi(X') - f_i(X')] + [f_i(X') - f_i(X)] \leqslant$$
$$\leqslant \frac{1}{4}\gamma + \varepsilon + \frac{1}{4}\gamma \leqslant a_{\bar{k}+1}(X) - \frac{1}{2}\gamma.$$

Noting that $R_{a_{\bar{k}+1}(X)-1/2\gamma}(X) = R_\varepsilon(X)$, we obtain $i \in R_\varepsilon(X)$. Consequently,

$$R_\varepsilon(X') \subset R_\varepsilon(X). \tag{8.6}$$

Now let $i \in R_\varepsilon(X)$. In view of the equality

$$R_\varepsilon(X) = R_{a_{\bar{k}}(X)+1/2\gamma}(X),$$

we see that for $\| X' - X \| < \delta$

$$\varphi(X') - f_i(X') =$$
$$= [\varphi(X') - \varphi(X)] + [\varphi(X) - f_i(X)] + [f_i(X) - f_i(X')] \leqslant$$
$$\leqslant \frac{1}{4}\gamma + a_{\bar{k}}(X) + \frac{1}{2}\gamma + \frac{1}{4}\gamma = a_{\bar{k}}(X) + \gamma \leqslant \varepsilon.$$

Thus $i \in R_\varepsilon(X')$, so that

$$R_\varepsilon(X) \subset R_\varepsilon(X'). \tag{8.7}$$

Combining (8.6) and (8.7), we complete the proof.

We now set

$$\Gamma_\gamma(X) = \Gamma'_\gamma(X) \cap [0, \bar{\varepsilon}],$$

where γ satisfies (8.5) and $\gamma < \bar{\varepsilon}$. It is obvious that $\Gamma_\gamma(X)$ is a nonempty closed bounded set.

We introduce the notation

$$D_\gamma(X, X') = \inf_{\varepsilon \in \Gamma_\gamma(X)} \varepsilon \psi_\varepsilon(X').$$

Lemma 8.3. For fixed X and γ, and any $\varepsilon^* > 0$, there exists $\delta > 0$ such that for all $X' \in E_n$, $\| X' - X \| < \delta$,

$$| D_\gamma(X, X') - D_\gamma(X, X) | < \varepsilon^*.$$

Proof. In view of Lemma 8.2, we have for all X', $\| X' - X \| < \delta_1$ (see also Appendix III, §2, inequality VIII)

$$| D_\gamma(X, X') - D_\gamma(X, X) | \leqslant \sup_{\varepsilon \in \Gamma_\gamma(X)} \varepsilon | \psi_\varepsilon(X') - \psi_\varepsilon(X) | \leqslant$$

$$\leqslant \bar{\varepsilon} \sup_{\varepsilon \in \Gamma_\gamma(X)} \left| \min_{\|g\|=1} \max_{i \in R_\varepsilon(X)} \left(\frac{\partial f_i(X')}{\partial X}, g \right) - \min_{\|g\|=1} \max_{i \in R_\varepsilon(X)} \left(\frac{\partial f_i(X)}{\partial X}, g \right) \right| \leqslant$$

$$\leqslant \bar{\varepsilon} \sup_{\varepsilon \in \Gamma_\gamma(X)} \max_{\|g\|=1} \max_{i \in R_\varepsilon(X)} \left| \left(\frac{\partial f_i(X')}{\partial X} - \frac{\partial f_i(X)}{\partial X}, g \right) \right| \leqslant$$
$$\leqslant \bar{\varepsilon} \max_{i \in [0:N]} \left\| \frac{\partial f_i(X')}{\partial X} - \frac{\partial f_i(X)}{\partial X} \right\|.$$

Now select $\delta > 0$, $\delta \leqslant \delta_1$, such that for all X', $\| X' - X \| < \delta$,

$$\bar{\varepsilon} \max_{i \in [0:N]} \left\| \frac{\partial f_i(X')}{\partial X} - \frac{\partial f_i(X)}{\partial X} \right\| < \varepsilon^*.$$

Then, for the same points X', we have

$$| D_\gamma(X, X') - D_\gamma(X, X) | < \varepsilon^*.$$

Q. E. D.

Lemma 8.4. *For any* $X' \in E_n$,

$$0 \leqslant D_\gamma(X, X') - D(X') \leqslant 3\gamma C_1(X'),$$

where $C_1(X') = \max_{i \in [0:N]} \left\| \frac{\partial f_i(X')}{\partial X} \right\|$.

Proof. If $\varepsilon \in \Gamma_\gamma(X)$, then

$$\varepsilon \psi_\varepsilon(X') \geqslant \inf_{\varepsilon \in \Gamma_\gamma(X)} \varepsilon \psi_\varepsilon(X') \stackrel{df}{=} D_\gamma(X, X'). \tag{8.8}$$

Now let $\varepsilon \in [0, \bar{\varepsilon}]$ but $\varepsilon \notin \Gamma_\gamma(X)$. We first consider the case

$$0 < \varepsilon < \gamma.$$

We have

$$\begin{aligned} \varepsilon \psi_\varepsilon(X') &= [\varepsilon \psi_\varepsilon(X') - \varepsilon \psi_\gamma(X')] + \\ &\quad + [\varepsilon \psi_\gamma(X') - \gamma \psi_\gamma(X')] + \gamma \psi_\gamma(X') \geqslant \\ &\geqslant \gamma \psi_\gamma(X') - |(\varepsilon - \gamma) \psi_\gamma(X')| - \varepsilon | \psi_\varepsilon(X') - \psi_\gamma(X') | \geqslant \\ &\geqslant \inf_{\varepsilon \in \Gamma_\gamma(X)} \varepsilon \psi_\varepsilon(X') - \gamma C_1(X') - \gamma(2C_1(X')) = \\ &= D_\gamma(X, X') - 3\gamma C_1(X'). \end{aligned} \tag{8.9}$$

If $a_{\bar{k}}(X) - \gamma < \varepsilon < a_{\bar{k}}(X) + \gamma$ then, setting

$$\hat{\varepsilon} = a_{\bar{k}}(X) - \gamma \in \Gamma_\gamma(X),$$

we obtain

$$\begin{aligned} \varepsilon \psi_\varepsilon(X') &\geqslant \varepsilon \psi_{\hat{\varepsilon}}(X') = \hat{\varepsilon} \psi_{\hat{\varepsilon}}(X') + (\varepsilon - \hat{\varepsilon}) \psi_{\hat{\varepsilon}}(X') \geqslant \\ &\geqslant D_\gamma(X, X') - |(\varepsilon - \hat{\varepsilon}) \psi_{\hat{\varepsilon}}(X')| \geqslant D_\gamma(X, X') - 2\gamma C_1(X'). \end{aligned} \tag{8.10}$$

Thus, by (8.8), (8.9) and (8.10), it follows that for all $\varepsilon \in [0, \bar{\varepsilon}]$

$$\varepsilon\psi_\varepsilon(X') \geqslant D_\gamma(X, X') - 3\gamma C_1(X').$$

Hence $D_\gamma(X, X') - D(X') \leqslant 3\gamma C_1(X')$. The inequality $D_\gamma(X, X') - D(X') \geqslant 0$ is obvious.

Theorem 8.2. The function $D(X)$ is continuous on E_n.

Proof. Fix $X \in E_n$ and choose arbitrary $\varepsilon^* > 0$ and $\Delta > 0$. Denote

$$C_1 = \max_{\|X'-X\| \leqslant \Delta} \max_{i \in [0:N]} \left\| \frac{\partial f_i(X')}{\partial X} \right\|.$$

Now choose γ so that

$$3\gamma C_1 < \frac{1}{3}\varepsilon^*. \tag{8.11}$$

By Lemma 8.3, there exists δ, $0 < \delta < \Delta$, such that for all X', $\|X' - X\| < \delta$,

$$|D_\gamma(X, X') - D_\gamma(X, X)| < \frac{1}{3}\varepsilon^*. \tag{8.12}$$

Using Lemma 8.4 and inequalities (8.11) and (8.12), we finally see that for all X', $\|X' - X\| < \delta$,

$$\begin{aligned} |D(X') - D(X)| \leqslant |D(X') - D_\gamma(X, X')| + \\ + |D_\gamma(X, X') - D_\gamma(X, X)| + |D_\gamma(X, X) - D(X)| < \\ < \frac{1}{3}\varepsilon^* + \frac{1}{3}\varepsilon^* + \frac{1}{3}\varepsilon^* = \varepsilon^*. \end{aligned}$$

This completes the proof.

3. We can now describe a third successive approximation procedure for determining stationary points of $\varphi(X)$, utilizing the D-function.

Let $X_0 \in E_n$ be an arbitrary initial approximation. Assume that the set $M(X_0) = \{X \mid \varphi(X) \leqslant \varphi(X_0)\}$ is bounded.

Suppose we have already found the k-th approximation $X_k \in M(X_0)$. If $D(X_k) = 0$, it follows from Lemma 8.1 that X_k is a stationary point and the procedure terminates.

Suppose, then, that $D(X_k) < 0$. In view of (8.2), we can write

$$D(X_k) = \varepsilon'_k \psi_{\varepsilon''_k}(X_k), \tag{8.13}$$

where $(\varepsilon'_k, \varepsilon''_k)$ is one of the pairs

$$(a_1(X_k), a_0(X_k)), \ldots, (a_{\bar{m}}(X_k), a_{\bar{m}-1}(X_k)), \quad (\bar{\varepsilon}, a_{\bar{m}}(X_k)).$$

Since by assumption $D(X_k) < 0$, we have $\psi_{\varepsilon''_k}(X_k) < 0$.

Let $g_k = g_{\varepsilon''_k}(X_k)$ be the direction of ε''_k-steepest descent of the function $\varphi(X)$ at the point X_k. Consider the ray

$$X = X_k(\alpha) \overset{\text{df}}{=} X_k + \alpha g_k, \quad \alpha \geqslant 0,$$

and let $X_k(\alpha_k)$ denote a point at which

$$\varphi(X_k(\alpha_k)) = \min_{\alpha \in [0, \infty)} \varphi(X_k(\alpha)).$$

We now define the $(k+1)$-th approximation to be the point $X_k(\alpha_k)$:

$$X_{k+1} = X_k(\alpha_k).$$

As usual, one proves that $\varphi(X_{k+1}) < \varphi(X_k)$, so that, in particular $X_{k+1} \in M(X_0)$.

If the sequence $\{X_k\}$ thus obtained is finite, its last term is by construction a stationary point of $\varphi(X)$. Otherwise:

Theorem 8.3. Any limit point of the sequence $\{X_k\}$ is a stationary point of $\varphi(X)$.

Proof. Let

$$X_{k_j} \xrightarrow[k_j \to \infty]{} X^*.$$

We assert that $D(X^*) = 0$. If not,

$$D(X^*) = -b < 0.$$

By the continuity of $D(X)$ and by (8.13), we have for sufficiently large $k_j > K_1$,

$$D\left(X_{k_j}\right) = \varepsilon'_{k_j} \psi_{\varepsilon''_{k_j}}\left(X_{k_j}\right) \leqslant -\frac{b}{2}.$$

Now, for any $\varepsilon \in \left[\varepsilon''_{k_j}, \varepsilon'_{k_j}\right)$, we have

$$\psi_{\varepsilon}\left(X_{k_j}\right) = \psi_{\varepsilon''_{k_j}}\left(X_{k_j}\right),$$

and so, setting $\tilde{\varepsilon}''_{k_j} = \left(\varepsilon'_{k_j} + \varepsilon''_{k_j}\right)/2$, we obtain

$$\varepsilon'_{k_j} \psi_{\tilde{\varepsilon}''_{k_j}}\left(X_{k_j}\right) \leqslant -\frac{b}{2}, \tag{8.14}$$

with

$$\frac{\varepsilon'_{k_j}}{2} \leqslant \tilde{\varepsilon}''_{k_j} \leqslant \varepsilon'_{k_j}. \tag{8.15}$$

Without loss of generality, we may assume that

$$\varepsilon'_{k_j} \to \varepsilon', \quad \tilde{\varepsilon}''_{k_j} \to \tilde{\varepsilon}'',$$

where $\varepsilon' > 0$ and, by (8.15), $\tilde{\varepsilon}'' > 0$.

Choose $K \geqslant K_1$ such that for $k_j > K$

$$\varepsilon'_{k_j} \leqslant 2\varepsilon', \quad \bar{\varepsilon}''_{k_j} \geqslant \frac{\bar{\varepsilon}''}{2}.$$

By (8.14), we have for the same values of k_j

$$\psi_{\bar{\varepsilon}''/2}(X_{k_j}) \leqslant -\frac{b}{4\varepsilon'}.$$

Combined with the relations

$$R_{\bar{\varepsilon}''/2}(X_{k_j}) \subset R_{\bar{\varepsilon}''_{k_j}}(X_{k_j}), \quad R_{\bar{\varepsilon}''_{k_j}}(X_{k_j}) = R_{\varepsilon''_{k_j}}(X_{k_j})$$

this inequality leads to a contradiction, proving the theorem.

Remark 1. The advantage of the D-function as a tool for determination of stationary points X^* of $\varphi(X)$ $(D(X^*)=0)$ over the ψ-function $(\psi(X^*)\geqslant 0)$ is that $D(X)$ is continuous throughout E_n, whereas $\psi(X)$ may experience discontinuities at the stationary points (see remark at the end of §5).

Remark 2. A more general definition of $D(X)$ is

$$D(X) = \inf_{\varepsilon \in [0, \bar{\varepsilon}]} h(\varepsilon)\psi_\varepsilon(X),$$

where $h(\varepsilon)$ is a strictly increasing function on $[0, \bar{\varepsilon}]$, vanishing at $\varepsilon = 0$.

In this general case too the function $D(X)$ is continuous and the successive approximations based on this new D-function will converge.

4. Let us use the D-method to find a stationary point of the function $\varphi(X)$ of Example 2 (§5). Recall that

$$X = (x_1, x_2) \in E_2,$$
$$f_1(X) = -5x_1 + x_2, \; f_2(X) = 4x_2 + x_1^2 + x_2^2, \; f_3(X) = 5x_1 + x_2,$$
$$\varphi(X) = \max_{i \in [1:3]} f_i(X).$$

Set $\bar{\varepsilon} = 2$. As initial approximation, we take $X_0 = (-0{,}1; 0{,}2)$. The results of the computations are presented in Table 3. It is noteworthy that X_1 lies on a ray which, if utilized as the basis for a steepest-descent procedure, leads to a nonstationary point $\bar{X} = (0, 0)$.

Thus, the D-method not only enables one to avoid the point $\bar{X}$, which was a major obstacle in the method of steepest descent, but also, in only four steps, yields the stationary point $X^* = (0, -3)$, accurate up to six decimal places.

Figure 24 illustrates the procedure geometrically.

5. In order to examine the meaning of the D-method, let us assume that the functions $f_i(X)$ are linear:

$$f_i(X) = (A_i, X) + b_i, \quad i \in [0:N].$$

TABLE 3.

k	$X_k=\begin{pmatrix}x_1^{(k)}\\x_2^{(k)}\end{pmatrix}$	$f_1(X_k)$ $f_2(X_k)$ $f_3(X_k)$	$a_1(X_k)$ $a_2(X_k)$	$R(X_k)$ $R_{a_1}(X_k)$ $R_{a_2}(X_k)$	$\psi(X_k)$ $\psi_{a_1}(X_k)$ $\psi_{a_2}(X_k)$	$a_1\psi(X_k)$ $a_2\psi_{a_1}(X_k)$ $\bar{\varepsilon}\psi_{a_2}(X_k)$	$D(X_k)$	$R_{\varepsilon_k''}(X_k)$	$g_k=\begin{pmatrix}g_1^{(k)}\\g_2^{(k)}\end{pmatrix}$	α_k
0	−0.100000 0.200000	0.700000 0.850000 −0.300000	0.150000 1.150000	{2} {1, 2} {1, 2, 3}	−4.404500 −3.700000 −1.000000	−0.660675 −4.255000 −2.000000	−4.255000	{1, 2}	0.578017 −0.816025	0.206321
1	0.019257 0.031648	−0.064636 0.127962 0.127932	0.000030 0.192598	{2} {2, 3} {1, 2, 3}	−4.063962 −3.477632 −1.000000	−0.000122 −0.669785 −2.000000	−2.000000	{1, 2, 3}	0.000000 −1.000000	3.063275
2	0.019257 −3.031627	−3.127911 −2.935374 −2.935343	0.000031 0.192568	{3} {2, 3} {1, 2, 3}	−5.099020 −1.775831 1.000000	−0.000158 −0.341968 2.000000	−0.341968	{2, 3}	−0.525345 0.850889	0.036909
3	−0.000133 −3.000222	−2.999557 −2.999554 −3.000885	0.000003 0.001333	{2} {1, 2} {1, 2, 3}	−2.000445 −1.715513 1.000000	−0.000006 −0.002287 2.000000	−0.002287	{1, 2}	0.514590 0.857437	0.000258
4	0.000000 −3.000000	−3.000000 −3.000000 −3.000000	—	$R(X_4)=$ $=\{1, 2, 3\}$	$\psi(X_4)=$ $=1.000000$	—	0.000000	—	—	—

Then $\frac{\partial f_i(X)}{\partial X} = A_i$. Let $\max_{i\in[0:N]} \|A_i\| = M$.

Let X_k be a point for which $D(X_k) < 0$. For each $a_s = a_s(X_k)$, $0 = a_0 < a_1 < \ldots < a_m$, we can construct a vector $g_k^{(s)} = g_{a_s}(X_k)$ such that $\|g_k^{(s)}\| = 1$ and

$$\psi_{a_s}(X_k) = \max_{i\in R_{a_s}(X_k)} (A_i, g_k^{(s)}). \qquad (8.16)$$

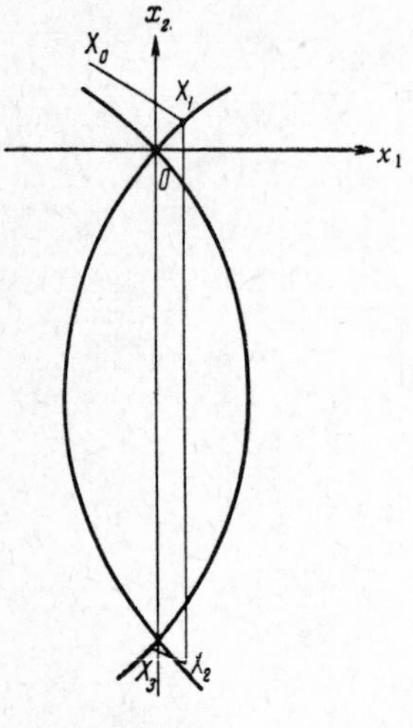

FIGURE 24.

Now, the D-method singles out a vector $g_k^{(s)}$ for which the product $a_{s+1}\psi_{a_s}(X_k)$ is minimal ($s \in [0:\bar{m}]$, $a_{\bar{m}+1} = \bar{\varepsilon}$). This choice is justified as follows.

Since

$$f_i(X_k + \alpha g_k^{(s)}) = f_i(X_k) + \alpha(A_i, g_k^{(s)}),$$

it follows that for $i \in R(X_k)$

$$f_i(X_k + \alpha g_k^{(s)}) \geqslant \varphi(X_k) - \alpha M.$$

Hence it follows that

$$\max_{i\in R_{a_s}(X_k)} f_i(X_k + \alpha g_k^{(s)}) \geqslant \varphi(X_k) - \alpha M. \quad (8.17)$$

Consider $i \notin R_{a_s}(X_k)$. In view of the fact that for $\varepsilon \in [a_s, a_{s+1})$

$$R_\varepsilon(X_k) = R_{a_s}(X_k),$$

we see that

$$f_i(X_k + \alpha g_k^{(s)}) \leqslant \varphi(X_k) - \varepsilon + \alpha M.$$

Since ε may be chosen arbitrarily close to a_{s+1}, it follows that for $i \notin R_{a_s}(X_k)$

$$f_i(X_k + \alpha g_k^{(s)}) \leqslant \varphi(X_k) - a_{s+1} + \alpha M.$$

Hence

$$\max_{i\notin R_{a_s}(X_k)} f_i(X_k + \alpha g_k^{(s)}) \leqslant \varphi(X_k) - (a_{s+1} - \alpha M). \qquad (8.18)$$

Now the expressions on the right of (8.17) and (8.18) are equal if $\alpha_0^{(s)} = \frac{a_{s+1}}{2M}$. Thus, if $\alpha \in [0, \alpha_0^{(s)}]$, the inequality

$$\max_{i\in R_{a_s}(X_k)} f_i(X_k + \alpha g_k^{(s)}) \geqslant \max_{i\notin R_{a_s}(X_k)} f_i(X_k + \alpha g_k^{(s)})$$

is guaranteed. In particular,

$$\varphi(X_k + \alpha_0^{(s)} g_k^{(s)}) = \max_{i \in R_{a_s}(X_k)} f_i(X_k + \alpha_0^{(s)} g_k^{(s)}).$$

Next, in view of (8.16), we obtain

$$\max_{i \in R_{a_s}(X_k)} f_i(X_k + \alpha_0^{(s)} g_k^{(s)}) \leqslant \varphi(X_k) + \alpha_0^{(s)} \max_{i \in R_{a_s}(X_k)} (A_i, g_k^{(s)}) =$$
$$= \varphi(X_k) + \frac{1}{2M} a_{s+1} \psi_{a_s}(X_k).$$

Finally,

$$\varphi(X_k + \alpha_0^{(s)} g_k^{(s)}) \leqslant \varphi(X_k) + \frac{1}{2M} a_{s+1} \psi_{a_s}(X_k).$$

Now if $a_{s+1}\psi_{a_s}(X_k)$, is negative, its absolute value (disregarding the factor $1/2M$ indicates by how much one can expect the function $\varphi(X)$ to decrease if the search sets out from X_k in the direction $g_k^{(s)}$. It is natural to select the vector $g_k^{(s)}$ in such a way as to minimize the product $a_{s+1}\psi_{a_s}(X_k)$. Note that in this case it follows from (8.2) that

$$\varphi(X_k + \alpha_0^{(s)} g_k^{(s)}) \leqslant \varphi(X_k) + \frac{1}{2M} D(X_k).$$

It is this idea that is utilized in the D-method.

§9. CONCLUDING REMARKS

1. In §§6–8 we described three successive approximation procedures for determination of stationary points of the maximum function $\varphi(X)$. The first is mainly of theoretical interest, as a direct generalization of the method of steepest descent. The third method is highly effective for small n and N. The most convenient for practical use, however, is the second method. We shall therefore devote closer attention to the basic component of this method – the determination of ε-stationary points ($\varepsilon > 0$).

At each step of the computation procedure described in §7.1, we have to solve three auxiliary problems:

I. Does the origin lie inside the polyhedron $L_\varepsilon(X_k)$?

II. If $\mathbf{0} \notin L_\varepsilon(X_k)$, find $g_k = g_\varepsilon(X_k)$, i. e., the direction of ε-steepest descent of $\varphi(X)$ at X_k.

III. Find α_k – the minimum point of the function $\varphi(X_k + \alpha g_k)$ on the half-axis $\alpha \in [0, \infty)$.

The first problem is essentially a linear programming problem. Indeed, let k be fixed and let $Z_1, \ldots, Z_s$ be points of $H_\varepsilon(X_k)$. Any point of $L_\varepsilon(X_k)$ may be expressed in the form

$$V = \sum_{i=1}^{s} \alpha_i Z_i; \quad \alpha_i \geqslant 0, \quad \sum_{i=1}^{s} \alpha_i = 1.$$

The question whether the origin lies in $L_\varepsilon(X_k)$ is equivalent to whether the system

$$\sum_{i=1}^{s} \alpha_i Z_i = 0$$

is solvable under the additional conditions $\sum_{i=1}^{s} \alpha_i = 1$; $\alpha_i \geqslant 0$, $i \in [1:s]$. This suggests the following linear programming problem in the space of vectors $W = (\alpha_1, \ldots, \alpha_s, u) \in E_{s+1}$: find min u, under the constraints

$$\sum_{i=1}^{s} \alpha_i z_k^{(i)} - u \leqslant 0, \quad k \in [1:n];$$

$$-\sum_{i=1}^{s} \alpha_i z_k^{(i)} - u \leqslant 0, \quad k \in [1:n];$$

$$\sum_{i=1}^{s} \alpha_i = 1; \quad \alpha_i \geqslant 0, \quad i \in [1:s]; \quad u \geqslant 0.$$

Let $W_0 = (\alpha_1^{(0)}, \ldots, \alpha_s^{(0)}, u^{(0)})$ denote a solution of this problem. It is readily seen that $\mathbf{0} \in L_\varepsilon(X_k)$ if and only if $u^{(0)} = 0$. If $u^{(0)} = 0$, then X_k is an ε-stationary point of $\varphi(X)$.

Let $u^{(0)} > 0$. Then $\mathbf{0} \notin L_\varepsilon(X_k)$, and we may proceed to the second auxiliary problem. We shall need the vector $\alpha_0 = (\alpha_1^{(0)}, \ldots, \alpha_s^{(0)})$, whose coordinates are the first s coordinates of the vector W_0.

The second auxiliary problem reduces to finding $Z_\varepsilon(X_k)$, the point of $L_\varepsilon(X_k)$ nearest the origin, since (see §6.2)

$$g_\varepsilon(X_k) = -Z_\varepsilon(X_k)/\| Z_\varepsilon(X_k) \|.$$

To determine $Z_\varepsilon(X_k)$ one can use any simple iterative method (see Appendix IV), letting the initial approximation be

$$Z_0 = \sum_{i=1}^{s} \alpha_i^{(0)} Z_i,$$

where $(\alpha_1^{(0)}, \ldots, \alpha_s^{(0)})$ is the vector obtained by solving the first auxiliary problem.

Note that the direction g_k of the search need not necessarily be $g_\varepsilon(X_k)$. Fixing the parameter $\theta_1 \in (0, 1)$ (the same for all k), it suffices to take any point V_k in $L_\varepsilon(X_k)$ for which

$$\frac{\min\limits_{Z \in H_\varepsilon(X_k)} (Z, V_k)}{(V_k, V_k)} \geqslant \theta_1 \tag{9.1}$$

(in Figure 25, the quotient on the left of (9.1) is equal to $\frac{B_k O}{V_k O}$), and then set

$$g_k = -\frac{V_k}{\| V_k \|}.$$

It can be shown that Theorem 7.1 remains valid with this choice of direction.

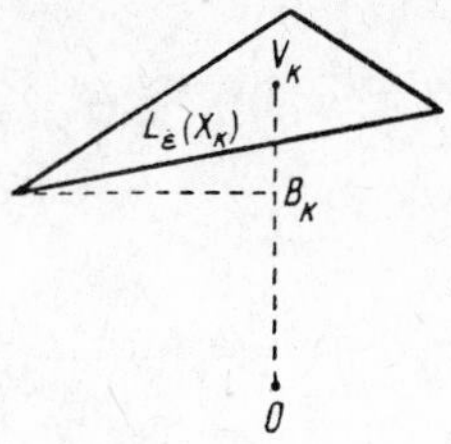

FIGURE 25.

Now for the third auxiliary problem. For fixed k, set

$$\Phi_k(\alpha) = \varphi(X_k + \alpha g_k), \quad \alpha \geqslant 0.$$

The function $\Phi_k(\alpha)$ is continuous and has continuous left and right derivatives at any point $\alpha \in (0, \infty)$, with

$$\Phi_k'(\alpha + 0) = \max_{i \in R(X_k(\alpha))} \left(\frac{\partial f_i(X_k(\alpha))}{\partial X}, g_k\right), \tag{9.2}$$

$$\Phi_k'(\alpha - 0) = \min_{i \in R(X_k(\alpha))} \left(\frac{\partial f_i(X_k(\alpha))}{\partial X}, g_k\right). \tag{9.3}$$

Here $X_k(\alpha) = X_k + \alpha g_k$. Since $\mathbf{0} \notin L_\varepsilon(X_k)$,

$$\Phi_k'(+0) = \max_{i \in R(X_k)} \left(\frac{\partial f_i(X_k)}{\partial X}, g_k\right) < 0. \tag{9.4}$$

We observe moreover that the set

$$M_k = \{\alpha \in [0, \infty) \mid \Phi_k(\alpha) \leqslant \Phi_k(0)\}$$

is bounded, since the set $M(X_0)$ is bounded and $X_k \in M(X_0)$.

We have to find α_k — the minimum point of $\Phi_k(\alpha)$ on M_k. However, Theorem 7.1 will remain in force if instead we let α_k be any point of $[0, \infty)$ for which

$$\frac{\varphi(X_k) - \varphi(X_k(\alpha_k))}{\varphi(X_k) - \min\limits_{\alpha \geqslant 0} \varphi(X_k(\alpha))} \geqslant \theta_2, \tag{9.5}$$

where $\theta_2 \in (0, 1)$ is fixed, the same for all k. The proof of this statement is left to the reader.

Let $\Phi_k(\alpha)$ be convex on $[0, \infty)$. We shall now see how to find a number α_k satisfying (9.5), in finitely many trials. We consider three number sequences $\{\alpha_{kj}^{(1)}\}$, $\{\alpha_{kj}^{(2)}\}$, $\{\alpha_{kj}^{(3)}\}$ with the properties:

I. $0 \leqslant \alpha_{kj}^{(1)} < \alpha_{kj}^{(2)} < \alpha_{kj}^{(3)}$.

II. $\Phi_k(\alpha_{kj}^{(2)}) \leqslant \Phi_k(\alpha_{kj}^{(1)})$, $\Phi_k(\alpha_{kj}^{(2)}) \leqslant \Phi_k(\alpha_{kj}^{(3)})$.

III. Each of the three sequences tends to the same limit α_k^* as $j \to \infty$ (the construction of these sequences will be explained later).

Set

$$\Delta_{kj} = \max\{\Phi'_k(\alpha_{kj}^{(2)}+0)(\alpha_{kj}^{(2)}-\alpha_{kj}^{(3)}),\ \Phi'_k(\alpha_{kj}^{(2)}-0)(\alpha_{kj}^{(2)}-\alpha_{kj}^{(1)}),\ 0\}.$$

Clearly (Figure 26),

$$0 \leqslant \Phi_k(\alpha_{kj}^{(2)}) - \min_{\alpha \geqslant 0} \Phi_k(\alpha) \leqslant \Delta_{kj}. \tag{9.6}$$

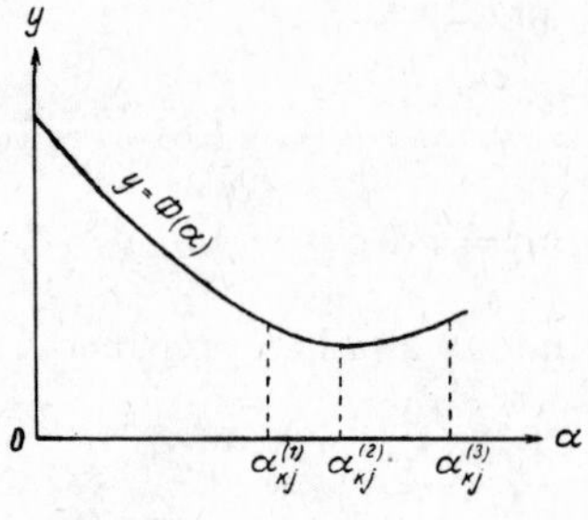

FIGURE 26.

By (9.2), (9.3) and III,

$$\Delta_{kj} \xrightarrow[j\to\infty]{} 0. \tag{9.7}$$

Letting $j \to \infty$ in (9.6), we get

$$\Phi_k(\alpha_k^*) = \min_{\alpha \geqslant 0} \Phi_k(\alpha).$$

We now define α_k to be the first term of the sequence $\{\alpha_{kj}^{(2)}\}$ for which

$$(\Phi_k(0) - \Phi_k(\alpha_{kj}^{(2)}))(1-\theta_2) \geqslant \theta_2 \Delta_{kj}$$

(the existence of such a term follows from (9.4) and (9.7). Then, as is readily verified, inequality (9.5) will hold.

It remains to show how to construct sequences $\{\alpha_{kj}^{(1)}\}$, $\{\alpha_{kj}^{(2)}\}$, $\{\alpha_{kj}^{(3)}\}$ satisfying conditions I, II, III.

We first point out a simple geometric lemma. Let B be a point on a segment AC (see Figure 27) such that

$$\frac{BC}{AC} = \frac{\sqrt{5}-1}{2} > \frac{1}{2},$$

and D the point symmetric to B relative to the midpoint of AC. Then

$$\frac{DC}{BC} = \frac{AB}{AD} = \frac{\sqrt{5}-1}{2}. \tag{9.8}$$

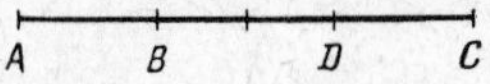

FIGURE 27.

Now let $\alpha_0 > 0$ be an arbitrary number. There are two possibilities:

$$\Phi_k(\alpha_0) < \Phi_k(0) \quad \text{and} \quad \Phi_k(\alpha_0) \geqslant \Phi_k(0).$$

In the first case, we examine the sequence $\alpha_i = q^i \alpha_0$, $i = 1, 2, \ldots$, where $q = \frac{3+\sqrt{5}}{2}$, until we find a term α_i such that $\Phi_k(\alpha_i) \geqslant \Phi_k(\alpha_{i-1})$ (the existence of α_i follows from the fact that M_k is bounded). Set

$$\alpha_{k0}^{(1)} = 0, \quad \alpha_{k0}^{(2)} = \alpha_{i-1}, \quad \alpha_{k0}^{(3)} = \alpha_i.$$

In the second case, we examine the sequence $\alpha_{-i} = q^{-i}\alpha_0$, $i = 1, 2, \ldots$, until we find a term α_{-i} such that $\Phi_k(\alpha_{-i}) \leqslant \Phi_k(0)$ (the existence of α_{-i} follows from (9.4)). Set

$$\alpha_{k0}^{(1)} = 0, \quad \alpha_{k0}^{(2)} = \alpha_{-i}, \quad \alpha_{k0}^{(3)} = \alpha_{-i+1}.$$

Thus, in either case we have constructed $\alpha_{k0}^{(1)}, \alpha_{k0}^{(2)}, \alpha_{k0}^{(3)}$, with $0 = \alpha_{k0}^{(1)} < \alpha_{k0}^{(2)} < \alpha_{k0}^{(3)}$,

$$\Phi_k(\alpha_{k0}^{(2)}) \leqslant \Phi_k(\alpha_{k0}^{(1)}), \quad \Phi_k(\alpha_{k0}^{(2)}) \leqslant \Phi_k(\alpha_{k0}^{(3)}),$$
$$\frac{\alpha_{k0}^{(3)} - \alpha_{k0}^{(2)}}{\alpha_{k0}^{(3)} - \alpha_{k0}^{(1)}} = \frac{\sqrt{5}-1}{2}.$$

Let α_{k0} denote the point symmetric to $\alpha_{k0}^{(2)}$ relative to the midpoint of the segment $[\alpha_{k0}^{(1)}, \alpha_{k0}^{(3)}]$. If $\Phi_k(\alpha_{k0}) < \Phi_k(\alpha_{k0}^{(2)})$, we set

$$\alpha_{k1}^{(1)} = \alpha_{k0}^{(2)}, \quad \alpha_{k1}^{(2)} = \alpha_{k0}, \quad \alpha_{k1}^{(3)} = \alpha_{k0}^{(3)}.$$

Otherwise,

$$\alpha_{k1}^{(1)} = \alpha_{k0}^{(1)}, \quad \alpha_{k1}^{(2)} = \alpha_{k0}^{(2)}, \quad \alpha_{k1}^{(3)} = \alpha_{k0}.$$

By (9.8), we have

$$\frac{\alpha_{k1}^{(3)} - \alpha_{k1}^{(1)}}{\alpha_{k0}^{(3)} - \alpha_{k0}^{(1)}} = \frac{\sqrt{5}-1}{2}$$

and the ratio of the length of the greater of the segments $[\alpha_{k1}^{(1)}, \alpha_{k1}^{(2)}]$, $[\alpha_{k1}^{(2)}, \alpha_{k1}^{(3)}]$ to the length of the whole segment $[\alpha_{k1}^{(1)}, \alpha_{k1}^{(3)}]$ is $\frac{\sqrt{5}-1}{2}$.

This procedure may be continued: at the $(j+1)$-th step one considers the point α_{kj} symmetric to $\alpha_{kj}^{(2)}$ relative to the midpoint of $[\alpha_{kj}^{(1)}, \alpha_{kj}^{(3)}]$. The ultimate result of this procedure are three sequences $\{\alpha_{kj}^{(1)}\}$, $\{\alpha_{kj}^{(2)}\}$, $\{\alpha_{kj}^{(3)}\}$, which, by construction, satisfy conditions I and II. Since the segments $[\alpha_{kj}^{(1)}, \alpha_{kj}^{(3)}]$ are nested and

$$\alpha_{kj}^{(3)} - \alpha_{kj}^{(1)} = \left(\frac{\sqrt{5}-1}{2}\right)^{j} (\alpha_{k0}^{(3)} - \alpha_{k0}^{(1)}),$$

all three sequences $\{\alpha_{kj}^{(1)}\}$, $\{\alpha_{kj}^{(2)}\}$, $\{\alpha_{kj}^{(3)}\}$ tend to the same limit as $j \to \infty$. Thus condition III is also fulfilled.

The construction is thus complete.

2. The ε-stationarity concept considered in §7 is most naturally termed *absolute* ε-stationarity. *Relative* ε-stationarity may be defined as follows.

Let $\mu = \inf_{X \in E_n} \varphi(X) > 0$. Set

$$R'_\varepsilon(X) = \{i \in [0:N] \mid \varphi(X) - f_i(X) \leqslant \varepsilon\varphi(X)\}, \quad \varepsilon \geqslant 0.$$

A point X^* is said to be *relatively* ε-*stationary* if

$$\min_{\|g\|=1} \max_{i \in R'_\varepsilon(X^*)} \left(\frac{\partial f_i(X^*)}{\partial X}, g\right) \geqslant 0.$$

If X^* is a relatively ε-stationary point and the functions $f_i(X^*)$, $i \in R'_\varepsilon(X^*)$, are convex on E_n, then

$$0 \leqslant \frac{\varphi(X^*) - \mu}{\varphi(X^*)} \leqslant \varepsilon.$$

Thus, in this case $\varphi(X^*)$ is an approximation to μ, with relative error at most ε.

Relatively ε-stationary points ($\varepsilon > 0$) may be determined by a modification of any method for absolute ε-stationary points (see §7.1). It can be shown that the analogs of Theorems 7.1 and 7.3 are valid.

Chapter IV

THE DISCRETE MINIMAX PROBLEM WITH CONSTRAINTS

§1. STATEMENT OF THE PROBLEM

Let $f_i(X)$, $i \in [0:N]$, $X = (x_1, \ldots, x_n)$ be functions defined and continuously differentiable on some open set $\Omega' \subset E_n$. Let Ω be a convex closed (not necessarily bounded) subset of Ω'.

Our problem is to find a point $X^* \in \Omega$ such that

$$\max_{i \in [0:N]} f_i(X^*) = \inf_{X \in \Omega} \max_{i \in [0:N]} f_i(X).$$

As in Chap. III, we shall consider the function

$$\varphi(X) = \max_{i \in [0:N]} f_i(X).$$

This function is defined on Ω'. The probem we have set is precisely to minimize the function $\varphi(X)$ on the set Ω. As proved in Chap. III, §2, $\varphi(X)$ is differentiable in all directions at any point of Ω', therefore, in particular, throughout Ω. We shall thus be able to utilize the technique developed in that chapter.

In this chapter we shall derive necessary conditions for $\varphi(X)$ to have a minimum on Ω; these conditions will then be interpreted in the geometric framework. We shall also deduce sufficient conditions for a local minimax.

We shall consider a method of successive approximations for finding stationary points — points at which the necessary condition for a minimum of $\varphi(X)$ on Ω is satisfied.

§2. NECESSARY CONDITIONS FOR A MINIMAX

1. Let $f_i(X)$, $i \in [0:N]$ be continuously differentiable functions on an open set $\Omega' \subset E_n$. As usual, we shall use the notation

$$\varphi(X) = \max_{i \in [0:N]} f_i(X),$$
$$R(X) = \{i \in [0:N] \mid f_i(X) = \varphi(X)\}.$$

Let Ω be a convex closed subset of Ω', and consider the problem of minimizing $\varphi(X)$ on Ω.

Theorem 2.1. A necessary condition for a point $X^* \in \Omega$ to be a minimum point of $\varphi(X)$ on Ω is that

$$\inf_{Z \in \Omega} \max_{i \in R(X^*)} \left(\frac{\partial f_i(X^*)}{\partial X}, Z - X^* \right) = 0. \tag{2.1}$$

If $\varphi(X)$ is convex, this condition is also sufficient.

Proof. Necessity. Let $X^* \in \Omega$ be a minimum point of $\varphi(X)$ on Ω and suppose that (2.1) fails to hold. Then there is a point $Z_1 \in \Omega$ such that

$$\max_{i \in R(X^*)} \left(\frac{\partial f_i(X^*)}{\partial X}, \; Z_1 - X^* \right) = -\rho < 0 \tag{2.2}$$

(the expression on the left of (2.1) cannot be positive). Clearly, $Z_1 \neq X^*$. We set

$$g_1 = \frac{Z_1 - X^*}{\|Z_1 - X^*\|}; \qquad \|g_1\| = 1.$$

By formula (2.7) in Chap. III,

$$\varphi(X^* + \alpha g_1) = \varphi(X^*) + \alpha \frac{\partial \varphi(X^*)}{\partial g_1} + o(g_1; \alpha). \tag{2.3}$$

It follows from (2.2) and the formula for the directional derivative that

$$\frac{\partial \varphi(X^*)}{\partial g_1} = -\frac{\rho}{\|Z_1 - X^*\|}. \tag{2.4}$$

Thus, for sufficiently small α, we have from (2.3) and (2.4)

$$\varphi(X^* + \alpha g_1) \leqslant \varphi(X^*) - \frac{\alpha\rho}{2\|Z_1 - X^*\|} < \varphi(X^*). \tag{2.5}$$

Inequality (2.5) contradicts the assumption that X^* is a minimum point of $\varphi(X)$ on Ω, since for all $\alpha \in [0, \|Z_1 - X^*\|]$ the points $X^* + \alpha g_1$ are in Ω.

Sufficiency. Let $\varphi(X)$ be convex on Ω and suppose that (2.1) is satisfied at a point $X^* \in \Omega$. We wish to show that X^* is a minimum point of $\varphi(X)$ on the set Ω. Suppose the contrary. Then there exists a point $Z_1 \in \Omega$ for which

$$\varphi(Z_1) < \varphi(X^*). \tag{2.6}$$

It is obvious that $Z_1 \neq X^*$. Setting $g_1 = \frac{Z_1 - X^*}{\|Z_1 - X^*\|}$, $\|g_1\| = 1$, we evaluate

$$\frac{\partial \varphi(X^*)}{\partial g_1} \overset{df}{=} \lim_{\alpha \to +0} \frac{1}{\alpha} [\varphi(X^* + \alpha g_1) - \varphi(X^*)]. \tag{2.7}$$

Since $\varphi(X)$ is convex, it follows that for $\beta \in [0, 1]$

$$\varphi(X^* + \beta(Z_1 - X^*)) = \varphi(\beta Z_1 + (1-\beta)X^*) \leqslant \\ \leqslant \beta\varphi(Z_1) + (1-\beta)\varphi(X^*) = \varphi(X^*) + \beta[\varphi(Z_1) - \varphi(X^*)].$$

Hence, for $\alpha \in (0, \| Z_1 - X^* \|)$,

$$\frac{1}{\alpha}[\varphi(X^* + \alpha g_1) - \varphi(X^*)] = \\ = \frac{1}{\alpha}\left[\varphi\left(X^* + \frac{\alpha}{\|Z_1 - X^*\|}(Z_1 - X^*)\right) - \varphi(X^*)\right] \leqslant \\ \leqslant \frac{1}{\|Z_1 - X^*\|}[\varphi(Z_1) - \varphi(X^*)]. \qquad (2.8)$$

Thus, by (2.6), (2.7) and (2.8),

$$\frac{\partial \varphi(X^*)}{\partial g_1} \leqslant \frac{\varphi(Z_1) - \varphi(X^*)}{\|Z_1 - X^*\|} < 0.$$

It remains to observe that

$$\max_{i \in R(X^*)} \left(\frac{\partial f_i(X^*)}{\partial X}, \ Z_1 - X^*\right) = \| Z_1 - X^* \| \frac{\partial \varphi(X^*)}{\partial g_1} \leqslant \\ \leqslant \varphi(Z_1) - \varphi(X^*) < 0.$$

This contradicts (2.1), completing the proof of the theorem.

. Remark 1. It is not hard to see that condition (2.1) may be written equivalently as

$$\inf_{\substack{Z \in \Omega \\ \|Z - X^*\| \leqslant 1}} \max_{i \in R(X^*)} \left(\frac{\partial f_i(X^*)}{\partial X}, \ Z - X^*\right) = 0. \qquad (2.1')$$

We saw in Chap. III, §3, that for fixed X the function

$$\chi(g) = \max_{i \in R(X)} \left(\frac{\partial f_i(X)}{\partial X}, \ g\right)$$

is continuous in g throughout E_n. Since the set

$$\Omega(X) = \{g = Z - X \mid Z \in \Omega, \ \| Z - X \| \leqslant 1\},$$

is closed and bounded for any fixed X, this implies that $\chi(g)$ achieves its minimum on $\Omega(X)$.

In view of this remark, we conclude that the infimum on the left of (2.1') is achieved, so that we may rewrite (2.1') thus:

$$\min_{\substack{Z \in \Omega \\ \|Z - X^*\| \leqslant 1}} \max_{i \in R(X^*)} \left(\frac{\partial f_i(X^*)}{\partial X}, \ Z - X^*\right) = 0.$$

Remark 2. If $N = 0$, Theorem 2.1 implies a necessary condition for a continuously differentiable function $f_0(X)$ to have a minimum on a convex set Ω: for $f_0(X)$ to have a minimum on Ω at a point $X^* \in \Omega$ it is necessary and, if $f_0(X)$ is convex on Ω, also sufficient that

$$\inf_{Z \in \Omega} \left(\frac{\partial f_0(X^*)}{\partial X}, \ Z - X^* \right) = 0. \tag{2.9}$$

Definition. A point $X^* \in \Omega$ at which condition (2.1) holds is called a stationary point of $\varphi(X)$ on Ω.

2. Let X be any point of Ω, fixed from now on.

Consider the cone $\Gamma(X)$ defined by

$$\Gamma(X) = \{V = \lambda(Z - X) \mid \lambda > 0, \ Z \in \Omega\}.$$

It is clear that $0 \in \Gamma(X)$. We shall call the closure of $\Gamma(X)$ the cone of feasible directions of Ω at X, and denote it by $\overline{\Gamma}(X)$. The properties of $\overline{\Gamma}(X)$ are discussed in detail in Appendix II.

Theorem 2.2. Condition (2.1) is equivalent to the inequality

$$\inf_{\substack{g \in \overline{\Gamma}(X^*) \\ \|g\|=1}} \max_{i \in R(X^*)} \left(\frac{\partial f_i(X^*)}{\partial X}, \ g \right) \geqslant 0. \tag{2.10}$$

Proof. We first show that (2.1) implies (2.10). If this is not true, we may suppose that (2.1) holds but there is a vector $g_1 \in \overline{\Gamma}(X^*)$, $\| g_1 \| = 1$, such that

$$\max_{i \in R(X^*)} \left(\frac{\partial f_i(X^*)}{\partial X}, \ g_1 \right) = -\rho < 0.$$

By the definition of $\overline{\Gamma}(X^*)$, there is a vector $V_1 \in \Gamma(X^*)$ for which

$$\max_{i \in R(X^*)} \left(\frac{\partial f_i(X^*)}{\partial X}, \ V_1 \right) \leqslant -\frac{1}{2}\rho. \tag{2.11}$$

Since $V_1 \in \Gamma(X^*)$, we have

$$V_1 = \lambda_1(Z_1 - X^*); \quad \lambda_1 > 0, \ Z_1 \in \Omega.$$

Hence, by (2.11), we have

$$\max_{i \in R(X^*)} \left(\frac{\partial f_i(X^*)}{\partial X}, \ Z_1 - X^* \right) \leqslant -\frac{\rho}{2\lambda_1} < 0,$$

contradicting (2.1). We have thus shown that (2.10) follows from (2.1).

We now prove that if (2.10) is satisfied at a point $X^* \in \Omega$, then so is (2.1). If not, there exists a point $Z_1 \in \Omega$ such that

$$\max_{i \in R(X^*)} \left(\frac{\partial f_i(X^*)}{\partial X}, \ Z_1 - X^* \right) = -\rho < 0. \tag{2.12}$$

It is clear that $Z_1 \neq X^*$. By (2.12),

$$\max_{i \in R(X^*)} \left(\frac{\partial f_i(X^*)}{\partial X}, \frac{Z_1 - X^*}{\|Z_1 - X^*\|} \right) = -\frac{\rho}{\|Z_1 - X^*\|} < 0,$$

which contradicts (2.10), because the vector

$$g_1 = \frac{1}{\|Z_1 - X^*\|}(Z_1 - X^*)$$

is in $\Gamma(X^*)$ and a fortiori in $\overline{\Gamma}(X^*)$, and $\|g_1\| = 1$.

This completes the proof.

Remark 1. Since the set

$$\{g \mid g \in \overline{\Gamma}(X^*),\ \|g\| = 1\}$$

is closed and bounded, the infimum on the left of (2.10) is attained and so we may rewrite (2.10) as

$$\min_{\substack{g \in \overline{\Gamma}(X^*) \\ \|g\|=1}} \max_{i \in R(X^*)} \left(\frac{\partial f_i(X^*)}{\partial X}, g \right) \geqslant 0. \tag{2.10'}$$

Remark 2. Consider the case $\Omega = E_n$. For any point $X \in E_n$,

$$\overline{\Gamma}(X) = \Gamma(X) = E_n,$$

so that condition (2.10') becomes

$$\min_{\|g\|=1} \max_{i \in R(X^*)} \left(\frac{\partial f_i(X^*)}{\partial X}, g \right) \geqslant 0,$$

which is simply condition (3.1) of Chap. III.

§3. GEOMETRIC INTERPRETATION OF THE NECESSARY CONDITIONS

1. Fix $X \in \Omega$, and let $\overline{\Gamma}(X)$ be the cone of feasible directions of Ω at X. We define the *dual cone* $\Gamma^+(X)$ to be

$$\Gamma^+(X) = \{Z \in E_n \mid (Z, V) \geqslant 0 \quad \text{for all} \quad V \in \overline{\Gamma}(X)\}.$$

The properties of dual cones are presented in detail in Appendix II, §2. In particular, it is shown there that $\Gamma^+(X)$ is closed and bounded.

As before, let

$$L(X) = \left\{ Z = \sum_{i \in R(X)} \alpha_i \frac{\partial f_i(X)}{\partial X} \,\middle|\, \alpha_i \geqslant 0,\ \sum_{i \in R(X)} \alpha_i = 1 \right\}.$$

Theorem 3.1. Condition (2.1) (or, equivalently, condition (2.10)) is equivalent to

$$\Gamma^{+}(X^{*}) \cap L(X^{*}) \neq \varnothing. \tag{3.1}$$

Proof. We first show that if condition (2.10) holds at some point $X^{*} \in \Omega$, then so does (3.1). Suppose the contrary. Then, by the separation theorem (Appendix II, Theorem 2.1), there exists a vector $W_0 \in \Gamma^{++}(X^{*})$ such that

$$\max_{Z \in L(X^{*})} (W_0, Z) < 0. \tag{3.2}$$

In view of Appendix II, Lemma 2.3, we have

$$\Gamma^{++}(X^{*}) = \overline{\Gamma}(X^{*}),$$

so that $W_0 \in \overline{\Gamma}(X^{*})$. In addition, by (3.2), $\| W_0 \| > 0$. Set $g_0 = W_0/\| W_0 \|$. Since

$$\max_{Z \in L(X^{*})} (g_0, Z) = \max_{i \in R(X^{*})} \left(\frac{\partial f_i(X^{*})}{\partial X}, g_0 \right), \tag{3.3}$$

it follows from (3.2) that

$$\max_{i \in R(X^{*})} \left(\frac{\partial f_i(X^{*})}{\partial X}, g_0 \right) = \frac{1}{\| W_0 \|} \max_{Z \in L(X^{*})} (W_0, Z) < 0,$$

which contradicts (2.10), because $g_0 \in \overline{\Gamma}(X^{*})$ and $\| g_0 \| = 1$.

We have thus shown that (2.10) implies (3.1).

Conversely, suppose that (3.1) is true but not (2.10). Then there exists a vector $g_0 \in \overline{\Gamma}(X^{*})$, $\| g_0 \| = 1$ for which

$$\max_{i \in R(X^{*})} \left(\frac{\partial f_i(X^{*})}{\partial X}, g_0 \right) < 0.$$

By (3.3), this inequality may be written

$$\max_{Z \in L(X^{*})} (g_0, Z) < 0,$$

with $g_0 \in \Gamma^{++}(X^{*})$, since $\overline{\Gamma}(X^{*}) = \Gamma^{++}(X^{*})$.

By Appendix II, Theorem 2.1, this implies that

$$\Gamma^{+}(X^{*}) \cap L(X^{*}) = \varnothing,$$

contradicting (3.1).

This completes the proof.

Remark 1. If $\Omega = E_n$, then $\Gamma^{+}(X) = \{0\}$ for any $X \in E_n$. Thus condition (3.1) reduces to

$$0 \in L(X^{*}). \tag{3.4}$$

This is precisely the necessary condition derived in Chap. III, §3, for the existence of a minimax on the entire space E_n.

Remark 2. Let $N=0$. Then $L(X)=\left\{\frac{\partial f_0(X)}{\partial X}\right\}$ for any $X\in\Omega$. Using (3.1) we get the following geometric condition for a continuously differentiable function to have a minimum on Ω:

A necessary condition for a continuously differentiable function $f_0(X)$ to attain its minimum on Ω at a point $X^*\in\Omega$ is that

$$\frac{\partial f_0(X^*)}{\partial X}\in\Gamma^+(X^*).$$

If $f_0(X)$ is convex, this condition is also sufficient.

2. Now let X_0 be any point of Ω. Construct the polyhedron $L(X_0)$ and cone $\Gamma^+(X_0)$. Set

$$\rho(X_0)=\inf_{\substack{Z\in L(X_0)\\ Y\in\Gamma^+(X_0)}}\|Z-Y\|. \tag{3.5}$$

Since $L(X_0)$ and $\Gamma^+(X_0)$ are closed sets and $(L(X_0))$ is also bounded, the infimum in (3.5) is achieved. Thus there exist points $Z(X_0)\in L(X_0)$ and $Y(X_0)\in\Gamma^+(X_0)$ such that

$$\|Z(X_0)-Y(X_0)\|=\rho(X_0). \tag{3.6}$$

If $\rho(X_0)=0$, X_0 is a stationary point of $\varphi(X)$ on Ω, since then condition (3.1) is clearly satisfied.

Let $\rho(X_0)>0$, i. e., X_0 is not a stationary point. It is readily shown that then the vector $Z(X_0)-Y(X_0)$ satisfying (3.6) is unique (although the points $Z(X_0)$ and $Y(X_0)$ need not be unique; see Figure 28). We introduce the notation

$$g(X_0)=\frac{Y(X_0)-Z(X_0)}{\|Y(X_0)-Z(X_0)\|},\quad \|g(X_0)\|=1.$$

Definition. A vector $\bar g\in E_n$, $\|\bar g\|=1$ is called a direction of steepest descent for $\varphi(X)$ on Ω at X_0 if

$$\frac{\partial\varphi(X_0)}{\partial\bar g}=\min_{\substack{g\in\overline{\Gamma}(X_0)\\ \|g\|=1}}\frac{\partial\varphi(X_0)}{\partial g}.$$

Theorem 3.2. If $\rho(X_0)>0$, then $g(X_0)$ is a direction of steepest descent for $\varphi(X)$ on Ω at X_0, and

$$\frac{\partial\varphi(X_0)}{\partial g(X_0)}=-\rho(X_0). \tag{3.7}$$

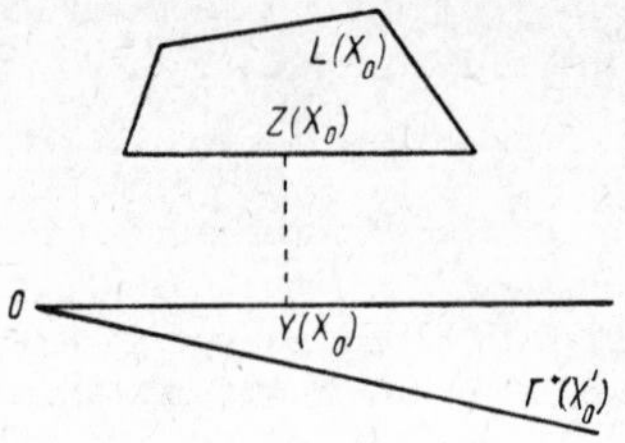

FIGURE 28.

For the proof, we need the following
Lemma 3.1.

$$(g(X_0),\ Y(X_0))=0. \tag{3.8}$$

Proof. We shall use the notation

$$Y_0=Y(X_0),\quad Z_0=Z(X_0),\quad g_0=g(X_0),\quad \rho_0=\rho(X_0).$$

Since $Y_0 \in \Gamma^+(X_0)$, it follows that for all $\lambda>0$

$$\lambda Y_0 \in \Gamma^+(X_0).$$

Next,

$$\begin{aligned}(\lambda Y_0-Z_0,\ \lambda Y_0-Z_0)&=\\ &=((\lambda-1)Y_0+(Y_0-Z_0),\ (\lambda-1)Y_0+(Y_0-Z_0))=\\ &=(\lambda-1)[(\lambda-1)(Y_0,Y_0)+2(Y_0-Z_0,\ Y_0)]+\rho_0^2.\end{aligned}$$

If we assume that $(Y_0-Z_0,\ Y_0)\neq 0$, it follows at once that for some $\lambda_0>0$

$$(\lambda_0 Y_0-Z_0,\ \lambda_0 Y_0-Z_0)<\rho_0^2.$$

But this inequality contradicts (3.5), so that $(Y_0-Z_0,\ Y_0)=0$, which is equivalent to (3.8).

Proof of Theorem 3.2. We observe that by Appendix II, Lemma 1.4 for any $Y\in\Gamma^+(X_0)$ and $Z\in L(X_0)$

$$(Y-Z,\ Y_0-Z_0)\geqslant(Y_0-Z_0,\ Y_0-Z_0). \tag{3.9}$$

We first prove that $g_0\in\overline{\Gamma}(X_0)$ or, equivalently, that $g_0\in\Gamma^{++}(X_0)$. It will suffice to show that for all $Y\in\Gamma^+(X_0)$

$$(Y,\ g_0)\geqslant 0. \tag{3.10}$$

By (3.8) and the definition of ρ_0,

$$(Z_0,\ Y_0-Z_0)=-\rho_0^2. \tag{3.11}$$

In view of (3.9) and (3.11), we obtain

$$(Y, Y_0 - Z_0) = \\ = (Y - Z_0, Y_0 - Z_0) + (Z_0, Y_0 - Z_0) \geqslant \rho_0^2 - \rho_0^2 = 0,$$

which is equivalent to (3.10).

We have thus shown that $g_0 \in \overline{\Gamma}(X_0)$.

To evaluate $\frac{\partial\varphi(X_0)}{\partial g(X_0)}$, we use (3.8) and (3.9):

$$\frac{\partial\varphi(X_0)}{\partial g(X_0)} = \max_{i \in R(X_0)} \left(\frac{\partial f_i(X_0)}{\partial X},\ g_0\right) = \max_{Z \in L(X_0)} (Z, g_0) = \\ = \frac{1}{\|Y_0 - Z_0\|} \max_{Z \in L(X_0)} (Z - Y_0, Y_0 - Z_0) = \\ = -\frac{1}{\|Y_0 - Z_0\|} (Y_0 - Z_0, Y_0 - Z_0) = -\rho_0.$$

This proves (3.7).

It remains to show that g_0 is a direction of steepest descent for $\varphi(X)$ on Ω at X_0.

We consider any vector $\bar{g} \in \overline{\Gamma}(X_0)$, $\|\bar{g}\| = 1$, and show that

$$\frac{\partial\varphi(X_0)}{\partial\bar{g}} \geqslant -\rho(X_0). \tag{3.12}$$

Since $\bar{g} \in \Gamma^{++}(X_0)$, it follows that for all $Y \in \Gamma^{+}(X_0)$

$$(Y, \bar{g}) \geqslant 0.$$

Fix $Y \in \Gamma^{+}(X_0)$ and $Z \in L(X_0)$. Then

$$\|Y - Z\| \geqslant (Y - Z, \bar{g}) \geqslant -(Z, \bar{g}) \geqslant -\max_{Z' \in L(X_0)} (Z', \bar{g}) = -\frac{\partial\varphi(X_0)}{\partial\bar{g}}.$$

Hence it follows that

$$\rho_0 \geqslant -\frac{\partial\varphi(X_0)}{\partial\bar{g}},$$

which is equivalent to (3.12), proving the theorem.

R e m a r k 1. The direction of steepest descent of $\varphi(X)$ on Ω at X_0 is unique, provided $\rho(X_0) > 0$. The proof of this statement is analogous to that of Lemma 3.4 in Chap. III.

R e m a r k 2. Theorem 3.2 asserts only that $g(X_0) \in \overline{\Gamma}(X_0)$. It may happen that for any $\alpha > 0$ we have $X_0 + \alpha g(X_0) \notin \Omega$, so that it is impossible to move along the direction of steepest descent.

3. As an example, we consider the function (see Chap. III, §5)

$$\varphi(X) = \max_{i \in [1:3]} f_i(X),$$

where

$$f_1(X) = -5x_1 + x_2, \quad f_2(X) = 4x_2 + x_1^2 + x_2^2,$$
$$f_3(X) = 5x_1 + x_2.$$

Let $X_0 = (0, 0)$. As shown previously, $L(X_0)$ is the triangle with vertices at the points $(-5,1)$, $(0,4)$, $(5,1)$. Since $0 \notin L(X_0)$, X_0 is not a stationary point for $\varphi(X)$ on E_2.

We now introduce a constraint

$$\Omega_1 = \{X \in E_2 \mid 2x_1 + x_2 \geqslant 0\}.$$

It is readily verified that in this case

$$\Gamma^+(X_0) = \{V = \alpha(2, 1) \mid \alpha \geqslant 0\}.$$

Since the ray $\Gamma^+(X_0)$ cuts the set $L(X_0)$ (Figure 29), X_0 is a stationary point for $\varphi(X)$ on the set Ω_1, and since $\varphi(X)$ is a convex function it follows that X_0 is a minimum point for $\varphi(X)$ on Ω_1.

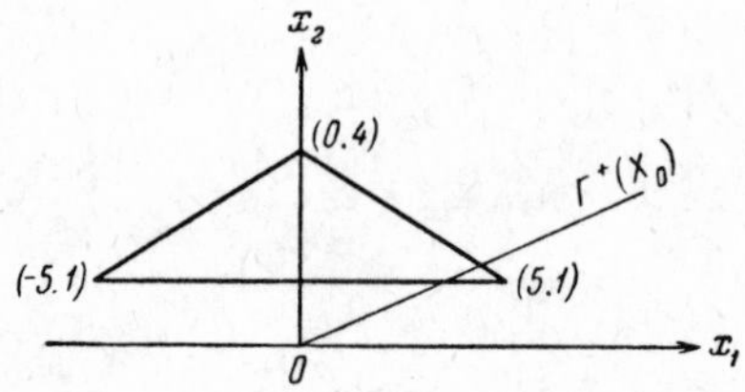

FIGURE 29.

Now consider another constraint:

$$\Omega_2 = \{X \in E_2 \mid 10x_1 + x_2 \geqslant 0\}.$$

In this case, $\Gamma^+(X_0) = \{V = \alpha(10, 1) \mid \alpha \geqslant 0\}$. Since $L(X_0)$ and $\Gamma^+(X_0)$ are disjoint, (Figure 30), we see that $X_0 = (0, 0)$ is not a stationary point of $\varphi(X)$ on Ω_2. It is not difficult to find the direction of steepest descent for $\varphi(X)$ at X_0 on Ω_2:

$$g(X_0) = \frac{(1, -10)}{\sqrt{101}}.$$

Note that $X_0 + \alpha g(X_0) \in \Omega_2$ for all $\alpha \geqslant 0$.

Finally, let

$$\Omega_4 = \{X \in E_2 \mid (x_1 - 1)^2 + x_2^2 \leqslant 1\}.$$

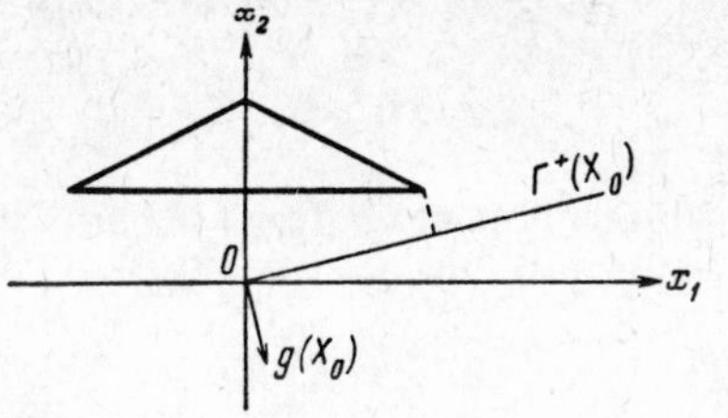

FIGURE 30.

It is readily shown that in this case

$$\Gamma^+(X_0) = \{V = \alpha(1, 0) \mid \alpha \geqslant 0\}, \quad L(X_0) \cap \Gamma^+(X_0) = \varnothing,$$
$$g(X_0) = (0, -1), \quad X_0 + \alpha g(X_0) \notin \Omega_3, \quad \text{if} \quad \alpha > 0.$$

Thus $X_0 = (0, 0)$ is not a stationary point of $\varphi(X)$ on Ω_3. At the same time, no motion in the direction of steepest descent is possible, for the vector $g(X_0)$ points out of the region Ω_3.

§4. SUFFICIENT CONDITIONS FOR A LOCAL MINIMAX WITH CONSTRAINTS

1. It was shown in §2 that if $\varphi(X)$ is a convex function then any stationary point of $\varphi(X)$ on Ω is a minimum point (on Ω). We shall now establish sufficient conditions for the existence of a local minimax.

Let $X \in \Omega$. Consider the function

$$\psi(X) \overset{df}{=} \psi(\Omega, X) = \min_{\substack{g \in \overline{\Gamma}(X) \\ \|g\| = 1}} \max_{i \in R(X)} \left(\frac{\partial f_i(X)}{\partial X}, g \right).$$

In terms of this function, the necessary condition (2.10) may be written

$$\psi(X^*) \geqslant 0.$$

Definition. A point $X^* \in \Omega$ is called a *local minimum point* of $\varphi(X)$ on the set Ω if there exists $\delta > 0$ such that, for $X \in \Omega$, $\| X - X^* \| < \delta$,

$$\varphi(X) \geqslant \varphi(X^*).$$

Under the same assumptions, if

$$\varphi(X) > \varphi(X^*)$$

for $X \in \Omega$, $\| X - X^* \| < \delta$, $X \neq X^*$, then X^* is called a *strict local minimum point* for $\varphi(X)$ on Ω.

Theorem 4.1. Let $X^* \in \Omega$ be a stationary point of $\varphi(X)$ on Ω. If

$$\psi(X^*) > 0, \tag{4.1}$$

then X^* is a strict local minimum point for $\varphi(X)$ on Ω.

Proof. If the assertion is false, we can find a sequence of points $\{X_k\}$, $X_k \in \Omega$, $k = 1, 2, \ldots$, $X_k \neq X^*$, such that

$$\| X_k - X^* \| \xrightarrow[k\to\infty]{} 0 \quad \text{and} \quad \varphi(X_k) \leqslant \varphi(X^*). \tag{4.2}$$

We may assume without loss of generality that

$$g_k \stackrel{\mathrm{df}}{=} \frac{X_k - X^*}{\|X_k - X^*\|} \xrightarrow[k\to\infty]{} g_0.$$

Clearly, $g_0 \in \bar{\Gamma}(X^*)$, $\| g_0 \| = 1$.

Take any $i \in R(X^*)$. Since $\varphi(X^*) = f_i(X^*)$, we see via (4.2) that

$$f_i(X_k) \leqslant \varphi(X_k) \leqslant \varphi(X^*) = f_i(X^*),$$

i.e.,

$$f_i(X_k) \leqslant f_i(X^*). \tag{4.3}$$

Next,

$$f_i(X_k) = f_i(X^* + \| X_k - X^* \| g_k) =$$
$$= f_i(X^*) + \| X_k - X^* \| \left(\frac{\partial f_i(X^* + \theta_k g_k)}{\partial X}, \; g_k \right),$$

where $\theta_k \in (0, \| X_k - X^* \|)$.

Hence, by (4.3), it follows that for all k

$$\left(\frac{\partial f_i(X^* + \theta_k g_k)}{\partial X}, \; g_k \right) \leqslant 0.$$

Since $\theta_k \xrightarrow[k\to\infty]{} 0$, we obtain

$$\left(\frac{\partial f_i(X^*)}{\partial X}, \; g_0 \right) \leqslant 0. \tag{4.4}$$

But (4.4) is valid for all $i \in R(X^*)$, and so

$$\max_{i \in R(X^*)} \left(\frac{\partial f_i(X^*)}{\partial X}, \; g_0 \right) \leqslant 0,$$

contradicting (4.1) and completing the proof.

We now interpret condition (4.1) geometrically.

Theorem 4.2. Condition (4.1) holds if and only if the set $L(X^*)$ and cone $\Gamma^+(X^*)$ cannot be separated; that is to say, there exist no nonzero vector $W \in E_n$ and number α such that

$$(W, Z) \leqslant \alpha \quad \text{for all} \quad Z \in L(X^*), \tag{4.5}$$
$$(W, V) \geqslant \alpha \quad \text{for all} \quad V \in \Gamma^+(X^*). \tag{4.6}$$

Proof. We first show that if $L(X^*)$ and $\Gamma^+(X^*)$ cannot be separated then $\psi(X^*) > 0$. If this is false, there exists a vector $g_0 \in \overline{\Gamma}(X^*)$, $\|g_0\| = 1$ for which

$$\max_{i \in R(X^*)} \left(\frac{\partial f_i(X^*)}{\partial X}, g_0 \right) = \max_{Z \in L(X^*)} (Z, g_0) \leqslant 0.$$

Since $\overline{\Gamma}(X^*) = \Gamma^{++}(X^*)$, it follows that $g_0 \in \Gamma^{++}(X^*)$. Thus $(g_0, V) \geqslant 0$ for all $V \in \Gamma^+(X^*)$. But then $L(X^*)$ and $\Gamma^+(X^*)$ are separable, for the vector $W = g_0$ and number $\alpha = 0$ satisfy conditions (4.5) and (4.6). Thus, it follows that $\psi(X^*) > 0$.

Now let $\psi(X^*) > 0$. We claim that $L(X^*)$ and $\Gamma^+(X^*)$ cannot be separated. Otherwise, there exist $W_0 \in E_n$ and α_0 such that

$$(W_0, Z) \leqslant \alpha_0 \quad \text{for all} \quad Z \in L(X^*), \tag{4.7}$$
$$(W_0, V) \geqslant \alpha_0 \quad \text{for all} \quad V \in \Gamma^+(X^*). \tag{4.8}$$

Without loss of generality, we may assume that $\|W_0\| = 1$. Since $\Gamma^+(X^*)$ is a cone containing the origin, it follows from (4.8) that $\alpha_0 \leqslant 0$.

Hence, by (4.7), we have

$$\max_{Z \in L(X^*)} (W_0, Z) \leqslant 0. \tag{4.9}$$

We claim that

$$(W_0, V) \geqslant 0 \quad \text{for all} \quad V \in \Gamma^+(X^*).$$

Supposing the contrary, we can find a vector $V' \in \Gamma^+(X^*)$ such that $(W_0, V') = -b < 0$. Since $\Gamma^+(X^*)$ is a cone, we must have $\beta V' \in \Gamma^+(X^*)$ for any $\beta > 0$. By (4.8),

$$(W_0, \beta V') = -\beta b \geqslant \alpha_0.$$

But α_0 is bounded and $\beta b \xrightarrow[\beta \to \infty]{} \infty$ — contradiction. This shows that indeed

$$(W_0, V) \geqslant 0 \quad \text{for all} \quad V \in \Gamma^+(X^*). \tag{4.10}$$

It follows from (4.10) that $W_0 \in \Gamma^{++}(X^*)$, and since $\Gamma^{++}(X^*) = \overline{\Gamma}(X^*)$, we have $W_0 \in \overline{\Gamma}(X^*)$.

Thus, $W_0 \in \overline{\Gamma}(X^*)$ and $\|W_0\| = 1$. In view of (4.9), we get

$$\psi(X^*) = \min_{\substack{g \in \overline{\Gamma}(X^*) \\ \|g\| = 1}} \max_{Z \in L(X^*)} (Z, g) \leqslant \max_{Z \in L(X^*)} (W_0, Z) \leqslant 0,$$

contrary to assumption. This completes the proof.

Corollary. If $\Omega = E_n$ then $\Gamma^+(X^*) = \{\mathbf{0}\}$, and the nonseparability of $L(X^*)$ and $\Gamma^+(X^*)$ means that $\mathbf{0}$ is an interior point of $L(X^*)$.

This is precisely the sufficient condition established in Chapter III.

2. The example presented below will show that a stationary point may not be a local minimum point if condition (4.1) fails to hold.

Example. Let

$$X = (x_1, x_2) \in E_2,$$
$$\varphi(X) = \max_{i \in [1:3]} f_i(X),$$

where

$$f_1(X) = -5x_1 + x_2,$$
$$f_2(X) = 4x_2 + x_1^2 + x_2^2,$$
$$f_3(X) = 5x_1 + x_2 - x_1^2.$$

Set $\Omega = \{X \in E_2 \mid 5x_1 + x_2 \geqslant 0\}$, and let $X^* = (0, 0)$. It is clear that $R(X^*) = \{1, 2, 3\}$; $L(X^*)$ is the triangle with vertices at $(-5, 1)$, $(0, 4)$, $(5, 1)$; $\Gamma^+(X^*) = \{V = \alpha(5, 1) \mid \alpha \geqslant 0\}$. It is readily shown (Figure 31) that

$$L(X^*) \cap \Gamma^+(X^*) \neq \varnothing,$$

so that $X^* = (0, 0)$ is a stationary point for $\varphi(X)$ on Ω.

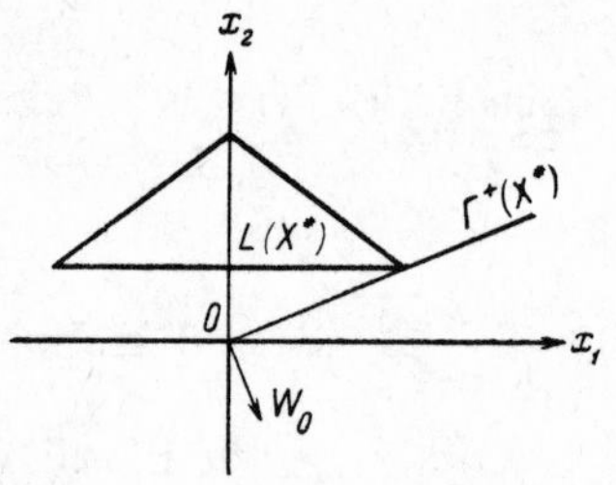

FIGURE 31.

On the other hand, the condition $\psi(X^*) > 0$ is not satisfied, for $L(X^*)$ and $\Gamma^+(X^*)$ may be separated by the straight line $(X, W_0) = 0$, where $W_0 = (1, -5)$. Indeed,

$$(W_0, Z) \leqslant 0 \quad \text{for all} \quad Z \in L(X^*),$$
$$(W_0, V) = 0 \quad \text{for all} \quad V \in \Gamma^+(X^*).$$

Thus $\psi(X^*) = 0$.

We claim that X^* is not a local minimum point of $\varphi(X)$ on Ω. Consider the point $X_\alpha = \alpha W_0$, $\alpha > 0$. Clearly, $X_\alpha \in \Omega$. We have

$$f_1(X_\alpha) = -10\alpha, \quad f_2(X_\alpha) = -20\alpha + 26\alpha^2, \quad f_3(X_\alpha) = -\alpha^2.$$

Hence it follows that for small $\alpha>0$

$$\varphi(X_\alpha)<0=\varphi(X^*),$$

so that $X^*=(0,0)$ is not a local minimum point of $\varphi(X)$ on Ω.

The reason for this situation is that $f_3(X)$ is not a convex function, as a result of which $\varphi(X)$ is also nonconvex.

3*. We end the section with another sufficient condition for the existence of a local minimum.

Theorem 4.3. Let $X^*\in\Omega$ be a stationary point of $\varphi(X)$ on Ω, such that $\psi(X^*)=0$. Assume that the functions $f_i(X)$, $i\in[0:N]$, are twice continuously differentiable in some neighborhood $S_\delta(X^*)$, $\delta>0$, of X^*. If there exists $\gamma>0$ such that

$$\min_{g\in G_\gamma}\ \max_{i\in R_2(X^*,g)}\left(\frac{\partial^2 f_i(X^*)}{\partial X^2}g,\ g\right)>0,$$

where

$$G_\gamma=\left\{g \,\Big|\, \|g\|=1,\ g\in\overline{\Gamma}(X^*),\ 0\leqslant\frac{\partial\varphi(X^*)}{\partial g}\leqslant\gamma\right\},$$

$$R_2(X^*,g)=\left\{i \,\Big|\, i\in R(X^*),\ \left(\frac{\partial f_i(X^*)}{\partial X},\ g\right)=\frac{\partial\varphi(X^*)}{\partial g}\right\},$$

then X^* is a strict local minimum point of $\varphi(X)$ on Ω.

The proof, which is analogous to that of Theorem 4.2 in Chap. III, is left to the reader.

§5. SOME ESTIMATES

1. *Theorem 5.1. Suppose that for some point $X_0\in\Omega$ and some index set $Q\subset[0:N]$*

$$\inf_{Z\in\Omega}\ \max_{i\in Q}\left(\frac{\partial f_i(X_0)}{\partial X},\ Z-X_0\right)\overset{\mathrm{df}}{=}-a\leqslant 0, \tag{5.1}$$

and that the functions $f_i(X)$ are convex on Ω for all $i\in Q$. Then

$$\min_{i\in Q} f_i(X_0)-a\leqslant\inf_{X\in\Omega}\varphi(X)\leqslant\varphi(X_0). \tag{5.2}$$

Proof. The right-hand inequality in (5.2) is trivial. To prove the other inequality, suppose that it is false. Then there is a point $\overline{X}\in\Omega$ such that

$$\min_{i\in Q} f_i(X_0)-\varphi(\overline{X})\overset{\mathrm{df}}{=}a_1>a. \tag{5.3}$$

Note that for all $i \in Q$

$$f_i(X_0) - f_i(\bar{X}) \geqslant \min_{i \in Q} f_i(X_0) - \varphi(\bar{X}) = a_1. \tag{5.4}$$

In view of the convexity of the functions $f_i(X), i \in Q$, Lemma 3.2 of Appendix II and (5.4), we obtain

$$\left(\frac{\partial f_i(X_0)}{\partial X}, \bar{X} - X_0\right) \leqslant f_i(\bar{X}) - f_i(X_0) \leqslant -a_1.$$

This inequality is valid for all $i \in Q$, so that

$$\max_{i \in Q}\left(\frac{\partial f_i(X_0)}{\partial X}, \bar{X} - X_0\right) \leqslant -a_1.$$

It now follows from this inequality and from (5.1) that $-a \leqslant -a_1$, or $a \geqslant a_1$, contradicting (5.3) and proving the theorem.

Corollary 1. *Under the assumptions of Theorem 5.1, if $Q \subset R(X_0)$, then*

$$\varphi(X_0) - a \leqslant \inf_{X \in \Omega} \varphi(X) \leqslant \varphi(X_0).$$

Indeed, in this case

$$\min_{i \in Q} f_i(X_0) = \varphi(X_0).$$

Corollary 2. *Under the assumptions of Theorem 5.1, if $a = 0$, then*

$$\min_{i \in Q} f_i(X_0) \leqslant \inf_{X \in \Omega} \varphi(X) \leqslant \varphi(X_0).$$

It is easy to see that $a = 0$ if and only if

$$\Gamma^+(X_0) \cap L(Q, X_0) \neq \varnothing,$$

where $L(Q, X_0)$ is the convex hull of the points $\frac{\partial f_i(X_0)}{\partial X}$, $i \in Q$ (see the proof of Theorem 3.1).

Corollary 3. *Under the assumptions of Theorem 5.1, if $Q \subset R(X_0)$ and $a = 0$, then*

$$\inf_{X \in \Omega} \varphi(X) = \varphi(X_0),$$

i.e., X_0 is a minimum point of $\varphi(X)$ on the closed convex set Ω.

Remark. Suppose that Ω is a bounded set, and let $d > 0$ be an upper bound for its diameter. Suppose moreover that at $X_0 \in \Omega$ we have

$$\min_{\substack{g \in \bar{\Gamma}(X_0) \\ \|g\|=1}} \max_{i \in Q}\left(\frac{\partial f_i(X_0)}{\partial X}, g\right) \stackrel{df}{=} -a \leqslant 0, \tag{5.5}$$

where $Q \subset [0:N]$. Suppose further that the functions $f_i(X)$ are convex on Ω for all $i \in Q$. Then

$$\min_{i \in Q} f_i(X_0) - ad \leqslant \min_{X \in \Omega} \varphi(X) \leqslant \varphi(X_0). \tag{5.6}$$

We first prove that

$$\inf_{Z \in \Omega} \max_{i \in Q} \left(\frac{\partial f_i(X_0)}{\partial X}, Z - X_0 \right) \geqslant -ad. \tag{5.7}$$

It will suffice to show that for all $Z \in \Omega$

$$\max_{i \in Q} \left(\frac{\partial f_i(X_0)}{\partial X}, Z - X_0 \right) \geqslant -ad.$$

This is trivial for $Z = X_0$, so we may assume that $Z \neq X_0$. Then, by (5.5),

$$\begin{aligned} \max_{i \in Q} \left(\frac{\partial f_i(X_0)}{\partial X}, Z - X_0 \right) = \\ = \| Z - X_0 \| \max_{i \in Q} \left(\frac{\partial f_i(X_0)}{\partial X}, \frac{Z - X_0}{\| Z - X_0 \|} \right) \geqslant \\ \geqslant -a \| Z - X_0 \| \geqslant -ad. \end{aligned}$$

This proves inequality (5.7), and (5.6) now follows from (5.7) and Theorem 5.1.

2. *Theorem 5.2. Suppose that for some point $X_0 \in \Omega$ there exists an index set $Q \subset [0:N]$ such that*

$$\min_{\substack{g \in \overline{\Gamma}(X_0) \\ \|g\|=1}} \max_{i \in Q} \left(\frac{\partial f_i(X_0)}{\partial X}, g \right) \stackrel{\text{df}}{=} -a \leqslant 0. \tag{5.8}$$

Suppose further that the functions $f_i(X)$ are twice continuously differentiable on Ω' for all $i \in Q$, and for all $i \in Q$, $X \in \Omega$ and $V \in E_n$

$$\left(\frac{\partial^2 f_i(X)}{\partial X^2} V, V \right) \geqslant m \| V \|^2, \tag{5.9}$$

where $m > 0$ is independent of i, X, V. Then

$$\min_{i \in Q} f_i(X_0) - \frac{a^2}{2m} \leqslant \inf_{X \in \Omega} \varphi(X) \leqslant \varphi(X_0). \tag{5.10}$$

Proof. Since the right-hand inequality in (5.10) is trivial, we confine attention to the left-hand one.

For any $Z \in \Omega$ and all $i \in [0:N]$,

$$\begin{aligned} f_i(Z) = f_i(X_0 + (Z - X_0)) = f_i(X_0) + \left(\frac{\partial f_i(X_0)}{\partial X}, Z - X_0 \right) + \\ + \frac{1}{2} \left(\frac{\partial^2 f_i(X_0 + \theta_i(Z - X_0))}{\partial X^2} (Z - X_0), Z - X_0 \right), \end{aligned} \tag{5.11}$$

where $\theta_i \in (0, 1)$.

By (5.8), for any $Z \in \Omega$ there exists $i(Z) \in Q$ such that

$$\left(\frac{\partial f_{i(Z)}(X_0)}{\partial X},\ Z - X_0\right) \geqslant -\|Z - X_0\| a. \tag{5.12}$$

In view of (5.11), (5.12), (5.9), we obtain

$$f_{i(Z)}(Z) \geqslant f_{i(Z)}(X_0) - a\|Z - X_0\| + \frac{1}{2} m \|Z - X_0\|^2 \geqslant$$
$$\geqslant \min_{i \in Q} f_i(X_0) - a\|Z - X_0\| + \frac{1}{2} m\|Z - X_0\|^2.$$

Since

$$\frac{1}{2} m\|Z - X_0\|^2 - a\|Z - X_0\| \geqslant \min_{-\infty<\beta<\infty}\left[\frac{1}{2} m\beta^2 - a\beta\right] = -\frac{a^2}{2m},$$

it follows that

$$\varphi(Z) \geqslant f_{i(Z)}(Z) \geqslant \min_{i \in Q} f_i(X_0) - \frac{a^2}{2m}.$$

Hence $\inf_{Z \in \Omega} \varphi(Z) \geqslant \min_{i \in Q} f_i(X_0) - \frac{a^2}{2m}$, and the proof is complete.

Corollary. Under the assumptions of Theorem 5.2, if $Q \subset R(X_0)$, then

$$0 \leqslant \varphi(X_0) - \inf_{Z \in \Omega} \varphi(Z) \leqslant \frac{a^2}{2m}.$$

§6. METHOD OF SUCCESSIVE APPROXIMATIONS

1. In this section we shall make systematic use of a special norm: if $V = (v_1, \ldots, v_n)$, then

$$\|V\|_I = \max_{i \in [1:n]} |v_i|.$$

Clearly, $\|V\| \leqslant \sqrt{n}\|V\|_I$.

As usual, we assume that Ω is a closed convex (not necessarily bounded) subset of E_n. For fixed $X \in \Omega$, we set

$$\Omega(X) = \{Z \in \Omega \mid \|Z - X\|_I \leqslant 1\}.$$

We introduce the functions

$$\psi_1(X) = \min_{Z \in \Omega(X)} \max_{i \in R(X)} \left(\frac{\partial f_i(X)}{\partial X},\ Z - X\right),$$
$$\psi_{1\varepsilon}(X) = \min_{Z \in \Omega(X)} \max_{i \in R_\varepsilon(X)} \left(\frac{\partial f_i(X)}{\partial X},\ Z - X\right),$$

where

$$R(X) = \{i \in [0:N] \mid f_i(X) = \varphi(X)\},$$
$$R_\varepsilon(X) = \{i \in [0:N] \mid \varphi(X) - f_i(X) \leqslant \varepsilon\}.$$

It is readily seen that $X^* \in \Omega$ is a stationary point of $\varphi(X)$ on Ω if and only if $\psi_1(X^*) = 0$. We observe moreover that for fixed $X \in \Omega$ the function $\psi_{1\varepsilon}(X)$ is a nondecreasing and piecewise constant function of ε, $0 \leqslant \varepsilon < \infty$, and there exists $a_1(X) > 0$ such that

$$\psi_{1\varepsilon}(X) = \psi_1(X) \quad \text{for} \quad \varepsilon \in [0, a_1(X)). \tag{6.1}$$

We now describe a successive approximation procedure for finding stationary points of $\varphi(X)$ on Ω.

Fix $\varepsilon_0 > 0$ and $\rho_0 > 0$ and select an initial approximation $X_0 \in \Omega$. We shall assume that the set $M(X_0) = \{X \in \Omega \mid \varphi(X) \leqslant \varphi(X_0)\}$ is bounded (it is clearly also closed).

Suppose we have already determined the k-th approximation $X_k \in M(X_0)$. If $\psi_1(X_k) = 0$ X_k is a stationary point and we are done.

If $\psi_1(X_k) < 0$, we examine the sequence of numbers

$$\varepsilon_\nu = \frac{\varepsilon_0}{2^\nu}, \quad \nu = 0, 1, 2, \ldots,$$

to find a term for which

$$\psi_{1\varepsilon_\nu}(X_k) \leqslant -\frac{\rho_0}{\varepsilon_0} \varepsilon_\nu \tag{6.2}$$

(this may of course be the case for $\nu = 0$). Denote the first ν satisfying (6.2) by ν_k. That ν_k is finite follows from the fact that the right-hand side of inequality (6.2) tends to zero as $\nu \to \infty$, while the left-hand side tends to $\psi_1(X_k) < 0$ by (6.1).

Thus,

$$\psi_{1\varepsilon_{\nu_k}}(X_k) \leqslant -\frac{\rho_0}{\varepsilon_0} \varepsilon_{\nu_k},$$

and if $\nu_k > 0$ then

$$\psi_{1\varepsilon_{\nu_k - 1}}(X_k) > -\frac{\rho_0}{\varepsilon_0} \varepsilon_{\nu_k - 1}.$$

We shall use the notation

$$\hat{\varepsilon}_k = \varepsilon_{\nu_k}.$$

Let $Z_k \in \Omega(X_k)$ be a point such that

$$\psi_{1\hat{\varepsilon}_k}(X_k) = \max_{i \in R_{\hat{\varepsilon}_k}(X_k)} \left(\frac{\partial f_i(X_k)}{\partial X}, Z_k - X_k \right).$$

Consider the segment

$$X = X_k(\alpha) \overset{df}{=} X_k + \alpha(Z_k - X_k), \quad 0 \leqslant \alpha \leqslant 1.$$

Clearly, $X_k(\alpha) \in \Omega$ for $\alpha \in [0, 1]$. Determine $\alpha_k \in [0, 1]$ for which

$$\varphi(X_k(\alpha_k)) = \min_{\alpha \in [0, 1]} \varphi(X_k(\alpha)).$$

Now set

$$X_{k+1} = X_k(\alpha_k).$$

It is clear that $X_{k+1} \in \Omega$. It is also easily seen that

$$\varphi(X_{k+1}) < \varphi(X_k)$$

(see the proof of inequality (6.6) in Chap. III). Hence it follows in particular that $X_{k+1} \in M(X_0)$.

Continuing the procedure as described, we obtain a sequence $\{X_k\}$ and auxiliary sequences $\{\hat{\varepsilon}_k\}$, $\{Z_k\}$, $\{\alpha_k\}$, with $X_k \in M(X_0)$, $Z_k \in \Omega(X_k)$, $k = 0, 1, 2, \ldots$, and

$$\varphi(X_0) > \varphi(X_1) > \ldots > \varphi(X_k) > \ldots \tag{6.3}$$

If the sequence $\{X_k\}$ is finite, its last term is a stationary point of $\varphi(X)$ on Ω.

Suppose, then, that the sequence $\{X_k\}$ is infinite. Since the set $M(X_0)$ is bounded and closed and the function $\varphi(X)$ is continuous on $M(X_0)$, it follows that $\varphi(X)$ is bounded below on $M(X_0)$. Hence, by (6.3), it follows that there exists a limit

$$\lim_{k \to \infty} \varphi(X_k) = \varphi^*,$$

such that for all $k = 0, 1, 2, \ldots$

$$\varphi(X_k) > \varphi^*. \tag{6.4}$$

2. *Lemma 6.1.* $\hat{\varepsilon}_k \to 0$ *as* $k \to \infty$.

Proof. Suppose the contrary. Then there exist $\varepsilon^* > 0$ and a subsequence $\{\hat{\varepsilon}_{k_j}\}$ such that

$$\hat{\varepsilon}_{k_j} \geqslant \varepsilon^*.$$

Since $R_{\varepsilon^*}(X_{k_j}) \subset R_{\hat{\varepsilon}_{k_j}}(X_{k_j})$, we see that for $i \in R_{\varepsilon^*}(X_{k_j})$

$$\begin{aligned} f_i(X_{k_j}(\alpha)) = f_i(X_{k_j} + \alpha(Z_{k_j} - X_{k_j})) &\leqslant \\ \leqslant f_i(X_{k_j}) + \alpha\psi_{1\hat{\varepsilon}_{k_j}}(X_{k_j}) + o_{ik_j}(\alpha) &\leqslant \\ \leqslant \varphi(X_{k_j}) - \alpha\frac{\rho_0}{\varepsilon_0}\hat{\varepsilon}_{k_j} + o_{ik_j}(\alpha) \leqslant \varphi(X_{k_j}) - \alpha\rho^* &+ o_{ik_j}(\alpha), \end{aligned}$$

where $\rho^* = \frac{\rho_0}{\varepsilon_0}\varepsilon^*$.

Hence, since $X_{k_j} \in M(X_0)$ and $\|Z_{kj} - X_{kj}\|_l \leqslant 1$, it follows that there is a number α_0, $0 < \alpha_0 < 1$, such that for $i \in R_{\varepsilon^*}(X_{k_j})$ and $\alpha \in [0, \alpha_0]$

$$f_i(X_{k_j}(\alpha)) \leqslant \varphi(X_{k_j}) - \frac{1}{2}\alpha\rho^* \tag{6.5}$$

uniformly in k_j.

Set

$$C_1 = \max_{X \in M(X_0)} \max_{i \in [0:N]} \left\| \frac{\partial f_i(X)}{\partial X} \right\|.$$

Since $X_{k_j} \in M(X_0)$, it follows that for $i \notin R_{\varepsilon^*}(X_{k_j})$

$$f_i(X_{k_j}(\alpha)) = f_i(X_{k_j} + \alpha(Z_{k_j} - X_{k_j})) \leqslant f_i(X_{k_j}) + \\ + \sqrt{n}\,\alpha C_1 + o_{ik_j}(\alpha) \leqslant \varphi(X_{k_j}) - \varepsilon^* + \sqrt{n}\,\alpha C_1 + o_{ik_j}(\alpha).$$

Hence there exists a number α_1, $0 < \alpha_1 \leqslant \alpha_0$, such that for $i \notin R_{\varepsilon^*}(X_{k_j})$ and $\alpha \in [0, \alpha_1]$

$$f_i(X_{k_j}(\alpha)) \leqslant \varphi(X_{k_j}) - \frac{1}{2}\varepsilon^* \tag{6.6}$$

uniformly in k_j.

Now fix any $\alpha' \in (0, \alpha_1)$. In view of (6.5) and (6.6), we obtain, for all k_j,

$$\varphi(X_{k_j}(\alpha')) \leqslant \varphi(X_{k_j}) - \beta, \tag{6.7}$$

where $\beta = \min\left\{\frac{1}{2}\alpha'\rho^*, \frac{1}{2}\varepsilon^*\right\}$.

Since $\varphi(X_k) \xrightarrow[k\to\infty]{} \varphi^*$, it follows that $\varphi(X_{k_j}) \leqslant \varphi^* + \frac{1}{2}\beta$ for sufficiently large k_j. Then, by (6.7),

$$\varphi(X_{k_j}(\alpha')) \leqslant \varphi^* - \frac{1}{2}\beta.$$

A fortiori,

$$\varphi(X_{k_j+1}) = \min_{\alpha \in [0, 1]} \varphi(X_{k_j}(\alpha)) \leqslant \varphi(X_{k_j}(\alpha')) \leqslant \varphi^* - \frac{1}{2}\beta,$$

contradicting (6.4) and proving the lemma.

3. If the sequence $\{X_k\}$ constructed above is infinite, then, in view of the fact that $X_k \in M(X_0)$, $k = 0, 1, 2, \ldots$, and $M(X_0)$ is a bounded closed set, it must have at least one limit point.

Theorem 6.1. Any limit point of the sequence $\{X_k\}$ *is a stationary point of* $\varphi(X)$ *on the set* Ω.

Proof. Let $X_{k_j} \xrightarrow[k_j\to\infty]{} X^*$. It is clear that $X^* \in \Omega$. We must show that $\psi_1(X^*) = 0$. Suppose this is false:

$$\psi_1(X^*) \stackrel{df}{=} \min_{Z \in \Omega(X^*)} \max_{i \in R(X^*)} \left(\frac{\partial f_i(X^*)}{\partial X}, Z - X^*\right) = -b < 0. \tag{6.8}$$

Let $\bar{Z} \in \Omega(X^*)$ denote a point for which

$$\psi_1(X^*) = \max_{i \in R(X^*)} \left(\frac{\partial f_i(X^*)}{\partial X}, \ \bar{Z} - X^* \right). \tag{6.9}$$

By the definition of $\Omega(X^*)$ the point $\bar{Z}$ is in Ω, and

$$\| \bar{Z} - X^* \|_l \leqslant 1. \tag{6.10}$$

By Lemma 6.1 of Chap. III, there exists $\gamma^* > 0$ such that for sufficiently large $k_j > K$ and $\varepsilon \in [0, \gamma^*]$

$$R_\varepsilon(X_{k_j}) \subset R(X^*). \tag{6.11}$$

Increasing K if necessary, we may assume that for all $k_j > K$ it is also true that

$$\| X_{k_j} - X^* \|_l \leqslant 1, \tag{6.12}$$

$$\left| \max_{i \in R(X^*)} \left(\frac{\partial f_i(X^*)}{\partial X}, \ \bar{Z} - X^* \right) - \right.$$
$$\left. - \max_{i \in R(X^*)} \left(\frac{\partial f_i(X_{k_j})}{\partial X}, \ \bar{Z} - X_{k_j} \right) \right| < \frac{b}{2}. \tag{6.13}$$

We claim that for $k_j > K$ and $\varepsilon \in [0, \gamma^*]$

$$\psi_{1\varepsilon}(X_{k_j}) \leqslant -\frac{b}{4}. \tag{6.14}$$

By (6.11) and (6.13), we have

$$\max_{i \in R(X^*)} \left(\frac{\partial f_i(X^*)}{\partial X}, \ \bar{Z} - X^* \right) \geqslant$$
$$\geqslant \max_{i \in R(X^*)} \left(\frac{\partial f_i(X_{k_j})}{\partial X}, \ \bar{Z} - X_{k_j} \right) - \frac{b}{2} \geqslant$$
$$\geqslant \max_{i \in R_\varepsilon(X_{k_j})} \left(\frac{\partial f_i(X_{k_j})}{\partial X}, \ \bar{Z} - X_{k_j} \right) - \frac{b}{2}.$$

Hence, by (6.9), we obtain

$$\max_{i \in R_\varepsilon(X_{k_j})} \left(\frac{\partial f_i(X_{k_j})}{\partial X}, \ \bar{Z} - X_{k_j} \right) \leqslant -\frac{b}{2}. \tag{6.15}$$

Let $\tilde{Z}_{k_j} = \frac{1}{2}(\bar{Z} + X_{k_j})$. It is clear that $\tilde{Z}_{k_j} \in \Omega$ and

$$\tilde{Z}_{k_j} - X_{k_j} = \frac{1}{2}(\bar{Z} - X_{k_j}). \tag{6.16}$$

In view of (6.10) and (6.12), we obtain

$$\|\tilde{Z}_{k_j} - X_{k_j}\|_l = \frac{1}{2}\|\bar{Z} - X_{k_j}\|_l \leqslant$$
$$\leqslant \frac{1}{2}(\|\bar{Z} - X^*\|_l + \|X^* - X_{k_j}\|_l) \leqslant 1.$$

Thus $\tilde{Z}_{k_j} \in \Omega(X_{k_j})$. Next, by (6.16) and the definition of $\psi_{l\varepsilon}(X_{k_j})$ we have

$$\max_{i \in R_\varepsilon(X_{k_j})} \left(\frac{\partial f_i(X_{k_j})}{\partial X}, \bar{Z} - X_{k_j}\right) =$$
$$= 2 \max_{i \in R_\varepsilon(X_{k_j})} \left(\frac{\partial f_i(X_{k_j})}{\partial X}, \tilde{Z}_{k_j} - X_{k_j}\right) \geqslant 2\psi_{l\varepsilon}(X_{k_j}). \tag{6.17}$$

Combining (6.15) and (6.17), we get (6.14), as required.
Now set

$$\varepsilon' = \min\left\{\gamma^*, \frac{\varepsilon_0}{\rho_0} \cdot \frac{b}{4}\right\}.$$

Let ε^* denote the largest term of the sequence $\varepsilon_\nu = \varepsilon_0/2^\nu$, $\nu = 0, 1, 2, \ldots$, smaller than ε'. Then for all $k_j > K$

$$\psi_{l\varepsilon^*}(X_{k_j}) \leqslant -\frac{b}{4} \leqslant -\frac{\rho_0}{\varepsilon_0}\varepsilon' \leqslant -\frac{\rho_0}{\varepsilon_0}\varepsilon^*.$$

Recalling that $\hat{\varepsilon}_{k_j}$ is the first term of the sequence $\{\varepsilon_\nu\}$ satisfying the analogous inequality, we see that for the same values $k_j > K$

$$\hat{\varepsilon}_{k_j} \geqslant \varepsilon^*,$$

contradicting Lemma 6.1.

4. Each step of our successive approximation procedure involves finding $\hat{\varepsilon}_k$, Z_k and α_k. To find α_k we must minimize a function of one variable, namely $\Phi_k(\alpha) = \varphi(X_k(\alpha))$, on the interval $[0, 1]$. This raises no difficulties (see Chap. III, §9).

Successful determination of $\hat{\varepsilon}_k$ and Z_k depends in the final analysis on correct solution of the following minimax problem for fixed X:

$$\max_{i \in R_\varepsilon(X)} \left(\frac{\partial f_i(X)}{\partial X}, Z - X\right) \to \min_{Z \in \Omega(X)}. \tag{6.18}$$

If $L_\varepsilon(X)$ denotes the convex hull of the points $\partial f_i(X)/\partial X$, $i \in R_\varepsilon(X)$, we may rewrite (6.18) as

$$\max_{V \in L_\varepsilon(X)} (V, Z - X) \to \min_{Z \in \Omega(X)}.$$

This problem is generally far from trivial. Some relevant considerations will be presented in Chap. VI, §5.

For the moment, we only note that if the set Ω is defined by a system of linear inequalities, problem (6.18) becomes a linear programming problem. Indeed, let $\Omega=\{Z\in E_n \mid (A_j, Z)+b_j\leqslant 0$ for all $j\in[0:N_1]\}$. Let $W^*=(z_1^*, \ldots, z_n^*, u^*)$ be a vector minimizing the linear form $\mathscr{L}(W)=u$ under the constraints

$$\begin{aligned} \left(\frac{\partial f_i(X)}{\partial X}, Z-X\right) &\leqslant u, & i&\in R_\varepsilon(X); \\ (A_j, Z)+b_j &\leqslant 0, & j&\in[0:N_1]; \\ -1\leqslant z_k-x_k &\leqslant 1, & k&\in[1:n]. \end{aligned}$$

Then it is clear that $Z^*=(z_1^*, \ldots, z_n^*)$ is a solution of the minimax problem (6.18).

5. The method of this section is also applicable to solving minimax problems when there are no constraints on the parameters: $\Omega=E_n$. In that case the result is not one of the methods of Chapter III, but a new method with the special feature that determination of the direction of descent $g_k=Z_k-X_k$ involves solution of a linear programming problem.

Chapter V

THE GENERALIZED PROBLEM OF NONLINEAR PROGRAMMING

§1. STATEMENT OF THE PROBLEM

By the generalized problem of nonlinear (mathematical) programming we mean the problem of minimizing the maximum function $\varphi(X)$ on a set Ω defined by a system of inequalities

$$\Omega=\{X\in E_n \mid h_j(X)\leqslant 0 \quad \text{for all} \quad j\in[0:N_1]\}.$$

Below we shall demonstrate how to apply the results of Chapter IV to the solution of such problems. In particular, we shall present new versions of the necessary condition for a minimax, which take into consideration the specific properties of the set Ω, and develop new successive approximation procedures for finding the stationary points of $\varphi(X)$ on Ω.

A separate section is devoted to the method of penalty functions, in the case that the constraints are linear equalities.

§2. PROPERTIES OF SETS DEFINED BY INEQUALITIES

1. Let $h_j(X)$, $j\in[0:N_1]$, be functions defined on E_n. We shall assume that the functions $h_j(X)$ are convex and continuously differentiable on E_n. Consider the set

$$\Omega=\{X\in E_n \mid h_j(X)\leqslant 0 \quad \text{for all} \quad j\in[0:N_1]\}. \tag{2.1}$$

The set Ω is clearly convex and closed.

We introduce the notation

$$\varphi_1(X)=\max_{j\in[0:N_1]} h_j(X).$$

We shall say that the Slater condition holds if

$$\inf_{X\in E_n} \varphi_1(X)<0. \tag{2.2}$$

If this condition holds, Ω has interior points. We define

$$\partial\Omega=\{X\in E_n \mid \varphi_1(X)=0\}, \quad Q(X)=\{j\in[0:N_1] \mid h_j(X)=0\}.$$

If $X \in \partial\Omega$, then clearly $Q(X) \neq \varnothing$. Conversely, if $X \in \Omega$ and $Q(X) \neq \varnothing$, then $X \in \partial\Omega$.

Lemma 2.1. If $X_0 \in \partial\Omega$ and the Slater condition (2.2) is satisfied, then

$$\min_{\|g\|=1} \max_{j \in Q(X_0)} \left(\frac{\partial h_j(X_0)}{\partial X}, g\right) < 0. \tag{2.3}$$

Proof. Suppose the contrary:

$$\min_{\|g\|=1} \max_{j \in Q(X_0)} \left(\frac{\partial h_j(X_0)}{\partial X}, g\right) \geqslant 0.$$

Since $X_0 \in \partial\Omega$, we have

$$Q(X_0) = \{j \in [0:N_1] \mid h_j(X_0) = \varphi_1(X_0)\}.$$

By Chap. III, Theorem 3.1, and our assumptions on the functions $h_j(X)$,

$$\min_{X \in E_n} \varphi_1(X) = \varphi_1(X_0).$$

But $\varphi_1(X_0) = 0$, and so

$$\min_{X \in E_n} \varphi_1(X) = 0,$$

contradicting (2.2).

Remark. Inequality (2.3) is equivalent to the relation

$$\mathbf{0} \notin L_1(X_0),$$

where $L_1(X_0) = \operatorname{co} H_1(X_0)$,

$$H_1(X_0) = \left\{\frac{\partial h_j(X_0)}{\partial X} \,\middle|\, j \in Q(X_0)\right\}.$$

This follows from Theorem 3.2 in Chap. III.

Lemma 2.2. Suppose that the Slater condition (2.2) holds, and that $h_{j_0}(X_0) = 0$ for some point $X_0 \in \Omega$ and some $j_0 \in [0:N_1]$. Then

$$\frac{\partial h_{j_0}(X_0)}{\partial X} \neq \mathbf{0}.$$

Proof. We first observe that under the assumptions of the lemma $X_0 \in \partial\Omega$ and $j_0 \in Q(X_0)$.

The assumption that $\partial h_{j_0}(X_0)/\partial X = \mathbf{0}$ now leads to a contradiction with inequality (2.3). This proves the lemma.

Lemma 2.3. Suppose that the Slater condition (2.2) holds and that Ω is bounded. Then there exists $a_1 > 0$ such that for all $X \in \partial\Omega$

$$\min_{j \in Q(X)} \left\| \frac{\partial h_j(X)}{\partial X} \right\| \geqslant a_1. \tag{2.4}$$

Proof. If this is false, there exist a sequence $\{X_k\}$, $X_k \in \partial\Omega$, $k = 1, 2, \ldots$, and a corresponding index sequence $\{j(k)\}$, $j(k) \in [0 : N_1]$, $j(k) \in Q(X_k)$, such that

$$\left\| \frac{\partial h_{j(k)}(X_k)}{\partial X} \right\| \xrightarrow[k \to \infty]{} 0. \tag{2.5}$$

We may assume without loss of generality that $j(k) \equiv j_0$ and $X_k \xrightarrow[k \to \infty]{} X^*$, $X^* \in \partial\Omega$.

Letting $k \to \infty$ in the equality $h_{j_0}(X_k) = 0$ and using (2.5), we obtain

$$h_{j_0}(X^*) = 0, \quad \frac{\partial h_{j_0}(X^*)}{\partial X} = \mathbf{0}.$$

But this contradicts Lemma 2.2.

2*. The results of this subsection are auxiliary in nature and will not be used till §6.

Define the sets

$$\partial_\mu \Omega = \{X \mid -\mu \leqslant \varphi_1(X) \leqslant 0\},$$

$$Q_\mu(X) = \{j \in [0 : N_1] \mid -\mu \leqslant h_j(X) \leqslant 0\}.$$

If $X \in \partial_\mu\Omega$, it is obvious that $Q_\mu(X) \neq \varnothing$. Conversely, if $X \in \Omega$ and $Q_\mu(X) \neq \varnothing$, then $X \in \partial_\mu\Omega$.

Lemma 2.4. *Suppose that the set Ω is bounded and let the Slater condition (2.2) hold. Then for any $a' \in (0, a_1)$ (where a_1 is the number whose existence was proved in Lemma 2.3), there exists $\bar{\mu} > 0$ such that for all $\mu \in [0, \bar{\mu}]$*

$$\min_{X \in \partial_\mu \Omega} \min_{j \in Q_\mu(X)} \left\| \frac{\partial h_j(X)}{\partial X} \right\| \geqslant a'.$$

Proof. Suppose that for some $a' \in (0, a_1)$ no such number $\bar{\mu}$ exists. Then there exist sequences $\{\mu_k\}$, $\mu_k \geqslant 0$, $\mu_k \xrightarrow[k \to \infty]{} 0$, and $\{X_k\}$, $\{j(k)\}$, $X_k \in \partial_{\mu_k}\Omega$, $j(k) \in Q_{\mu_k}(X_k)$, such that

$$\| \partial h_{j(k)}(X_k)/\partial X \| < a'. \tag{2.6}$$

Without loss of generality, we may assume that $j(k) \equiv j_0$ and $X_k \xrightarrow[k \to \infty]{} X^*$, where $X^* \in \partial\Omega$ and $j_0 \in Q(X^*)$. Letting $k \to \infty$ in (2.6), we obtain

$$\left\| \frac{\partial h_{j_0}(X^*)}{\partial X} \right\| \leqslant a' < a_1,$$

which contradicts (2.4), proving the lemma.

We now consider the sets

$$\partial'_\mu \Omega = \{X \mid -\mu \leqslant \max_{j \in [0:N_1]} t_j(X) h_j(X) \leqslant 0\},$$
$$Q'_\mu(X) = \{j \in [0:N_1] \mid -\mu \leqslant t_j(X) h_j(X) \leqslant 0\},$$

where $t_j(X) = \| \partial h_j(X)/\partial X \|^{-1}$ $\left(\text{by convention, we set } \frac{-a}{0} = -\infty \text{ if } a > 0\right)$.

Lemma 2.5. Suppose that the Slater condition (2.2) holds and let Ω *be bounded. For any* $a' \in (0, a_1)$ *there exists* $\mu' > 0$ *such that for all* $\mu \in [0, \mu']$

$$\min_{X \in \partial'_\mu \Omega} \min_{j \in Q'_\mu(X)} \left\| \frac{\partial h_j(X)}{\partial X} \right\| \geqslant a'. \tag{2.7}$$

Proof. The lemma follows immediately from Lemma 2.4, with $\mu' = \bar{\mu}/a'$ in the notation of the latter. Indeed, let $\mu \in [0, \mu']$, $X \in \partial'_\mu \Omega$, $j \in Q'_\mu(X)$. Then

$$-\mu \leqslant t_j(X) h_j(X) \leqslant 0.$$

Since $\mu \leqslant \bar{\mu}/a'$, we may write

$$\frac{-\bar{\mu}}{a'} \leqslant t_j(X) h_j(X) \leqslant 0. \tag{2.8}$$

If $-\bar{\mu} \leqslant h_j(X) \leqslant 0$, then $X \in \partial_{\bar{\mu}} \Omega$, $j \in Q_{\bar{\mu}}(X)$ and therefore, by Lemma 2.4,

$$\left\| \frac{\partial h_j(X)}{\partial X} \right\| \geqslant a'. \tag{2.9}$$

But if $h_j(X) < -\bar{\mu}$, it follows from (2.8) that necessarily

$$t_j(X) < \frac{1}{a'},$$

or, equivalently,

$$\left\| \frac{\partial h_j(X)}{\partial X} \right\| > a'. \tag{2.10}$$

Inequality (2.7) now follows from (2.9) and (2.10).

Remark. We assumed in Lemmas 2.3 through 2.5 that the set Ω is bounded. Suppose now that Ω is not bounded and let $M \subset \Omega$ be a bounded closed subset. Then Lemmas 2.3 – 2.5 remain valid if we replace the sets $\partial\Omega$, $\partial_\mu \Omega$, $\partial'_\mu \Omega$ throughout by $(\partial\Omega) \cap M$, $(\partial_\mu \Omega) \cap M$, $(\partial'_\mu \Omega) \cap M$.

Lemma 2.6. Let M *be a bounded closed subset of* Ω, *and* μ *some positive number. Then there exists* $\delta > 0$ *such that*

$$h_j(X) \leqslant -\delta$$

uniformly in $X \in M$ *and* $j \notin Q'_\mu(X)$.

Proof. The set M_j of all X in M such that $t_j(X) h_j(X) \leqslant -\mu$ bounded and closed, and therefore $h_j(X)$ assumes its maximum value $-\delta_j < 0$ there. It remains to set $\delta = \min \delta_j$ and note that $X \in M$, $j \notin Q'_\mu(X)$ imply $X \in M_j$.

3. As before, let

$$\Omega=\{X \mid h_j(X)\leqslant 0,\quad j\in[0:N_1]\},$$

and suppose that the Slater condition (2.2) is satisfied, i. e., there exists a point $\overline{X}\in\Omega$ such that

$$\max_{j\in[0:N_1]} h_j(\overline{X})<0. \tag{2.11}$$

Fix $X_0\in\Omega$ and consider the set

$$\Gamma(X_0)=\{V\in E_n \mid h_j(X_0+\alpha V)\leqslant 0 \quad \text{for all} \quad j\in[0:N_1] \text{ and all } \alpha\in[0,\alpha_0(V)], \text{ where } \alpha_0(V)>0\}.$$

Let $\overline{\Gamma}(X_0)$ denote the closure of the set $\Gamma(X_0)$. Obviously, $\overline{\Gamma}(X_0)$ is the cone of feasible directions of Ω at X_0 (see Appendix II, §2).

Consider the set

$$B(X_0)=\left\{V\in E_n \,\middle|\, \left(V,\frac{\partial h_j(X_0)}{\partial X}\right)\leqslant 0 \text{ for all } j\in Q(X_0)\right\},$$

where $Q(X_0)=\{j\in[0:N_1] \mid h_j(X_0)=0\}$. If $Q(X_0)=\varnothing$, we set $B(X_0)=E_n$ by definition.

Lemma 2.7. If the Slater condition (2.2) holds, then

$$\overline{\Gamma}(X_0)=B(X_0). \tag{2.12}$$

Proof. The statement is obvious if $Q(X_0)=\varnothing$. Indeed, in that case X_0 is an interior point of Ω and so $\Gamma(X_0)=\overline{\Gamma}(X_0)=E_n$. On the other hand, $B(X_0)=E_n$ and this proves (2.12).

Let $Q(X_0)\neq\varnothing$. Then $X_0\in\partial\Omega$. We shall first show that

$$\overline{\Gamma}(X_0)\subset B(X_0). \tag{2.13}$$

Let $V_0\in\overline{\Gamma}(X_0)$. Then there exists a sequence of vectors $\{V_i\}$ such that

$$V_0=\lim_{i\to\infty} V_i;\quad V_i\in\Gamma(X_0),\quad i=1,2,\ldots$$

Fix an arbitrary index i. For all $j\in[0:N_1]$ and all $\alpha\in(0,\alpha_0(V_i))$, it follows from the definition of $\Gamma(X_0)$ that

$$h_j(X_0+\alpha V_i)\leqslant 0.$$

Hence

$$h_j(X_0+\alpha V_i)=h_j(X_0)+\alpha\left(\frac{\partial h_j(X_0+\tau_i V_i)}{\partial X},\ V_i\right)\leqslant 0,$$

where $\tau_i\stackrel{df}{=}\tau_i(\alpha,j)\in(0,\alpha)$.

If $j \in Q(X_0)$, then $h_j(X_0) = 0$, and therefore

$$\left(\frac{\partial h_j(X_0 + \tau_i(\alpha, j) V_i)}{\partial X}, V_i\right) \leqslant 0.$$

This last inequality is true for all $\alpha \in (0, \alpha_0(V_i))$. Letting α tend to zero, we get

$$\left(\frac{\partial h_j(X_0)}{\partial X}, V_i\right) \leqslant 0.$$

Now let $i \to \infty$ in this inequality. This gives, for all $j \in Q(X_0)$,

$$\left(\frac{\partial h_j(X_0)}{\partial X}, V_0\right) \leqslant 0,$$

i. e., $V_0 \in \dot{B}(X_0)$ and (2.13) is proved.

We now prove that

$$B(X_0) \subset \overline{\Gamma}(X_0). \tag{2.14}$$

Let $V_0 \in B(X_0)$. Assume first that

$$\left(\frac{\partial h_j(X_0)}{\partial X}, V_0\right) < 0 \quad \text{for all} \quad j \in Q(X_0).$$

Since

$$h_j(X_0 + \alpha V_0) = h_j(X_0) + \alpha \left(\frac{\partial h_j(X_0)}{\partial X}, V_0\right) + o_j(\alpha),$$

there exists $\alpha_0 > 0$ such that for all $j \in Q(X_0)$ and $\alpha \in [0, \alpha_0]$

$$h_j(X_0 + \alpha V_0) \leqslant h_j(X_0) = 0.$$

Now, $h_j(X_0) < 0$ for $j \notin Q(X_0)$. Since the functions $h_j(X)$ are continuous, there exists $\alpha_1 > 0$, $\alpha_1 \leqslant \alpha_0$, such that for $j \notin Q(X_0)$ and $\alpha \in [0, \alpha_1]$

$$h_j(X_0 + \alpha V_0) < 0.$$

Thus $X_0 + \alpha V_0 \in \Omega$ for all $\alpha \in [0, \alpha_1]$, and so $V_0 \in \Gamma(X_0)$.

Now let V_0 be any vector in $B(X_0)$. Since $X_0 \in \partial\Omega$, it follows from Lemma 2.1 that there is a vector $g_0 \in E_n$, $\| g_0 \| = 1$, for which

$$\left(\frac{\partial h_j(X_0)}{\partial X}, g_0\right) < 0 \quad \text{for all} \quad j \in Q(X_0).$$

For any $\varepsilon > 0$ and $j \in Q(X_0)$, we have

$$\left(V_0 + \varepsilon g_0, \frac{\partial h_j(X_0)}{\partial X}\right) < 0,$$

so that, by what we have proved, $V_0+\varepsilon g_0 \in \Gamma(X_0)$. Letting ε tend to zero, we obtain $V_0 \in \bar{\Gamma}(X_0)$. This proves (2.14), and with it the lemma.

Thus,

$$\bar{\Gamma}(X_0)=\begin{cases}\left\{V \mid \left(V, \frac{\partial h_j(X_0)}{\partial X}\right)\leqslant 0 \text{ for all } j\in Q(X_0)\right\}, \\ \qquad\qquad\qquad\qquad \text{if } Q(X_0)\neq\varnothing; \\ E_n, \quad \text{if } Q(X_0)=\varnothing.\end{cases}$$

4. Let $K(X_0)$ denote the convex conical hull of the set $\tilde{H}_1(X_0)=\left\{V=-\frac{\partial h_j(X_0)}{\partial X}\,\middle|\, j\in Q(X_0)\right\}$:

$$K(X_0)=\left\{V=-\sum_{j\in Q(X_0)}\lambda_j\frac{\partial h_j(X_0)}{\partial X}\,\middle|\,\lambda_j\geqslant 0\right\}.$$

If $Q(X_0)=\varnothing$, we set $K(X_0)=\{0\}$ by definition. Let $Q(X_0)\neq\varnothing$, i. e., $X_0\in\partial\Omega$. By Appendix II, Lemma 2.6, $K(X_0)$ is a closed convex cone.

Lemma 2.8. If the Slater condition is satisfied, then

$$\Gamma^+(X_0)=K(X_0). \tag{2.15}$$

Proof. If $Q(X_0)=\varnothing$, the statement is obvious. Let $Q(X_0)\neq\varnothing$. We claim that

$$B(X_0)=K^+(X_0). \tag{2.16}$$

Let $V_0\in B(X_0)$. Then

$$\left(-\frac{\partial h_j(X_0)}{\partial X},\ V_0\right)\geqslant 0 \quad \text{for all} \quad j\in Q(X_0).$$

Hence it follows that

$$(Z,\ V_0)\geqslant 0 \quad \text{for all} \quad Z\in K(X_0),$$

i. e., by the definition of the dual cone, $V_0\in K^+(X_0)$. Thus,

$$B(X_0)\subset K^+(X_0). \tag{2.17}$$

Now let $V_0\in K^+(X_0)$. This means that

$$(Z,\ V_0)\geqslant 0 \quad \text{for all} \quad Z\in K(X_0).$$

In particular, for all $j\in Q(X_0)$,

$$\left(-\frac{\partial h_j(X_0)}{\partial X},\ V_0\right)\geqslant 0,$$

so that $V_0 \in B(X_0)$. Thus,

$$K^+(X_0) \subset B(X_0). \tag{2.18}$$

Combining (2.17) and (2.18), we get (2.16). Dualizing (2.16), we get

$$B^+(X_0) = K^{++}(X_0). \tag{2.19}$$

By Lemma 2.7,

$$B^+(X_0) = \Gamma^+(X_0). \tag{2.20}$$

On the other hand, since $K(X_0)$ is a convex closed cone, it follows via Appendix II, Lemma 2.3, that

$$K^{++}(X_0) = K(X_0). \tag{2.21}$$

Combining (2.19), (2.20) and (2.21), we obtain (2.15), proving the lemma.

Lemma 2.8 implies

Theorem 2.1. If $X_0 \in \Omega$ and the Slater condition (2.2) is satisfied, then

$$\Gamma^+(X_0) = \begin{cases} \left\{ V = -\sum_{j \in Q(X_0)} \lambda_j \frac{\partial h_j(X_0)}{\partial X} \,\middle|\, \lambda_j \geqslant 0 \right\}, & \\ & \textit{if} \quad Q(X_0) \neq \varnothing; \\ \{0\}, \quad \textit{if} \quad Q(X_0) = \varnothing. & \end{cases}$$

5. Suppose that some of the functions $h_j(X)$, $j \in [0 : N_1]$ are linear:

$$h_j(X) = (A_j, X) + b_j, \quad j \in [N' + 1 : N_1]. \tag{2.22}$$

We may then replace the Slater condition (2.2) by a weaker condition: it will suffice to demand the existence of a point $\overline{X}$ such that

$$\left. \begin{aligned} h_j(\overline{X}) &< 0 \quad \text{for} \quad j \in [0 : N'], \\ h_j(\overline{X}) &\leqslant 0 \quad \text{for} \quad j \in [N' + 1 : N_1]. \end{aligned} \right\} \tag{2.23}$$

Condition (2.23) will be called the generalized Slater condition.

We claim that Theorem 2.1 remains valid when the generalized Slater condition (2.23) is satisfied. To verify this we need only review the proof of the inclusion $B(X_0) \subset \overline{\Gamma}(X_0)$ in Lemma 2.7, for only there was the Slater condition used.

We first observe that (2.23) implies the existence of a vector W_0 such that

$$\left. \begin{aligned} \left(\frac{\partial h_j(X_0)}{\partial X}, W_0 \right) &< 0 \quad \text{for} \quad j \in [0 : N'] \cap Q(X_0), \\ \left(\frac{\partial h_j(X_0)}{\partial X}, W_0 \right) &\leqslant 0 \quad \text{for} \quad j \in [N' + 1 : N_1] \cap Q(X_0). \end{aligned} \right\} \tag{2.24}$$

Indeed, set $W_0 = \bar{X} - X_0$, where $\bar{X}$ is the point in (2.23). By Lemma 3.2 of Appendix II, if $j \in Q(X_0)$ then

$$\left(\frac{\partial h_j(X_0)}{\partial X},\ \bar{X} - X_0\right) \leqslant h_j(\bar{X}) - h_j(X_0) = h_j(\bar{X}).$$

Combining this with (2.23), we easily infer (2.24).

Now, if $V_0 \in B(X_0)$ is a vector such that

$$\left.\begin{aligned} &\left(\frac{\partial h_j(X_0)}{\partial X},\ V_0\right) < 0 \quad \text{for} \quad j \in [0:N'] \cap Q(X_0), \\ &\left(\frac{\partial h_j(X_0)}{\partial X},\ V_0\right) \leqslant 0 \quad \text{for} \quad j \in [N'+1:N_1] \cap Q(X_0), \end{aligned}\right\} \tag{2.25}$$

then $V_0 \in \Gamma(X_0)$. Indeed, if $j \in [0:N'] \cap Q(X_0)$ and $\alpha \geqslant 0$ is small, we see as before that

$$h_j(X_0 + \alpha V_0) \leqslant h_j(X_0) = 0.$$

If $j \in [N'+1:N_1] \cap Q(X_0)$, we infer from (2.22) and (2.25) that for all $\alpha \geqslant 0$

$$h_j(X_0 + \alpha V_0) = h_j(X_0) + \alpha\left(\frac{\partial h_j(X_0)}{\partial X},\ V_0\right) \leqslant 0.$$

Thus, for sufficiently small $\alpha \geqslant 0$ and $j \in Q(X_0)$,

$$h_j(X_0 + \alpha V_0) \leqslant 0.$$

On the other hand, for small $\alpha \geqslant 0$ and $j \notin Q(X_0)$,

$$h_j(X_0 + \alpha V_0) < 0.$$

We have thus shown that $h_j(X_0 + \alpha V_0) \leqslant 0$ for all $j \in [0:N_1]$ and $\alpha \in [0, \alpha_0(V_0)]$, where $\alpha_0(V_0) > 0$. But this is equivalent to $V_0 \in \Gamma(X_0)$.

After these observations, the proof of $B(X_0) \subset \bar{\Gamma}(X_0)$ proceeds exactly as in Lemma 2.7, and our assertion is thus proved.

Remark 1. If all the functions $h_j(X)$ are linear, condition (2.23) becomes

$$h_j(\bar{X}) \leqslant 0 \qquad \text{for} \qquad j \in [0:N_1].$$

In this case, therefore, the generalized Slater condition is equivalent to the requirement that the set Ω be nonempty.

Remark 2. Since any equality of the type

$$(A_j,\ X) + b_j = 0$$

may be replaced by two inequalities

$$\begin{aligned} (A_j,\ X) + b_j &\leqslant 0, \\ -(A_j,\ X) - b_j &\leqslant 0, \end{aligned}$$

we may allow the definition of Ω to include linear equalities. In this case, however, the Slater condition will never be satisfied, though its generalization may hold.

§3. NECESSARY CONDITIONS FOR A MINIMAX

1. Let Ω be a convex closed set defined as in (2.1), and let Ω' be some open set containing Ω. We set

$$\varphi(X) = \max_{i \in [0:N]} f_i(X),$$

where the functions $f_i(X)$, $i \in [0:N]$ are continuously differentiable on Ω'.

Let us consider the minimization of $\varphi(X)$ on Ω. The general problem was examined in Chapter IV. The results to be achieved presently will take the specific features of the set Ω into consideration.

For the reader's convenience, we recall the following theorem (see Chap. IV, §3):

In order that the function $\varphi(X)$ attain its minimum on Ω at a point $X^* \in \Omega$ it is necessary and, if $\varphi(X)$ is convex on Ω, also sufficient, that

$$\Gamma^+(X^*) \cap L(X^*) \neq \varnothing, \tag{3.1}$$

where

$$L(X^*) = \mathrm{co}\, H(X^*), \quad H(X^*) = \left\{ \frac{\partial f_i(X^*)}{\partial X} \,\middle|\, i \in R(X^*) \right\},$$

and $\Gamma^+(X^*)$ is the dual of the cone of feasible directions for Ω at X^*.

In the case that Ω is defined by relations of type (2.1) and the Slater condition (2.2) is fulfilled, the form of $\Gamma^+(X^*)$ is that established in Theorem 2.1.

The following theorem proves useful in constructing minimization algorithms for $\varphi(X)$.

Theorem 3.1. Let Ω be a set of type (2.1) and suppose that the Slater condition (2.2) is satisfied. Then (3.1) is equivalent to

$$\mathbf{0} \in \tilde{L}(X^*), \tag{3.2}$$

where

$$\tilde{L}(X^*) = \mathrm{co}\, \tilde{H}(X^*), \quad \tilde{H}(X^*) = H(X^*) \cup H_1(X^*),$$
$$H(X^*) = \left\{ \frac{\partial f_i(X^*)}{\partial X} \,\middle|\, i \in R(X^*) \right\},$$
$$H_1(X^*) = \left\{ \frac{\partial h_j(X^*)}{\partial X} \,\middle|\, j \in Q(X^*) \right\},$$
$$R(X^*) = \{ i \in [0:N] \mid f_i(X^*) = \varphi(X^*) \},$$
$$Q(X^*) = \{ j \in [0:N_1] \mid h_j(X^*) = 0 \}.$$

Proof. The assertion is obvious if $Q(X^*) = \varnothing$, for in that case $\tilde{L}(X^*) = L(X^*)$, and by Theorem 2.1

$$\Gamma^+(X^*) = \{0\}.$$

Thus both (3.1) and (3.2) reduce to the inclusion

$$0 \in L(X^*).$$

This justifies our assuming that $Q(X^*) \neq \varnothing$, i. e., $X^* \in \partial\Omega$.

Suppose that condition (3.2) holds at the point X^*. We shall show that condition (3.1) is also satisfied.

By (3.2),

$$0 = \sum_{i \in R(X^*)} \lambda_{1i} \frac{\partial f_i(X^*)}{\partial X} + \sum_{j \in Q(X^*)} \lambda_{2j} \frac{\partial h_j(X^*)}{\partial X}, \tag{3.3}$$

where $\lambda_{1i} \geqslant 0$, $\lambda_{2j} \geqslant 0$,

$$\sum_{i \in R(X^*)} \lambda_{1i} + \sum_{j \in Q(X^*)} \lambda_{2j} = 1.$$

In this case

$$a = \sum_{i \in R(X^*)} \lambda_{1i} > 0,$$

since otherwise we would have

$$\sum_{j \in Q(X^*)} \lambda_{2j} \frac{\partial h_j(X^*)}{\partial X} = 0; \quad \lambda_{2j} \geqslant 0, \quad \sum_{j \in Q(X^*)} \lambda_{2j} = 1,$$

i. e., $0 \in L_1(X^*)$, where $L_1(X^*) = \operatorname{co} H_1(X^*)$, contradicting Lemma 2.1.

Thus $a > 0$. We rewrite (3.3) as

$$0 = \sum_{i \in R(X^*)} \lambda'_{1i} \frac{\partial f_i(X^*)}{\partial X} + \sum_{j \in Q(X^*)} \lambda'_{2j} \frac{\partial h_j(X^*)}{\partial X}, \tag{3.4}$$

where

$$\lambda'_{1i} = \frac{\lambda_{1i}}{a}, \quad \lambda'_{2j} = \frac{\lambda_{2j}}{a}.$$

Clearly,

$$\lambda'_{1i} \geqslant 0, \quad \lambda'_{2j} \geqslant 0, \quad \sum_{i \in R(X^*)} \lambda'_{1i} = 1.$$

We now have

$$\sum_{i \in R(X^*)} \lambda'_{1i} \frac{\partial f_i(X^*)}{\partial X} \in L(X^*), \tag{3.5}$$

and, by (3.4) and Theorem 2.1,

$$\sum_{i \in R(X^*)} \lambda'_{1i} \frac{\partial f_i(X^*)}{\partial X} = - \sum_{j \in Q(X^*)} \lambda'_{2j} \frac{\partial h_j(X^*)}{\partial X} \in \Gamma^+(X^*). \tag{3.6}$$

Condition (3.1) now follows from (3.5) and (3.6).

Conversely, let condition (3.1) be satisfied. We must prove that $\mathbf{0} \in \tilde{L}(X^*)$. Let $Z_0 \in \Gamma^+(X^*) \cap L(X^*)$, i. e.,

$$Z_0 = - \sum_{j \in Q(X^*)} \alpha_j \frac{\partial h_j(X^*)}{\partial X}, \qquad \alpha_j \geqslant 0;$$

$$Z_0 = \sum_{i \in R(X^*)} \lambda_i \frac{\partial f_i(X^*)}{\partial X}, \qquad \lambda_i \geqslant 0, \quad \sum_{i \in R(X^*)} \lambda_i = 1.$$

We have

$$\sum_{i \in R(X^*)} \lambda_i \frac{\partial f_i(X^*)}{\partial X} + \sum_{j \in Q(X^*)} \alpha_j \frac{\partial h_j(X^*)}{\partial X} = \mathbf{0}. \tag{3.7}$$

We introduce the notation

$$c = \sum_{j \in Q(X^*)} \alpha_j; \quad \lambda_{1i} = \frac{\lambda_i}{1+c}, \quad i \in R(X^*);$$

$$\lambda_{2j} = \frac{\alpha_j}{1+c}, \quad j \in Q(X^*).$$

Dividing (3.7) by $1 + c$, we obtain

$$\sum_{i \in R(X^*)} \lambda_{1i} \frac{\partial f_i(X^*)}{\partial X} + \sum_{j \in Q(X^*)} \lambda_{2j} \frac{\partial h_j(X^*)}{\partial X} = \mathbf{0},$$

with

$$\lambda_{1i} \geqslant 0, \; \lambda_{2j} \geqslant 0, \quad \sum_{i \in R(X^*)} \lambda_{1i} + \sum_{j \in Q(X^*)} \lambda_{2j} = 1.$$

But this means that $\mathbf{0} \in \tilde{L}(X^*)$. This completes the proof.

Remark 1. Theorem 3.1 is no longer true if the Slater condition is replaced by the generalized Slater condition.

Example. Let

$$\Omega = \{X = (x_1, x_2) \mid x_1 + x_2 = 0\},$$

$$\varphi(X) = f_0(X) = (X, X).$$

The generalized Slater condition is fulfilled. Setting $A=(1, 1)$, we write the representation of Ω as

$$\Omega=\{X\in E_2 \mid (A, X)\leqslant 0, -(A, X)\leqslant 0\}.$$

It is not hard to show that for all $X\in\Omega$

$$\mathbf{0}\in\tilde{L}(X).$$

Indeed, $H_1(X)$ consists of exactly two vectors, A and $-A$, so that $\mathbf{0}\in \operatorname{co} H_1(X)$, and a fortiori $\mathbf{0}\in\tilde{L}(X)$.

On the other hand, $\Gamma^+(X)=\{\lambda A \mid \lambda \text{ arbitrary}\}$, $L(X)=\{2X\}$. For $X_0=(1, -1)\in\Omega$, say, we have

$$L(X_0)\cap\Gamma^+(X_0)=\varnothing.$$

Thus conditions (3.1) and (3.2) are not equivalent.

Remark 2. Let us partition the index set $[0:N_1]$ into two subsets $[0:N']$ and $[N'+1:N_1]$ arbitrarily, and set

$$Q_1(X)=\{j\in[0:N'] \mid h_j(X)=0\},$$
$$Q_2(X)=\{j\in[N'+1:N_1] \mid h_j(X)=0\}.$$

Clearly, $Q_1(X)\cup Q_2(X)=Q(X)$.

The following generalization of Theorem 3.1 may be proved in analogous fashion:

Suppose that the Slater condition (2.2) is satisfied and let $X^*\in\Omega$. Then condition (3.1) is equivalent to

$$\Gamma'^+(X^*)\cap\tilde{L}'(X^*)\neq\varnothing,$$

where

$$\tilde{L}'(X^*)=\operatorname{co}\tilde{H}'(X^*),\quad \tilde{H}'(X^*)=H(X^*)\cup H_1'(X^*);$$
$$H_1'(X^*)=\left\{\frac{\partial h_j(X^*)}{\partial X}\,\middle|\, j\in Q_1(X^*)\right\};$$
$$\Gamma'^+(X^*)\overset{df}{=}\begin{cases}\left\{V=-\sum\limits_{j\in Q_2(X^*)}\lambda_j\dfrac{\partial h_j(X^*)}{\partial X}\,\middle|\,\lambda_j\geqslant 0\right\}, & \text{if } Q_2(X^*)\neq\varnothing;\\ \{\mathbf{0}\}, & \text{if } Q_2(X^*)=\varnothing.\end{cases}$$

2. Definition. A point $X^*\in\Omega$ is called a stationary point of $\varphi(X)$ on the set Ω if condition (3.1) is fulfilled.

Let $X_0\in\Omega$ be a nonstationary point. Then

$$d(X_0)\overset{df}{=}\min_{\substack{Z\in L(X_0)\\ V\in\Gamma^+(X_0)}}\|Z-V\|=\|Z(X_0)-V(X_0)\|>0,$$

where $Z(X_0)\in L(X_0)$, $V(X_0)\in\Gamma^+(X_0)$.

In Chap. IV, §3, we showed that

$$g(X_0)=\frac{V(X_0)-Z(X_0)}{\|V(X_0)-Z(X_0)\|}, \quad \|g(X_0)\|=1,$$

is a direction of steepest descent for $\varphi(X)$ on Ω at the point X_0; we also showed that $g(X_0)\in\bar{\Gamma}(X_0)$.

It may happen that $g(X_0)\notin\Gamma(X_0)$, i. e., no point of the ray $X=X_0(\alpha)\overset{df}{=}X_0+\alpha g(X_0)$, $\alpha>0$ lies in the set Ω. This is why we cannot utilize directions of steepest descent directly in order to construct successive approximation procedures for finding stationary points.

Assume that the Slater condition (2.2) holds, and let $Q(X_0)\neq\varnothing$. Let us evaluate

$$\tilde{d}(X_0)\overset{df}{=}\min_{Z\in\tilde{L}(X_0)}\|Z\|=\|\tilde{Z}(X_0)\|. \tag{3.8}$$

Since X_0 is not a stationary point, it follows from Theorem 3.1 that $\tilde{d}(X_0)>0$ and $\tilde{Z}(X_0)\neq\mathbf{0}$. It follows from (3.8) that the direction

$$\tilde{g}(X_0)=-\frac{\tilde{Z}(X_0)}{\|\tilde{Z}(X_0)\|}, \quad \|\tilde{g}(X_0)\|=1, \tag{3.9}$$

enjoys the properties

$$\left(\frac{\partial f_i(X_0)}{\partial X},\ \tilde{g}(X_0)\right)\leqslant-\tilde{d}(X_0)<0, \quad i\in R(X_0); \tag{3.10}$$

$$\left(\frac{\partial h_j(X_0)}{\partial X},\ \tilde{g}(X_0)\right)\leqslant-\tilde{d}(X_0)<0, \quad j\in Q(X_0). \tag{3.11}$$

Consider the ray $X=X_0(\alpha)\overset{df}{=}X_0+\alpha\tilde{g}(X_0)$, $\alpha\geqslant0$. By (3.11), for sufficiently small $\alpha>0$,

$$h_j(X_0+\alpha\tilde{g}(X_0))<h_j(X_0)=0 \quad \text{for all} \quad j\in Q(X_0).$$

Since the functions $h_j(X)$ are continuous, it follows that for $j\notin Q(X_0)$ and small $\alpha>0$

$$h_j(X_0+\alpha\tilde{g}(X_0))<0.$$

Thus we have $X_0+\alpha\tilde{g}(X_0)\in\Omega$ for sufficiently small $\alpha>0$.

On the other hand, by (3.10), for sufficiently small $\alpha>0$,

$$\varphi(X_0+\alpha\tilde{g}(X_0))<\varphi(X_0).$$

Thus we see that $\tilde{g}(X_0)$ is a direction of decrease for $\varphi(X)$ at X_0, and moreover we can move along this direction without leaving the set Ω. We shall use this result extensively in the sequel.

It must be emphasized, however, that $\tilde{g}(X_0)$ is not necessarily a direction of steepest descent for $\varphi(X)$ on Ω at X_0.

§4*. DEPENDENCE OF DIRECTION OF DESCENT ON SPECIFIC FEATURES OF Ω

We now wish to ascertain how the direction of descent (3.9) depends on the mode of specification of the set Ω. Recall that Ω was defined by a system of inequalities (2.1). This is clearly equivalent to

$$\Omega = \{X \mid t_j h_j(X) \leqslant 0 \quad \text{for all} \quad j \in [0:N_1]\}, \tag{4.1}$$

where t_j are arbitrary positive numbers. Instead of the polyhedron $\tilde{L}(X)$ (see Theorem 3.1), we then have a polyhedron

$$\tilde{L}_t(X) = \text{co}\, \tilde{H}_t(X),$$

where $t = (t_0, \ldots, t_{N_1})$,

$$\tilde{H}_t(X) = H(X) \cup H_t(X),$$
$$H_t(X) = \left\{ t_j \frac{\partial h_j(X)}{\partial X} \,\middle|\, j \in Q(X) \right\},$$

and the necessary condition (3.2) becomes

$$0 \in \tilde{L}_t(X^*). \tag{4.2}$$

Of course, conditions (3.2) and (4.2) are equivalent, so it is immaterial which of them we use to check whether a given point is stationary.

Let $X_0 \in \Omega$ be a nonstationary point of $\varphi(X)$ on Ω. Then the direction

$$\tilde{g}_t(X_0) = -\frac{\tilde{Z}_t(X_0)}{\|\tilde{Z}_t(X_0)\|}, \quad \|\tilde{g}_t(X_0)\| = 1,$$

where

$$\tilde{d}_t(X_0) \overset{df}{=} \min_{Z \in \tilde{L}_t(X_0)} \|Z\| = \|\tilde{Z}_t(X_0)\|,$$

depends essentially on the coefficient vector t. In fact, for any positive t (i. e., a vector t all of whose components are positive), $\tilde{g}_t(X_0)$ is a direction of decrease for $\varphi(X)$ and there exists $\alpha_0(t, X_0) > 0$ such that $X_0 + \alpha \tilde{g}_t(X_0) \in \Omega$ for all $\alpha \in [0, \alpha_0(t, X_0)]$.

Denote

$$m(t) = \min_{j \in [0:N_1]} t_j.$$

Lemma 4.1. If $X_0 \in \Omega$ is not a stationary point of $\varphi(X)$ on Ω and the Slater condition (2.2) is fulfilled, then

$$\lim_{m(t) \to \infty} \tilde{d}_t(X_0) = d(X_0),$$

where (see §3.2)

$$d(X_0) = \min_{\substack{Z \in L(X_0) \\ V \in \Gamma^+(X_0)}} \| Z - V \|.$$

Proof. If $Q(X_0) = \varnothing$, then

$$\Gamma^+(X_0) = \{0\}, \quad \tilde{L}_t(X_0) = L(X_0),$$

so that $\tilde{d}_t(X_0) = d(X_0)$ for all positive t. We may therefore assume that $Q(X_0) \neq \varnothing$, i. e., $X_0 \in \partial\Omega$. By Lemma 2.1,

$$\psi_1(X_0) \overset{df}{=} \min_{\|g\|=1} \max_{j \in Q(X_0)} \left(\frac{\partial h_j(X_0)}{\partial X}, g \right) = -a < 0. \tag{4.3}$$

We claim that for any $\gamma_j \geqslant 0$, $j \in [0 : N_1]$,

$$\left\| \sum_{j \in Q(X_0)} \gamma_j \frac{\partial h_j(X_0)}{\partial X} \right\| \geqslant a \sum_{j \in Q(X_0)} \gamma_j. \tag{4.4}$$

Indeed, if $\sum_{j \in Q(X_0)} \gamma_j = 0$, this is trivially true. Let

$$\sum_{j \in Q(X_0)} \gamma_j = b > 0.$$

Then

$$\sum_{j \in Q(X_0)} \gamma_j \frac{\partial h_j(X_0)}{\partial X} = b \left[\sum_{j \in Q(X_0)} \gamma'_j \frac{\partial h_j(X_0)}{\partial X} \right],$$

where

$$\gamma'_j = \frac{\gamma_j}{b}, \quad \sum_{j \in Q(X_0)} \gamma'_j = 1.$$

Since (see Remark after Lemma 2.1)

$$W_0 \overset{df}{=} \sum_{j \in Q(X_0)} \gamma'_j \frac{\partial h_j(X_0)}{\partial X} \in L_1(X_0),$$

it follows from (4.3) that

$$\| W_0 \| \geqslant -\psi_1(X_0) = a, \tag{4.5}$$

because

$$-\psi_1(X_0) = \min_{Z \in L_1(X_0)} \| Z \|.$$

Inequality (4.4) now follows from (4.5).

We now recall that

$$\tilde{d}_t(X_0)=\min_{\lambda\in\Lambda(t)}\left\|\sum_{i\in R(X_0)}\lambda_{1i}\frac{\partial f_i(X_0)}{\partial X}+\sum_{j\in Q(X_0)}\lambda_{2j}t_j\frac{\partial h_j(X_0)}{\partial X}\right\|,$$

where

$$\Lambda(t)=\left\{\lambda\,|\,\lambda_{1i}\geqslant 0,\ \lambda_{2j}\geqslant 0,\ \sum_{i\in R(X_0)}\lambda_{1i}+\sum_{j\in Q(X_0)}\lambda_{2j}=1\right\},$$

and

$$d(X_0)=\min_{\lambda\in\Lambda}\left\|\sum_{i\in R(X_0)}\lambda_{1i}\frac{\partial f_i(X_0)}{\partial X}+\sum_{j\in Q(X_0)}\lambda_{2j}\frac{\partial h_j(X_0)}{\partial X}\right\|,$$

where

$$\Lambda=\left\{\lambda\,|\,\lambda_{1i}\geqslant 0,\ \lambda_{2j}\geqslant 0,\ \sum_{i\in R(X_0)}\lambda_{1i}=1\right\}.$$

Setting

$$c=\max_{i\in R(X_0)}\left\|\frac{\partial f_i(X_0)}{\partial X}\right\|,$$

we see from (4.4) that

$$\tilde{d}_t(X_0)=\left\|\sum_{i\in R(X_0)}\lambda_{1i}(t)\frac{\partial f_i(X_0)}{\partial X}+\sum_{j\in Q(X_0)}\lambda_{2j}(t)\,t_j\frac{\partial h_j(X_0)}{\partial X}\right\|\geqslant$$
$$\geqslant a\sum_{j\in Q(X_0)}\lambda_{2j}(t)\,t_j-c\geqslant am(t)\sum_{j\in Q(X_0)}\lambda_{2j}(t)-c. \qquad (4.6)$$

Now, on the one hand, since

$$\tilde{d}_t(X_0)\leqslant\max_{i\in R(X_0)}\left\|\frac{\partial f_i(X_0)}{\partial X}\right\|=c,$$

it follows from (4.6) that

$$\sum_{j\in Q(X_0)}\lambda_{2j}(t)\xrightarrow[m(t)\to\infty]{}0,$$
$$\mu(t)\overset{df}{=}\sum_{i\in R(X_0)}\lambda_{1i}(t)\xrightarrow[m(t)\to\infty]{}1. \qquad (4.7)$$

Next, for sufficiently large $m(t)$ we have

$$\tilde{d}_t(X_0)=\mu(t)\left\|\sum_{i\in R(X_0)}\lambda'_{1i}(t)\frac{\partial f_i(X_0)}{\partial X}+\sum_{j\in Q(X_0)}\lambda'_{2j}(t)\frac{\partial h_j(X_0)}{\partial X}\right\|,$$

where

$$\lambda'_{1i}(t)=\frac{\lambda_{1i}(t)}{\mu(t)},\qquad \lambda'_{2j}(t)=\frac{\lambda_{2j}(t)\,t_j}{\mu(t)}.$$

Since

$$\lambda'_{1i}(t)\geqslant 0,\quad \lambda'_{2j}(t)\geqslant 0,\qquad \sum_{i\in R(X_0)}\lambda'_{1i}(t)=1,$$

it follows by the definition of $d(X_0)$ that

$$\tilde{d}_t(X_0)\geqslant\mu(t)\,d(X_0). \tag{4.8}$$

We now show that $d(X_0)\leqslant\tilde{d}_t(X_0)$ for all positive t. Indeed, consider the chain of equalities

$$\begin{aligned} d(X_0)&=\left\|\sum_{i\in R(X_0)}\tilde{\lambda}_{1i}\frac{\partial f_i(X_0)}{\partial X}+\sum_{j\in Q(X_0)}\tilde{\lambda}_{2j}\frac{\partial h_j(X_0)}{\partial X}\right\|=\\ &=\left\|\sum_{i\in R(X_0)}\tilde{\lambda}_{1i}\frac{\partial f_i(X_0)}{\partial X}+\sum_{j\in Q(X_0)}\tilde{\lambda}'_{2j}t_j\frac{\partial h_j(X_0)}{\partial X}\right\|=\\ &=b\left\|\sum_{i\in R(X_0)}\lambda'_{1i}\frac{\partial f_i(X_0)}{\partial X}+\sum_{j\in Q(X_0)}\lambda'_{2j}t_j\frac{\partial h_j(X_0)}{\partial X}\right\|.\end{aligned}$$

Here

$$\tilde{\lambda}'_{2j}=\tilde{\lambda}_{2j}/t_j,$$

$$b=\sum_{i\in R(X_0)}\tilde{\lambda}_{1i}+\sum_{j\in Q(X_0)}\tilde{\lambda}'_{2j}t_j=1+\sum_{j\in Q(X_0)}\tilde{\lambda}_{2j}\geqslant 1,$$

$$\lambda'_{1i}=\tilde{\lambda}_{1i}/b,\qquad \lambda'_{2j}=\tilde{\lambda}'_{2j}/b.$$

It is clear that $\lambda'_{1i}\geqslant 0$, $\lambda'_{2j}\geqslant 0$, $\sum_{i\in R(X_0)}\lambda'_{1i}+\sum_{j\in Q(X_0)}\lambda'_{2j}=1$. Therefore, in view of the definition of $\tilde{d}_t(X_0)$, we get

$$d(X_0)\geqslant b\tilde{d}_t(X_0)\geqslant\tilde{d}_t(X_0). \tag{4.9}$$

From (4.8) and (4.9) we have

$$d(X_0)\geqslant\tilde{d}_t(X_0)\geqslant\mu(t)\,d(X_0).$$

It remains only to let $m(t)\to\infty$ in these inequalities and use (4.7). This completes the proof.

Theorem 4.1. If $X_0\in\Omega$ is not a stationary point of $\varphi(X)$ on Ω and the Slater condition (2.2) is satisfied, then

$$\tilde{g}_t(X_0)\xrightarrow[m(t)\to\infty]{}g(X_0),$$

where $g(X_0)$ is a direction of steepest descent for $\varphi(X)$ at X_0 on Ω.

Proof. By Lemma 4.1, we have

$$\tilde{d}_t(X_0) \xrightarrow[m(t)\to\infty]{} d(X_0). \tag{4.10}$$

But

$$\tilde{d}_t(X_0) = \min_{Z \in \tilde{L}_t(X_0)} \|Z\| = \|Z(t)\|,$$
$$d(X_0) = \min_{Z \in G(X_0)} \|Z\| = \|Z_0\|,$$

where

$$G(X_0) = L(X_0) - \Gamma^+(X_0) = \{Z - V \mid Z \in L(X_0),\ V \in \Gamma^+(X_0)\}.$$

Note that the set $G(X_0)$ is closed, since the sets $L(X_0)$ and $\Gamma^+(X_0)$ are closed and $L(X_0)$ is bounded.

It follows from the proof of Lemma 4.1 that

$$Z(t) = \mu(t) Z'(t), \tag{4.11}$$

where $Z'(t) \in G(X_0)$ and $\mu(t) \xrightarrow[m(t)\to\infty]{} 1$. By (4.10), we may assume that the set of vectors $\{Z'(t)\}$ is bounded for sufficiently large $m(t)$.

We claim that

$$Z'(t) \xrightarrow[m(t)\to\infty]{} Z_0. \tag{4.12}$$

If this is false, there exists a sequence $\{t_s\}$ such that $Z'(t_s) \xrightarrow[m(t_s)\to\infty]{} Z' \neq Z_0$. Since $Z'(t_s) \in G(X_0)$ and $G(X_0)$ is closed, we have $Z' \in G(X_0)$. Since Z_0 is the unique point of $G(X_0)$ nearest the origin, we obtain

$$\|Z'\| > \|Z_0\|.$$

On the other hand, by (4.10) and (4.11),

$$\|Z'\| = \|Z_0\|.$$

This contradiction proves our assertion concerning (4.12).

We now observe that

$$g_t(X_0) = -\frac{Z(t)}{\|Z(t)\|} = -\frac{Z'(t)}{\|Z'(t)\|}, \qquad g(X_0) = -\frac{Z_0}{\|Z_0\|}.$$

Hence, via (4.12), follows the assertion of the theorem.

§5. LAGRANGE MULTIPLIERS AND THE KUHN-TUCKER THEOREM

1. As before, let

$$\varphi(X) = \max_{i \in [0:N]} f_i(X),$$

where the functions $f_i(X)$, $i \in [0:N]$ are continuously differentiable on $\Omega' \subset E_n$, and let Ω be a subset of Ω' defined by

$$\Omega = \{X \mid \max_{j \in [0:N_1]} h_j(X) \leqslant 0\},$$

where the functions $h_j(X)$, $j \in [0:N_1]$ are convex and continuously differentiable on E_n. Assume moreover that the Slater condition

$$\inf_{X \in E_n} \max_{j \in [0:N_1]} h_j(X) < 0$$

(or the generalized Slater condition (2.23)) is satisfied.

Theorem 5.1. A necessary condition for the function $\varphi(X)$ to assume its minimum on Ω at a point $X^* \in \Omega$ is that there exist a vector

$$\lambda^* = (\lambda_{10}^*, \ldots, \lambda_{1N}^*, \lambda_{20}^*, \ldots, \lambda_{2N_1}^*) \in E_{N+N_1+2}$$

such that

$$\text{I.} \quad \sum_{i=0}^{N} \lambda_{1i}^* \frac{\partial f_i(X^*)}{\partial X} + \sum_{j=0}^{N_1} \lambda_{2j}^* \frac{\partial h_j(X^*)}{\partial X} = 0.$$

$$\text{II.} \quad \lambda_{1i}^* \geqslant 0, \quad \sum_{i=0}^{N} \lambda_{1i}^* = 1; \quad \lambda_{1i}^* = 0, \text{ если } f_i(X^*) < \varphi(X^*).$$

$$\text{III.} \quad \lambda_{2j}^* \geqslant 0; \quad \lambda_{2j}^* h_j(X^*) = 0, \quad j \in [0:N_1].$$

If $\varphi(X)$ is convex on Ω, this condition is also sufficient.

Proof. Necessity. Let X^* be a minimum point for $\varphi(X)$ on Ω. To fix ideas, we shall assume that the set

$$Q(X^*) = \{j \in [0:N_1] \mid h_j(X^*) = 0\}$$

is not empty (otherwise, the proof is only simplified). By Theorems 2.1 and 3.1 of Chap. IV, there exists a point Z_0 in the intersection of $L(X^*)$ and $\Gamma^+(X^*)$. Thus,

$$Z_0 = \sum_{i \in R(X^*)} \alpha_i \frac{\partial f_i(X^*)}{\partial X}; \qquad \alpha_i \geqslant 0, \quad \sum_{i \in R(X^*)} \alpha_i = 1;$$

$$Z_0 = -\sum_{j \in Q(X_0)} \beta_j \frac{\partial h_j(X^*)}{\partial X}; \quad \beta_j \geqslant 0, \quad j \in Q(X^*).$$

Hence it follows that

$$\sum_{i \in R(X^*)} \alpha_i \frac{\partial f_i(X^*)}{\partial X} + \sum_{j \in Q(X^*)} \beta_j \frac{\partial h_j(X^*)}{\partial X} = 0.$$

We now define the components of the vector λ^* as follows:

$$\lambda_{1i}^* = \alpha_i \quad \text{for} \quad i \in R(X^*), \quad \lambda_{1i}^* = 0 \quad \text{for} \quad i \notin R(X^*),$$

$$\lambda_{2j}^* = \beta_j \quad \text{for} \quad j \in Q(X^*), \quad \lambda_{2j}^* = 0 \quad \text{for} \quad j \notin Q(X^*).$$

This gives

$$\sum_{i=1}^{N} \lambda_{1i}^{*} \frac{\partial f_i(X^*)}{\partial X} + \sum_{j=0}^{N_1} \lambda_{2j}^{*} \frac{\partial h_j(X^*)}{\partial X} = 0.$$

Conditions II and III are also fulfilled, and we have proved necessity.

S u f f i c i e n c y. Let $\varphi(X)$ be convex on Ω, $X^* \in \Omega$ and assume that conditions I — III are satisfied. To fix ideas, we again assume that $Q(X^*) \neq \varnothing$. By II and III,

$$Z_0 \stackrel{\text{df}}{=} \sum_{i=0}^{N} \lambda_{1i}^{*} \frac{\partial f_i(X^*)}{\partial X} = \sum_{i \in R(X^*)} \lambda_{1i}^{*} \frac{\partial f_i(X^*)}{\partial X} \in L(X^*),$$

$$Z_0' \stackrel{\text{df}}{=} -\sum_{j=0}^{N_1} \lambda_{2j}^{*} \frac{\partial h_j(X^*)}{\partial X} = -\sum_{j \in Q(X^*)} \lambda_{2j}^{*} \frac{\partial h_j(X^*)}{\partial X} \in \Gamma^+(X^*).$$

Since $Z_0 = Z_0'$, we have

$$L(X^*) \cap \Gamma^+(X^*) \neq \varnothing.$$

Theorems 2.1 and 3.1 of Chap. IV now imply that $\varphi(X)$ assumes its minimum on Ω at the point X^*, Q. E. D.

2. We define a function

$$F(X, \lambda) = \sum_{i=0}^{N} \lambda_{1i} f_i(X) + \sum_{j=0}^{N_1} \lambda_{2j} h_j(X).$$

$F(X, \lambda)$ is known as the Lagrange function and the coefficients λ_{1i}, λ_{2j} as Lagrange multipliers. Suppose that $f_i(X)$, $i \in [0:N]$ and $h_j(X)$, $j \in [0:N_1]$ are defined throughout E_n, and let

$$\Lambda = \{\lambda = (\lambda_{10}, \ldots, \lambda_{1N}, \lambda_{20}, \ldots, \lambda_{2N_1}) \mid \lambda_{1i} \geqslant 0, \\ \sum_{i=0}^{N} \lambda_{1i} = 1; \quad \lambda_{2j} \geqslant 0, \quad j \in [0:N_1]\}.$$

D e f i n i t i o n. A point $[X^*, \lambda^*]$ is called a s a d d l e p o i n t of $F(X, \lambda)$ on $E_n \times \Lambda$ if

$$F(X^*, \lambda) \leqslant F(X^*, \lambda^*) \leqslant F(X, \lambda^*) \tag{5.1}$$

for all $X \in E_n$, $\lambda \in \Lambda$.

T h e o r e m 5.2 (Kuhn-Tucker). Let the functions $f_i(X)$, $i \in [0:N]$, and $h_j(X)$, $j \in [0:N_1]$, be convex and continuously differentiable on E_n and suppose that the Slater condition (or its generalized version) is fulfilled. The function $\varphi(X)$ assumes its minimum on Ω at the point X^* if and only if there exists a vector $\lambda^* \in \Lambda$ such that $[X^*, \lambda^*]$ is a saddle point of the function $F(X, \lambda)$ on $E_n \times \Lambda$.

Proof. Necessity. Let X^* be a minimum point of $\varphi(X)$ on Ω. Then by Theorem 5.1 there exists a vector $\lambda^* \in \Lambda$ satisfying conditions I—III. In particular,

$$\sum_{i=0}^{N} \lambda_{1i}^* \frac{\partial f_i(X^*)}{\partial X} + \sum_{j=0}^{N_1} \lambda_{2j}^* \frac{\partial h_j(X^*)}{\partial X} = 0. \tag{5.2}$$

But the expression on the left of (5.2) is precisely $\partial F(X^*, \lambda^*)/\partial X$, and since $F(X, \lambda)$ is convex in X on E_n for any $\lambda \in \Lambda$ we have

$$F(X^*, \lambda^*) = \min_{X \in E_n} F(X, \lambda^*).$$

This implies the right-hand inequality in (5.1).

Next, by II, III and the definition of $F(X, \lambda)$, we have

$$F(X^*, \lambda^*) = \sum_{i \in R(X^*)} \lambda_{1i}^* f_i(X^*) = \varphi(X^*) \sum_{i \in R(X^*)} \lambda_{1i}^* = \varphi(X^*). \tag{5.3}$$

Take any $\lambda \in \Lambda$. Since $X^* \in \Omega$,

$$\begin{aligned} F(X^*, \lambda) = \sum_{i=0}^{N} \lambda_{1i} f_i(X^*) + \sum_{j=0}^{N_1} \lambda_{2j} h_j(X^*) \leqslant \\ \leqslant \sum_{i=0}^{N} \lambda_{1i} f_i(X^*) \leqslant \varphi(X^*) \sum_{i=0}^{N} \lambda_{1i} = \varphi(X^*). \end{aligned} \tag{5.4}$$

Combining (5.3) and (5.4), we get the left-hand inequality in (5.1), proving necessity.

Sufficiency. Let $[X^*, \lambda^*] \in E_n \times \Lambda$ be a saddle point of $F(X, \lambda)$ on $E_n \times \Lambda$. We want to prove that

$$\varphi(X^*) = \min_{X \in \Omega} \varphi(X).$$

We first show that $X^* \in \Omega$. Indeed, otherwise, there would be some $j_0 \in [0 : N_1]$ for which $h_{j_0}(X^*) > 0$. We set

$$\lambda' = (\lambda_{10}^*, \ldots, \lambda_{1N}^*, \lambda_{20}^*, \ldots, \lambda_{2, j_0-1}^*, \lambda_{2j_0}^* + a, \lambda_{2, j_0+1}^*, \ldots, \lambda_{2N_1}^*) \in \Lambda,$$

where $a > 0$. Then

$$F(X^*, \lambda') = F(X^*, \lambda^*) + a h_{j_0}(X^*) > F(X^*, \lambda^*),$$

contradicting (5.1), and so $X^* \in \Omega$.

We now note that

$$\lambda_{2j}^* h_j(X^*) = 0 \quad \text{for all} \quad j \in [0 : N_1]. \tag{5.5}$$

Indeed, otherwise, for some j_0,

$$\lambda_{2j_0}^* h_{j_0}(X^*) < 0.$$

Setting

$$\lambda' = (\lambda_{10}^*, \ldots, \lambda_{1N}^*, \lambda_{20}^*, \ldots, \lambda_{2,j_0-1}^*, 0, \lambda_{2,j_0+1}^*, \ldots, \lambda_{2N_1}^*) \in \Lambda,$$

we obtain

$$F(X^*, \lambda') = F(X^*, \lambda^*) - \lambda_{2j_0}^* h_{j_0}(X^*) > F(X^*, \lambda^*),$$

which contradicts (5.1). This proves (5.5).

We now claim that

$$\lambda_{1i}^* = 0, \quad \text{if} \quad i \notin R(X^*). \tag{5.6}$$

Suppose the contrary. Then there exists $\lambda_{1i_0}^* > 0$, $i_0 \notin R(X^*)$. Let $i_1 \in R(X^*)$. Set

$$\lambda' = (\lambda_{10}', \ldots, \lambda_{1N}', \lambda_{20}', \ldots, \lambda_{2N_1}') \in \Lambda,$$

where

$$\lambda_{1i}' = \begin{cases} \lambda_{1i}^* & \text{if} \quad i \neq i_0,\ i \neq i_1; \\ 0 & \text{if} \quad i = i_0; \\ \lambda_{1i_1}^* + \lambda_{1i_0}^* & \text{if} \quad i = i_1; \end{cases}$$
$$\lambda_{2j}' = \lambda_{2j}^*, \qquad j \in [0 : N_1].$$

We then have

$$F(X^*, \lambda') = F(X^*, \lambda^*) - \lambda_{1i_0}^* [f_{i_1}(X^*) - f_{i_0}(X^*)] > F(X^*, \lambda^*),$$

which contradicts (5.1), proving (5.6). It follows from (5.5) and (5.6) that

$$F(X^*, \lambda^*) = \varphi(X^*). \tag{5.7}$$

It is now easy to show that

$$\varphi(X^*) = \min_{X \in \Omega} \varphi(X).$$

Indeed, if this is false there exists $X' \in \Omega$ for which $\varphi(X') < \varphi(X^*)$. Then

$$F(X', \lambda^*) = \sum_{i=0}^{N} \lambda_{1i}^* f_i(X') + \sum_{j=0}^{N_1} \lambda_{2j}^* h_j(X') \leqslant$$
$$\leqslant \sum_{i=0}^{N} \lambda_{1i}^* f_i(X') \leqslant \varphi(X') < \varphi(X^*). \tag{5.8}$$

In view of (5.7) and (5.8), we get

$$F(X', \lambda^*) < F(X^*, \lambda^*),$$

contradicting (5.1) and completing the proof.

Remark. The sufficiency proof used no special properties of the functions $f_i(X)$ and $h_j(X)$. In other words:

The existence of a saddle point $[X^*, \lambda^*]$ *of the Lagrange function* $F(X, \lambda)$ *on* $E_n \times \Lambda$ *is a sufficient condition for the truth of*

$$\max_{i\in[0:N]} f_i(X^*) = \min_{X\in\Omega} \max_{i\in[0:N]} f_i(X),$$

where $\Omega = \{X \mid h_j(X) \leqslant 0,\ j \in [0:N_1]\}$, *for arbitrary functions* $f_i(X)$ *and* $h_j(X)$ *defined on* E_n.

§6. FIRST METHOD OF SUCCESSIVE APPROXIMATIONS

1. The results of this section generalize those of Chap. III, §6, in a natural manner.

Let

$$\varphi(X) = \max_{i\in[0:N]} f_i(X), \quad \varphi_1(X) = \max_{j\in[0:N_1]} h_j(X),$$
$$\Omega = \{X \in E_n \mid h_j(X) \leqslant 0 \quad \text{for all} \quad j \in [0:N_1]\}.$$

As usual, we shall assume that the functions $f_i(X)$, $i \in [0:N]$, are continuously differentiable on some open set Ω' containing Ω, the functions $h_j(X)$, $j \in [0:N_1]$, are convex and continuously differentiable on E_n, and the Slater condition (2.2) is satisfied.

For fixed $X \in \Omega$, we set

$$R_\varepsilon(X) = \{i \in [0:N] \mid \varphi(X) - f_i(X) \leqslant \varepsilon\}, \quad \varepsilon \geqslant 0;$$
$$Q_\mu(X) = \{j \in [0:N_1] \mid -\mu \leqslant h_j(X) \leqslant 0\}, \quad \mu \geqslant 0.$$

It is clear that $R_0(X) = R(X)$, and if $\varepsilon' \geqslant \varepsilon$, then

$$R_{\varepsilon'}(X) \supset R_\varepsilon(X). \tag{6.1}$$

It is also clear that $Q_0(X) = Q(X)$, and if $\mu' \geqslant \mu$, then

$$Q_{\mu'}(X) \supset Q_\mu(X). \tag{6.2}$$

In Chap. III, §6, we worked out an explicit expression for $R_\varepsilon(X)$ as a function of ε. In particular, we saw that for some (finite or infinite) $a_1(X) > 0$,

$$R_\varepsilon(X) = R(X), \quad \text{if} \quad 0 \leqslant \varepsilon < a_1(X). \tag{6.3}$$

An analogous representation is valid for $Q_\mu(X)$. Arrange the numbers $h_j \overset{\text{df}}{=} h_j(X)$ ($X \in \Omega$ is fixed) in decreasing order:

$$h_{j_{01}} = \ldots = h_{j_{0q_0}} > h_{j_{11}} = \ldots = h_{j_{1q_1}} > \ldots > h_{j_{r1}} = \ldots = h_{j_{rq_r}}.$$

Clearly, $0 \leqslant r \leqslant N_1$, and $r \overset{\text{df}}{=} r(X)$ depends on X.

We denote

$$J_k(X)=\{j_{k1},\ \ldots,\ j_{kq_k}\},\quad k\in[0:r];$$
$$b_k(X)=-h_j(X),\quad j\in J_k(X).$$

Clearly,

$$0\leqslant b_0(X)<b_1(X)<\ \ldots\ <b_r(X).$$

We set $b_{r+1}(X)=\infty$ by definition. Let

$$Q^s(X)=\bigcup_{k=0}^{s} J_k(X),\quad s\in[0:r].$$

We can now write down the representation of $Q_\mu(X)$:

$$Q_\mu(X)=\begin{cases}\varnothing, & \text{if}\quad \mu\in[0,\ b_0(X)),\\ Q^s(X), & \text{if}\quad b_s(X)\leqslant\mu<b_{s+1}(X),\quad s\in[0:r].\end{cases}$$

Hence, in particular:

$$\left.\begin{aligned}Q_\mu(X)&=Q(X)\ \text{for}\ 0\leqslant\mu<b_0(X),\ \text{if}\ b_0(X)>0;\\ Q_\mu(X)&=Q(X)\ \text{for}\ 0\leqslant\mu<b_1(X),\ \text{if}\ b_0(X)=0.\end{aligned}\right\}\qquad(6.4)$$

Lemma 6.1. For any fixed $X_0\in\Omega$, *there exist numbers* $\gamma_0>0,\ \delta_0>0,\ \gamma_1>0,\ \delta_1>0$, *such that:*

I. *For any* $\varepsilon\in[0,\ \gamma_0]$ *and* $X'\in\Omega,\ \|X'-X_0\|<\delta_0$,

$$R_\varepsilon(X')\subset R(X_0).$$

II. *For any* $\mu\in[0,\ \gamma_1]$ *and* $X'\in\Omega,\ \|X'-X_0\|<\delta_1$,

$$Q_\mu(X')\subset Q(X_0).$$

The proof is omitted, since it is analogous to that of Lemma 6.1 in Chap. III.

2. We introduce the notation

$$\tilde L_{\varepsilon\mu}(X)=\operatorname{co}\tilde H_{\varepsilon\mu}(X),$$

where

$$X\in\Omega,\quad \tilde H_{\varepsilon\mu}(X)=H_\varepsilon(X)\cup H_{1\mu}(X),$$
$$H_\varepsilon(X)=\left\{\frac{\partial f_i(X)}{\partial X}\,\middle|\, i\in R_\varepsilon(X)\right\},$$
$$H_{1\mu}(X)=\left\{\frac{\partial h_j(X)}{\partial X}\,\middle|\, j\in Q_\mu(X)\right\}.$$

It is clear that $\tilde L_{00}(X)=\tilde L(X)$ (see §3).

By (6.3) and (6.4), we have

$$\tilde{L}_{\varepsilon\mu}(X) = \tilde{L}(X) \tag{6.5}$$

for $\varepsilon \in [0, a_1(X))$ and

$$\mu \in \begin{cases} [0, b_0(X)), & \text{if} \quad b_0(X) > 0, \\ [0, b_1(X)), & \text{if} \quad b_0(X) = 0. \end{cases}$$

Let

$$\tilde{d}_{\varepsilon\mu}(X) = \min_{Z \in \tilde{L}_{\varepsilon\mu}(X)} \| Z \|, \quad \tilde{d}(X) = \min_{Z \in \tilde{L}_{00}(X)} \| Z \|.$$

Definition 1. A point $X^* \in \Omega$ for which

$$0 \in \tilde{L}_{\varepsilon\mu}(X^*),$$

will be called an (ε, μ)-quasistationary point of $\varphi(X)$ on the set Ω.

Definition 2. Let $0 \notin \tilde{L}_{\varepsilon\mu}(\bar{X})$. The vector

$$\tilde{g}_{\varepsilon\mu}(\bar{X}) = -\tilde{Z}_{\varepsilon\mu}(\bar{X}) / \| \tilde{Z}_{\varepsilon\mu}(\bar{X}) \|,$$

where $\| \tilde{Z}_{\varepsilon\mu}(\bar{X}) \| = \min_{Z \in \tilde{L}_{\varepsilon\mu}(\bar{X})} \| Z \|$, will be called a direction of (ε, μ)-quasi-steepest descent for $\varphi(X)$ at the point $\bar{X}$.

3. We now proceed to describe our first method of successive approximations for finding stationary points of $\varphi(X)$ on Ω.

We fix three parameters $\varepsilon_0 > 0$, $\mu_0 > 0$, $\rho_0 > 0$. Let $X_0 \in \Omega$ be some initial approximation.

Assume that the set

$$M(X_0) = \{X \in \Omega \mid \varphi(X) \leqslant \varphi(X_0)\}$$

is bounded. Since $\varphi(X)$ is continuous and Ω is closed, $M(X_0)$ is also closed.

Suppose we have already determined the k-th approximation $X_k \in M(X_0)$. If $\tilde{d}(X_k) = 0$ (i. e., $0 \in \tilde{L}(X_k)$), then X_k is a stationary point of $\varphi(X)$ on Ω and the procedure terminates.

Suppose, therefore, that $\tilde{d}(X_k) > 0$. Let ν_k denote the smallest $\nu \in \{0, 1, 2, \ldots\}$ such that

$$\tilde{d}_{\varepsilon_\nu \mu_\nu}(X_k) \geqslant \rho_\nu, \tag{6.6}$$

where $\varepsilon_\nu = \varepsilon_0 / 2^\nu$, $\mu_\nu = \mu_0 / 2^\nu$, $\rho_\nu = \rho_0 / 2^\nu$. Note that ν_k is finite — this follows from the fact that $\rho_\nu \to 0$ as $\nu \to \infty$ while the left-hand side of (6.6) tends to $\tilde{d}(X_k) > 0$ by (6.5).

We shall use the notation

$$\hat{\varepsilon}_k = \varepsilon_{\nu_k}, \quad \hat{\mu}_k = \mu_{\nu_k}, \quad \hat{\rho}_k = \rho_{\nu_k}, \quad \tilde{d}_k = \tilde{d}_{\varepsilon_{\nu_k} \mu_{\nu_k}}(X_k).$$

We set

$$\tilde{g}_k = \tilde{g}_{\hat{\varepsilon}_k \hat{\mu}_k}(X_k),$$

where $\tilde{g}_{\hat{\varepsilon}_k \hat{\mu}_k}(X_k)$ is the direction of $(\hat{\varepsilon}_k, \hat{\mu}_k)$-quasisteepest descent of $\varphi(X)$ at X_k. Consider the ray

$$X = X_k(\alpha) \overset{df}{=} X_k + \alpha \tilde{g}_k, \quad \alpha \geqslant 0.$$

Let $\alpha_k \in [0, \infty)$ be such that

$$\varphi(X_k(\alpha_k)) = \min_{\substack{\alpha \in [0, \infty) \\ X_k(\alpha) \in \Omega}} \varphi(X_k(\alpha))$$

(that the minimum exists follows from the fact that $M(X_0)$ is bounded and closed).

Now set $X_{k+1} = X_k(\alpha_k)$. It is clear that $X_{k+1} \in M(X_0)$.

We first prove that

$$\tilde{g}_k \in \Gamma(X_k). \tag{6.7}$$

If $\varphi_1(X_k) < 0$, this is obvious. Let $\varphi_1(X_k) = 0$. Since $\max_{Z \in \tilde{L}_{\varepsilon\mu}(x)} (Z, \tilde{g}_{\varepsilon\mu}(X)) = -\tilde{d}_{\varepsilon\mu}(X)$, it follows that

$$\max_{j \in Q_\mu(X)} \left(\frac{\partial h_j(X)}{\partial X}, \tilde{g}_{\varepsilon\mu}(X) \right) \leqslant \max_{Z \in \tilde{L}_{\varepsilon\mu}(X)} (Z, \tilde{g}_{\varepsilon\mu}(X)) = -\tilde{d}_{\varepsilon\mu}(X). \tag{6.8}$$

In view of (6.8), we see that for sufficiently small $\alpha > 0$

$$\begin{aligned} \varphi_1(X_k + \alpha \tilde{g}_k) &= \varphi_1(X_k) + \alpha \frac{\partial \varphi_1(X_k)}{\partial \tilde{g}_k} + o_k(\alpha) \leqslant \\ &\leqslant \alpha \max_{j \in Q_{\hat{\mu}_k}(X_k)} \left(\frac{\partial h_j(X_k)}{\partial X}, \tilde{g}_{\hat{\varepsilon}_k \hat{\mu}_k}(X_k) \right) + o_k(\alpha) \leqslant \\ &\leqslant -\alpha \tilde{d}_k + o_k(\alpha) \leqslant -\frac{1}{2} \alpha \hat{\rho}_k < 0, \end{aligned}$$

which implies (6.7).

One proves in analogous fashion that for small $\alpha > 0$

$$\varphi(X_k + \alpha \tilde{g}_k) \leqslant \varphi(X_k) - \frac{1}{2} \alpha \hat{\rho}_k < \varphi(X_k). \tag{6.9}$$

It now follows from (6.7) and (6.9) that

$$\varphi(X_{k+1}) < \varphi(X_k).$$

Continuing the construction in this way, we obtain a sequence $\{X_k\}$ and together with it auxiliary sequences

$$\{\hat{\varepsilon}_k\}, \quad \{\hat{\mu}_k\}, \quad \{\hat{\rho}_k\}, \quad \{\tilde{g}_k\}, \quad \{\alpha_k\},$$

with $X_k \in M(X_0)$, $k=0, 1, 2, \ldots$, and

$$\varphi(X_0) > \varphi(X_1) > \ldots > \varphi(X_k) > \ldots \tag{6.10}$$

If the sequence $\{X_k\}$ is finite, its last term is by construction a stationary point of $\varphi(X)$ on Ω.

Now suppose that the sequence $\{X_k\}$ is infinite.

Lemma 6.2. As $k \to \infty$,

$$\hat{\varepsilon}_k \to 0, \quad \hat{\mu}_k \to 0, \quad \hat{\rho}_k \to 0. \tag{6.11}$$

Proof. We first observe that by construction the convergence relations (6.11) are either all true or all false. Suppose that they are false. Then there exist numbers $\varepsilon^* > 0$, $\mu^* > 0$, $\rho^* > 0$ and a sequence of indices $\{k_s\}$, $k_s \to \infty$, such that

$$\hat{\varepsilon}_{k_s} \geqslant \varepsilon^*, \quad \hat{\mu}_{k_s} \geqslant \mu^*, \quad \hat{\rho}_{k_s} \geqslant \rho^*.$$

Since

$$R_{\varepsilon^*}(X_{k_s}) \subset R_{\hat{\varepsilon}_{k_s}}(X_{k_s}), \quad Q_{\mu^*}(X_{k_s}) \subset Q_{\hat{\mu}_{k_s}}(X_{k_s}),$$

it follows that for $i \in R_{\varepsilon^*}(X_{k_s})$

$$f_i(X_{k_s} + \alpha \tilde{g}_{k_s}) = f_i(X_{k_s}) + \alpha\left(\frac{\partial f_i(X_{k_s})}{\partial X}, \tilde{g}_{k_s}\right) + o_{ik_s}(\alpha) \leqslant$$
$$\leqslant f_i(X_{k_s}) - \alpha \tilde{d}_{k_s} + o_{ik_s}(\alpha) \leqslant f_i(X_{k_s}) - \alpha\rho^* + o_{ik_s}(\alpha).$$

Similarly, for $j \in Q_{\mu^*}(X_{k_s})$, we have

$$h_j(X_{k_s} + \alpha \tilde{g}_{k_s}) \leqslant h_j(X_{k_s}) - \alpha\rho^* + o_{jk_s}(\alpha).$$

We now select $\alpha_0 > 0$, independent of i, j and k_s, such that for $i \in R_{\varepsilon^*}(X_{k_s})$, $j \in Q_{\mu^*}(X_{k_s})$, $\alpha \in [0, \alpha_0]$, we have

$$f_i(X_{k_s} + \alpha \tilde{g}_{k_s}) \leqslant \varphi(X_{k_s}) - \frac{1}{2}\alpha\rho^* \tag{6.12}$$

$$h_j(X_{k_s} + \alpha \tilde{g}_{k_s}) \leqslant -\frac{1}{2}\alpha\rho^*. \tag{6.13}$$

Next, let

$$C_1 = \max_{X \in M(X_0)} \max_{i \in [0:N]} \left\| \frac{\partial f_i(X)}{\partial X} \right\|,$$

$$C_2 = \max_{X \in M(X_0)} \max_{j \in [0:N_1]} \left\| \frac{\partial h_j(X)}{\partial X} \right\|.$$

Since $X_{k_s} \in M(X_0)$, it follows that for $i \notin R_{\varepsilon^*}(X_{k_s})$ and $j \notin Q_{\mu^*}(X_{k_s})$,

$$f_i(X_{k_s}+\alpha\tilde{g}_{k_s}) \leqslant f_i(X_{k_s})+\alpha C_1+o_{ik_s}(\alpha) \leqslant \\ \leqslant \varphi(X_{k_s})-\varepsilon^*+\alpha C_1+o_{ik_s}(\alpha);$$

$$h_j(X_{k_s}+\alpha\tilde{g}_{k_s}) \leqslant h_j(X_{k_s})+\alpha C_2+o_{jk_s}(\alpha) \leqslant \\ \leqslant -\mu^*+\alpha C_2+o_{jk_s}(\alpha).$$

It is clear that there exists $\alpha_1>0,\ 0<\alpha_1\leqslant\alpha_0$, independent of i, j, k_s, such that for $i\notin R_{\varepsilon^*}(X_{k_s})$, $j\notin Q_{\mu^*}(X_{k_s})$ and $\alpha\in[0,\alpha_1]$,

$$f_i(X_{k_s}+\alpha\tilde{g}_{k_s}) \leqslant \varphi(X_{k_s})-\frac{1}{2}\varepsilon^*; \tag{6.14}$$

$$h_j(X_{k_s}+\alpha\tilde{g}_{k_s}) \leqslant -\frac{1}{2}\mu^*. \tag{6.15}$$

It follows from (6.12)—(6.15), in particular, that

$$\varphi(X_{k_s}+\alpha_1\tilde{g}_{k_s}) \leqslant \varphi(X_{k_s})-\beta_1; \tag{6.16}$$

$$\varphi_1(X_{k_s}+\alpha_1\tilde{g}_{k_s}) \leqslant -\beta_2, \tag{6.17}$$

uniformly in k_s, where

$$\beta_1=\min\left\{\frac{1}{2}\alpha_1\rho^*,\ \frac{1}{2}\varepsilon^*\right\},$$

$$\beta_2=\min\left\{\frac{1}{2}\alpha_1\rho^*,\ \frac{1}{2}\mu^*\right\}.$$

We now easily derive a contradiction. It follows from (6.10) that

$$\varphi(X_k)\xrightarrow[k\to\infty]{}\varphi^*, \tag{6.18}$$

and for all $k=0, 1, 2, \ldots,$

$$\varphi(X_k)>\varphi^*. \tag{6.19}$$

By (6.17),

$$X_{k_s}(\alpha_1)\stackrel{df}{=}X_{k_s}+\alpha_1\tilde{g}_{k_s}\in\Omega,$$

and by (6.16)

$$\varphi(X_{k_s}(\alpha_1))\leqslant\varphi(X_{k_s})-\beta_1.$$

For sufficiently large k_s, it follows from (6.16) and (6.18) that

$$\varphi(X_{k_s}(\alpha_1))\leqslant\varphi^*-\frac{1}{2}\beta_1.$$

But

$$\varphi(X_{k_s+1})=\min_{\substack{\alpha\geqslant 0\\ X_{k_s}(\alpha)\in\Omega}}\varphi(X_{k_s}(\alpha))\leqslant\varphi(X_{k_s}(\alpha_1))\leqslant\varphi^*-\frac{1}{2}\beta_1,$$

which contradicts (6.19), proving the lemma.

Note that the above arguments remain valid if

$$Q_{\mu^*}(X_{k_s}) = \varnothing.$$

4. If the sequence $\{X_k\}$ constructed above is infinite, it has at least one limit point, since $X_k \in M(X_0)$, $k = 0, 1, 2, \ldots$, and by assumption the set $M(X_0)$ is bounded and closed.

Theorem 6.1. Any limit point of the sequence $\{X_k\}$ is a stationary point of $\varphi(X)$ on Ω.

Proof. Let

$$X_{k_s} \xrightarrow[k_s \to \infty]{} X^*.$$

It is clear that $X^* \in M(X_0)$ and, in particular, $X^* \in \Omega$. Our assertion that X^* is a stationary point of $\varphi(X)$ on Ω will be proved if we can show that $\tilde{d}(X^*) = 0$.

Suppose the contrary:

$$\tilde{d}(X^*) \stackrel{df}{=} \min_{Z \in \tilde{L}(X^*)} \| Z \| = d^* > 0.$$

By Lemma 6.1, there exist numbers $\gamma^* > 0$, $\delta^* > 0$, $K_1 > 0$, such that for $k_s \geqslant K_1$, $\varepsilon \in [0, \gamma^*]$, $\mu \in [0, \delta^*]$,

$$R_\varepsilon(X_{k_s}) \subset R(X^*), \qquad Q_\mu(X_{k_s}) \subset Q(X^*).$$

Hence

$$\tilde{d}_{\varepsilon\mu}(X_{k_s}) \geqslant \min_{Z \in L'(X_{k_s})} \| Z \|,$$

where

$$L'(X_{k_s}) = \operatorname{co} H'(X_{k_s}),$$

$$H'(X_{k_s}) = \left\{ \frac{\partial f_i(X_{k_s})}{\partial X} \,\middle|\, i \in R(X^*) \right\} \cup \left\{ \frac{\partial h_j(X_{k_s})}{\partial X} \,\middle|\, j \in Q(X^*) \right\}.$$

Obviously, we can find $K \geqslant K_1$ so large that for any $k_s \geqslant K$

$$\min_{Z \in L'(X_{k_s})} \| Z \| \geqslant \frac{1}{2} \tilde{d}(X^*) = \frac{d^*}{2}.$$

Thus, for $\varepsilon \in [0, \gamma^*]$, $\mu \in [0, \delta^*]$ and $k_s \geqslant K$

$$\tilde{d}_{\varepsilon\mu}(X_{k_s}) \geqslant \frac{d^*}{2}. \tag{6.20}$$

Let ν_1 denote the smallest nonnegative integer such that

$$\frac{\varepsilon_0}{2^{\nu_1}} \leqslant \gamma^*, \quad \frac{\mu_0}{2^{\nu_1}} \leqslant \delta^*, \quad \frac{\rho_0}{2^{\nu_1}} \leqslant \frac{d^*}{2}. \tag{6.21}$$

Then it follows from (6.20) and (6.21) that for all $k_s \geqslant K$

$$\tilde{d}_{\varepsilon_{\nu_1}\mu_{\nu_1}}(X_{k_s}) \geqslant \frac{d^*}{2} \geqslant \rho_{\nu_1}.$$

Hence

$$\hat{\varepsilon}_{k_s} \geqslant \varepsilon_{\nu_1}, \quad \hat{\mu}_{k_s} \geqslant \mu_{\nu_1}, \quad \hat{\rho}_{k_s} \geqslant \rho_{\nu_1},$$

contradicting Lemma 6.2.

Remark. At each step of the above successive approximation procedure one must determine $\hat{\varepsilon}_k$, $\hat{\mu}_k$, $\hat{\rho}_k$, $\tilde{g}_k$ and α_k. By Lemma 6.2, $\hat{\varepsilon}_k \to 0$, $\hat{\mu}_k \to 0$, $\hat{\rho}_k \to 0$, so that the number of trials (comparisons of $\tilde{d}_{\varepsilon_\nu\mu_\nu}(X_k)$ with ρ_ν, $\nu = 0, 1, 2, \ldots$; see (6.6)) required to select $\hat{\varepsilon}_k$, $\hat{\mu}_k$, $\hat{\rho}_k$ goes to infinity as $k \to \infty$.

As remarked in Chap. III, in actual computation one may resort to the following device: begin the trials not from ε_0, μ_0, ρ_0 but from $\hat{\varepsilon}_{k-1}$, $\hat{\mu}_{k-1}$, $\hat{\rho}_{k-1}$, i. e., the parameters computed at the preceding step. If

$$\tilde{d}_{\hat{\varepsilon}_{k-1}\hat{\mu}_{k-1}}(X_k) < \hat{\rho}_{k-1},$$

we examine the sequence

$$\hat{\varepsilon}_{k-1}/2^\nu, \quad \hat{\mu}_{k-1}/2^\nu, \quad \hat{\rho}_{k-1}/2^\nu, \quad \nu = 1, 2, \ldots,$$

until we find $\hat{\varepsilon}_k$, $\hat{\mu}_k$, $\hat{\rho}_k$.

But if

$$\tilde{d}_{\hat{\varepsilon}_{k-1}\hat{\mu}_{k-1}}(X_k) \geqslant \hat{\rho}_{k-1},$$

we look for the largest $\nu \in \left[0 : \log_2 \frac{\varepsilon_0}{\hat{\varepsilon}_{k-1}}\right]$ such that

$$\tilde{d}_{\varepsilon_{k\nu}\mu_{k\nu}}(X_k) \geqslant \rho_{k\nu},$$

where

$$\varepsilon_{k\nu} = 2^\nu \hat{\varepsilon}_{k-1}, \quad \mu_{k\nu} = 2^\nu \hat{\mu}_{k-1}, \quad \rho_{k\nu} = 2^\nu \hat{\rho}_{k-1}.$$

The quantities $\varepsilon_{k\nu}$, $\mu_{k\nu}$, $\rho_{k\nu}$ corresponding to this largest ν will be the required $\hat{\varepsilon}_k$, $\hat{\mu}_k$, $\hat{\rho}_k$.

5*. The first method of successive approximations may be modified as follows.

Set $\bar{\varepsilon}_{-1} = \varepsilon_0$, $\bar{\mu}_{-1} = \mu_0$, $\bar{\rho}_{-1} = \rho_0$ and select an initial approximation $X_0' \in \Omega$. We shall assume as usual that the set $M(X_0') = \{X \in \Omega \mid \varphi(X) \leqslant \varphi(X_0')\}$ is bounded.

Suppose we have already found $X_k' \in M(X_0')$, $\bar{\varepsilon}_{k-1}$, $\bar{\mu}_{k-1}$, $\bar{\rho}_{k-1}$, and let $\tilde{d}(X_k') > 0$. To construct X_{k+1}', we find the smallest $\nu \in \{0, 1, 2, \ldots\}$ for which

$$\tilde{d}_{\varepsilon_\nu'\mu_\nu'}(X_k') \geqslant \rho_\nu',$$

where $\varepsilon_\nu' = \bar{\varepsilon}_{k-1}/2^\nu$, $\mu_\nu' = \bar{\mu}_{k-1}/2^\nu$, $\rho_\nu' = \bar{\rho}_{k-1}/2^\nu$.

Denote this value by ν'_k. Set

$$\bar{\varepsilon}_k = \varepsilon'_{\nu'_k}, \quad \bar{\mu}_k = \mu'_{\nu'_k}, \quad \bar{\rho}_k = \rho'_{\nu'_k}.$$

We now proceed as in the original version: find $g'_k = \tilde{g}_{\bar{\varepsilon}_k \bar{\mu}_k}(X'_k)$ and define X'_{k+1} to be the minimum point of the function $\varphi(X)$ on the segment $X = X'_k(\alpha) \stackrel{df}{=} X'_k + \alpha g'_k$, $\alpha \geqslant 0$, $X'_k(\alpha) \in \Omega$.

It is clear that $X'_{k+1} \in M(X'_0)$, $\bar{\varepsilon}_k \leqslant \bar{\varepsilon}_{k-1}$, $\bar{\mu}_k \leqslant \bar{\mu}_{k-1}$, $\bar{\rho}_k \leqslant \bar{\rho}_{k-1}$ and $\varphi(X'_{k+1}) < \varphi(X'_k)$.

If the sequence $\{X'_k\}$ is finite, we are done — the last term is a stationary point of $\varphi(X)$ on Ω.

Otherwise:

Theorem 6.2. Let $\{X'_{k_p}\}$ be a subsequence of $\{X'_k\}$ such that

$$\bar{\varepsilon}_{k_p} < \bar{\varepsilon}_{k_p - 1}, \quad \bar{\mu}_{k_p} < \bar{\mu}_{k_p - 1}.$$

If the set $M(X'_0)$ is bounded, then any limit point of the subsequence $\{X'_{k_p}\}$ is a stationary point of $\varphi(X)$ on Ω.

The proof is analogous to that of Theorem 6.2, Chap. III. The constraints have the same effect as in the proof of Theorem 6.1.

6*. Yet another modification of the first method of successive approximation is as follows. Replace the set $Q_\mu(X)$ by

$$Q'_\mu(X) = \{j \in [0:N_1] \mid -\mu \leqslant t_j(X) h_j(X) \leqslant 0\},$$

and the set $H_{1\mu}(X)$ by

$$H'_{1\mu}(X) = \left\{ b_0 t_j(X) \frac{\partial h_j(X)}{\partial X} \,\middle|\, j \in Q'_\mu(X) \right\}.$$

Here $t_j(X) = \left\| \frac{\partial h_j(X)}{\partial X} \right\|^{-1}$, and b_0 is an arbitrary positive constant.

The set $\tilde{L}'_{\varepsilon\mu}(X)$ is in this case

$$\tilde{L}'_{\varepsilon\mu}(X) = \mathrm{co}\left(H_\varepsilon(X) \cup H'_{1\mu}(X)\right),$$

and the condition $\mathbf{0} \in \tilde{L}'_{00}(X)$ again defines the stationary points of $\varphi(X)$ on the set Ω. The formal description of the method in this case is unchanged. The convergence proof, however, will also make use of the following facts:

I. There exist $a > 0$ and $\bar{\mu} > 0$ such that for all $\mu \in [0, \bar{\mu}]$, $X \in (\partial_\mu \Omega) \cap M(X_0)$ and $j \in Q'_\mu(X)$

$$\left\| \frac{\partial h_j(X)}{\partial X} \right\| \geqslant a.$$

II. For every $\mu > 0$, there exists $c > 0$ such that for all $X \in M(X_0)$ and $j \notin Q'_\mu(X)$

$$h_j(X) \leqslant -c.$$

The parameter μ_0 must be taken no larger than $\bar{\mu}$. The details of the convergence proof are left to the reader.

7*. The following argument should explain the idea behind the use of $Q'_\mu(X)$. Suppose that all the constraints defining Ω are linear:

$$\Omega = \{Z \in E_n \mid h_j(Z) \overset{\text{df}}{=} (A_j, Z) + b_j \leqslant 0 \quad \text{for all } j \in [0:N_1]\}.$$

We claim that, for any point $X \in \Omega$, an index j is in $Q'_\mu(X)$ if and only if the distance of X from the plane $h_j(Z) = 0$ is at most μ (Figure 32).

Indeed, the distance ρ_j of X from the plane $(A_j, Z) + b_j = 0$ is (up to sign)

$$\beta_j(X) = \| A_j \|^{-1} ((A_j, X) + b_j).$$

Since $X \in \Omega$, i. e., $\beta_j(X) \leqslant 0$, we have

$$\rho_j = -\beta_j(X).$$

Thus the distance in question does not exceed μ if and only if

$$0 \geqslant \| A_j \|^{-1} ((A_j, X) + b_j) \geqslant -\mu,$$

or, equivalently,

$$-\mu \leqslant t_j(X)\, h_j(X) \leqslant 0, \tag{6.22}$$

where $t_j(X) \overset{\text{df}}{=} \left\| \frac{\partial h_j(X)}{\partial X} \right\|^{-1} = \| A_j \|^{-1}$.

On the other hand, inequality (6.22) is precisely the condition for the index j to be in the set $Q'_\mu(X)$, and our assertion is proved.

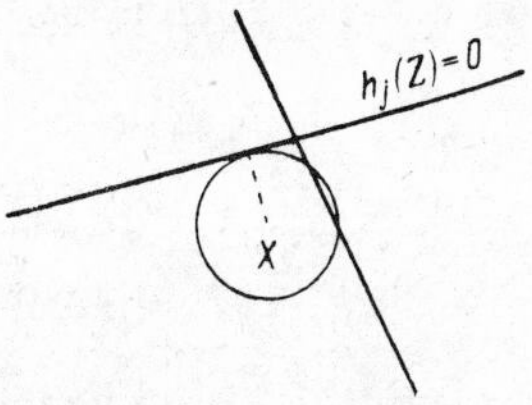

FIGURE 32.

As for the coefficient b_0 in the definition of $H'_{1\mu}(X)$, one must remember (see §4) that for $\varepsilon = 0$, $\mu = 0$ and large b_0 the vector $\tilde{g}_{00}(\bar{X})$ approximates the direction of steepest descent of $\varphi(X)$ at $\bar{X}$ on Ω (provided, of course, that $\tilde{d}(\bar{X}) > 0$).

§7. DETERMINATION OF (ε, μ)-QUASISTATIONARY POINTS. SECOND METHOD OF SUCCESSIVE APPROXIMATIONS

1. The definition of an (ε, μ)-quasistationary point was presented in the last section. We shall now describe a successive approximation procedure for finding (ε, μ)-quasistationary points of $\varphi(X)$ on Ω.

Fix $\varepsilon > 0$ and $\mu > 0$. Select an initial approximation $X_0 \in \Omega$. We shall assume that the set

$$M(X_0) = \{X \in \Omega \mid \varphi(X) \leqslant \varphi(X_0)\}$$

is bounded (it is of course also closed).

Suppose we have already determined the k-th approximation $X_k \in M(X_0)$. If $\tilde{d}_{\varepsilon\mu}(X_k) = 0$ (i. e., $\mathbf{0} \in \tilde{L}_{\varepsilon\mu}(X_k)$), then X_k is an (ε, μ)-quasistationary point of $\varphi(X)$ on Ω and the procedure terminates.

If $\tilde{d}_{\varepsilon\mu}(X_k) > 0$, we construct the next approximation X_{k+1} as follows. Set

$$g_k = \tilde{g}_{\varepsilon\mu}(X_k),$$

where $\tilde{g}_{\varepsilon\mu}(X_k)$ is the direction of (ε, μ)-quasisteepest descent of the function $\varphi(X)$ at X_k, and consider the ray $X = X_k(\alpha) \stackrel{\text{df}}{=} X_k + \alpha g_k$, $\alpha \geqslant 0$. Find $\alpha_k \in [0, \infty)$ such that

$$\varphi(X_k(\alpha_k)) = \min_{\substack{\alpha \geqslant 0 \\ X_k(\alpha) \in \Omega}} \varphi(X_k(\alpha))$$

(the infimum is achieved because $M(X_0)$ is closed and bounded). We now define the $(k+1)$-th approximation to be

$$X_{k+1} = X_k(\alpha_k).$$

It is clear that $X_{k+1} \in M(X_0)$. As in §6, one proves that

$$\varphi(X_{k+1}) < \varphi(X_k). \tag{7.1}$$

The procedure produces a sequence $\{X_k\}$ all of whose terms lie in $M(X_0)$. If it is finite, its last term is by construction an (ε, μ)-quasistationary point of $\varphi(X)$ on Ω. Otherwise:

Theorem 7.1. Any limit point of the sequence $\{X_k\}$ is an (ε, μ)-quasistationary point of $\varphi(X)$ on the set Ω.

We first prove the following

Lemma 7.1.

$$\lim_{k \to \infty} \tilde{d}_{\varepsilon\mu}(X_k) = 0. \tag{7.2}$$

Proof. Suppose the contrary. Then there exist a subsequence $\{X_{k_s}\}$ and a number $b > 0$ such that

$$\tilde{d}_{\varepsilon\mu}(X_{k_s}) \geqslant b.$$

Now, for $j \in Q_\mu(X_{k_s})$ and small α we have

$$h_j(X_{k_s} + \alpha g_{k_s}) \leqslant h_j(X_{k_s}) - \frac{1}{2}\alpha b,$$

uniformly in k_s, while for $j \notin Q_\mu(X_{k_s})$ and small α,

$$h_j(X_{k_s} + \alpha g_{k_s}) \leqslant -\frac{1}{2}\mu,$$

again uniformly in k_s; hence there exists $\alpha_1 > 0$ such that for all $\alpha \in [0, \alpha_1]$ and all k_s

$$\varphi_1(X_{k_s} + \alpha g_{k_s}) \leqslant 0,$$

i. e., $X_{k_s} + \alpha g_{k_s} \in \Omega$.

Analogous reasoning shows that there exists $\alpha' \in (0, \alpha_1)$ so small that

$$\varphi(X_{k_s} + \alpha' g_{k_s}) \leqslant \varphi(X_{k_s}) - \beta, \tag{7.3}$$

where $\beta = \min\left\{\frac{1}{2}\alpha' b, \frac{1}{2}\varepsilon\right\}$.

Thus the point $X_{k_s}(\alpha') = X_{k_s} + \alpha' g_{k_s}$ lies in Ω and inequality (7.3) holds.

We now readily derive a contradiction. It follows from (7.1) that

$$\varphi(X_k) \xrightarrow[k\to\infty]{} \varphi^* \geqslant \min_{X \in M(X_0)} \varphi(X),$$

with

$$\varphi(X_k) > \varphi^* \tag{7.4}$$

for all $k = 0, 1, 2, \ldots$ For sufficiently large k_s, we have

$$\varphi(X_{k_s}) \leqslant \varphi^* + \frac{1}{2}\beta,$$

so that by (7.3)

$$\varphi(X_{k_s+1}) \overset{df}{=} \min_{\substack{\alpha \geqslant 0 \\ X_{k_s}(\alpha) \in \Omega}} \varphi(X_{k_s}(\alpha)) \leqslant \varphi(X_{k_s}(\alpha')) \leqslant \varphi^* - \frac{1}{2}\beta,$$

which contradicts (7.4), proving the lemma.

Proof of Theorem 7.1. Let

$$X_{k_s} \xrightarrow[k_s\to\infty]{} X^*.$$

It is clear that $X^* \in M(X_0)$ and so $X^* \in \Omega$. We must show that $\tilde{d}_{\varepsilon\mu}(X^*) = 0$. If this is false,

$$\tilde{d}_{\varepsilon\mu}(X^*) \overset{df}{=} \min_{Z \in \tilde{L}_{\varepsilon\mu}(X^*)} \|Z\| = d^* > 0.$$

By Lemma 7.1, Chap. III, there exists $\delta>0$ such that for all $X \in \Omega$, $\| X - X^* \| < \delta$,

$$R_\varepsilon(X) \subset R_\varepsilon(X^*), \quad Q_\mu(X) \subset Q_\mu(X^*).$$

Hence it follows that for large k_s

$$\tilde{d}_{\varepsilon\mu}(X_{k_s}) \geqslant \frac{d^*}{2}$$

(see the proof of Lemma 7.2, Chap. III). But this contradicts (7.2). The proof is complete.

2. The method for (ε, μ)-quasistationary points yields a new method for finding stationary points of $\varphi(X)$ on Ω.

Let $\varepsilon_0 > 0$, $\mu_0 > 0$, $\rho_0 > 0$ be arbitrary, and select an initial approximation $X_0 \in \Omega$. Assume as usual that the set $M(X_0)$ is bounded.

If $\tilde{d}(X_0) = 0$, X_0 is a stationary point of $\varphi(X)$ on Ω. If $\tilde{d}(X_0) > 0$, finitely many steps of the above procedure for finding (ε, μ)-quasistationary points will yield a point $X_1 \in M(X_0)$ such that

$$\tilde{d}_{\varepsilon_0\mu_0}(X_1) \leqslant \rho_0.$$

Moreover, $\varphi(X_1) < \varphi(X_0)$. We now set

$$\varepsilon_1 = \frac{1}{2}\varepsilon_0, \quad \mu_1 = \frac{1}{2}\mu_0, \quad \rho_1 = \frac{1}{2}\rho_0$$

and repeat the procedure.

The result is a sequence $\{X_k\}$ such that

$$\tilde{d}_{\varepsilon_k\mu_k}(X_{k+1}) \leqslant \rho_k, \tag{7.5}$$

where $\varepsilon_k = \varepsilon_0/2^k$, $\mu_k = \mu_0/2^k$, $\rho_k = \rho_0/2^k$, $k = 0, 1, 2, \ldots$.

If this sequence is finite, its last term is by construction a stationary point of $\varphi(X)$ on Ω. Otherwise:

Theorem 7.2. Any limit point of the sequence $\{X_k\}$ is a stationary point of the function $\varphi(X)$ on Ω.

Proof. Let $X_{k_s} \xrightarrow[k_s\to\infty]{} X^*$. Obviously, $X^* \in \Omega$. We claim that $\tilde{d}(X^*) = 0$.

Indeed, if $\tilde{d}(X^*) > 0$, it follows from (6.3) and (6.4) that there exist $\varepsilon > 0$ and $\mu > 0$ for which

$$R_\varepsilon(X^*) = R(X^*), \quad Q_\mu(X^*) = Q(X^*).$$

Thus,

$$\tilde{d}_{\varepsilon\mu}(X^*) = \tilde{d}(X^*) > 0.$$

For sufficiently large k_s we have

$$\tilde{d}_{\varepsilon_{k_s}\mu_{k_s}}(X_{k_s+1}) \geqslant \frac{1}{2}\tilde{d}(X^*),$$

contradicting (7.5).

This proves the theorem.

Remark. Recall that the first and second methods of successive approximation may be used only provided the Slater condition (2.2) is satisfied, for only in this case is $\tilde{d}(X^*)=0$ equivalent to condition (3.1) (the definition of a stationary point).

§8. METHOD OF STEEPEST DESCENT. CASE OF LINEAR CONSTRAINTS

1. The successive approximation procedures of §§6 and 7 were based on the necessary condition for a minimax in the form (3.1).

Let $X_0 \in \Omega$ be a nonstationary point. Then

$$d(X_0) \stackrel{df}{=} \min_{\substack{Z \in L(X_0) \\ V \in \Gamma^+(X_0)}} \| Z - V \| = \| Z(X_0) - V(X_0) \| > 0.$$

The direction

$$g(X_0) = \frac{V(X_0) - Z(X_0)}{\| V(X_0) - Z(X_0) \|}, \qquad \| g(X_0) \| = 1,$$

is a direction of steepest descent for $\varphi(X)$ at X_0 on Ω.

In this section we consider the case that all the functions $h_j(X)$ are linear:

$$h_j(X) = (A_j, X) + b_j, \qquad j = [0 : N_1].$$

Supposing that $\| A_j \| \neq 0$, we may assume without loss of generality that $\| A_j \| = 1$, $j \in [0 : N_1]$. We shall also assume that the set

$$\Omega = \{X \in E_n \mid (A_j, X) + b_j \leqslant 0 \text{ for all } j \in [0 : N_1]\}$$

is not empty. Then, by the remark following Theorem 2.1, for any point $X_0 \in \Omega$

$$\Gamma^+(X_0) = \left\{ V = - \sum_{j \in Q(X_0)} \alpha_j A_j \,\middle|\, \alpha_j \geqslant 0 \right\}.$$

We introduce the notation

$$L_\varepsilon(X_0) = \operatorname{co} H_\varepsilon(X_0), \quad H_\varepsilon(X_0) = \left\{ \frac{\partial f_i(X_0)}{\partial X} \,\middle|\, i \in R_\varepsilon(X_0) \right\};$$
$$R_\varepsilon(X_0) = \{ i \in [0 : N] \mid \varphi(X_0) - f_i(X_0) \leqslant \varepsilon \}, \quad \varepsilon \geqslant 0;$$
$$\Gamma_\mu^+(X_0) = \left\{ V = - \sum_{j \in Q_\mu(X_0)} \alpha_j A_j \mid \alpha_j \geqslant 0 \right\};$$
$$Q_\mu(X_0) = \{ j \in [0 : N_1] \mid -\mu \leqslant h_j(X_0) \leqslant 0 \}, \quad \mu \geqslant 0.$$

If $Q_\mu(X_0) = \varnothing$, we set $\Gamma_\mu^+(X_0) = \{0\}$.

Definition 1. Let $\varepsilon \geqslant 0, \mu \geqslant 0$. A point $X_0 \in \Omega$ such that

$$d_{\varepsilon\mu}(X_0) \overset{df}{=} \min_{\substack{Z \in L_\varepsilon(X_0) \\ V \in \Gamma_\mu^+(X_0)}} \| Z - V \| = 0,$$

is called an (ε, μ)-stationary point of $\varphi(X)$ on the set Ω.

Definition 2. Let $X_0 \in \Omega$. If

$$d_{\varepsilon\mu}(X_0) = \| Z_{\varepsilon\mu}(X_0) - V_{\varepsilon\mu}(X_0) \| > 0,$$

where $Z_{\varepsilon\mu}(X_0) \in L_\varepsilon(X_0)$, $V_{\varepsilon\mu}(X_0) \in \Gamma_\mu^+(X_0)$, then

$$g_{\varepsilon\mu}(X_0) = \frac{V_{\varepsilon\mu}(X_0) - Z_{\varepsilon\mu}(X_0)}{\| X_{\varepsilon\mu}(X_0) - Z_{\varepsilon\mu}(X_0) \|} = \frac{V_{\varepsilon\mu}(X_0) - Z_{\varepsilon\mu}(X_0)}{d_{\varepsilon\mu}(X_0)}$$

is called a direction of (ε, μ)-steepest descent of $\varphi(X)$ at X_0 on Ω.

Clearly, $\| g_{\varepsilon\mu}(X_0) \| = 1$. As in Chap. IV, it is readily shown that the direction $g_{\varepsilon\mu}(X_0)$ is unique, although the pair of points $Z_{\varepsilon\mu}(X_0)$ and $V_{\varepsilon\mu}(X_0)$ need not be unique.

It follows from (6.3) and (6.4) that

$$L_\varepsilon(X_0) = L(X_0) \quad \text{for} \quad \varepsilon \in [0, a_1(X_0)),$$

$$\Gamma_\mu^+(X_0) = \Gamma^+(X_0) \text{ for } \mu \in \begin{cases} [0, b_0(X_0)), & \text{if} \quad b_0(X_0) > 0, \\ [0, b_1(X_0)), & \text{if} \quad b_0(X_0) = 0. \end{cases}$$

Therefore, for ε and μ as indicated,

$$d_{\varepsilon\mu}(X_0) = d(X_0). \tag{8.1}$$

Lemma 8.1. If $d_{\varepsilon\mu}(X_0) > 0$, then

$$\left(\frac{\partial f_i(X_0)}{\partial X}, g_{\varepsilon\mu}(X_0) \right) \leqslant - d_{\varepsilon\mu}(X_0), \quad i \in R_\varepsilon(X_0);$$

$$\left(\frac{\partial h_j(X_0)}{\partial X}, g_{\varepsilon\mu}(X_0) \right) = (A_j, g_{\varepsilon\mu}(X_0)) \leqslant 0, \quad j \in Q_\mu(X_0).$$

The proof follows immediately from the definitions of $d_{\varepsilon\mu}(X_0)$ and $g_{\varepsilon\mu}(X_0)$ (see the proof of Theorem 2.1 in Appendix II).

2. We can now present the steepest descent method for finding stationary points of $\varphi(X)$ on Ω.

As a preliminary, we fix three parameters $\varepsilon_0 > 0, \mu_0 > 0, \rho_0 > 0$. Select an initial approximation $X_0 \in \Omega$. We shall assume as always that the set

$$M(X_0) = \{X \in \Omega \mid \varphi(X) \leqslant \varphi(X_0)\}$$

is bounded.

Suppose we have already determined the k-th approximation $X_k \in M(X_0)$. If $d(X_k)=0$, then X_k is a stationary point and the procedure terminates.

Let $d(X_k)>0$. Determine the smallest $\nu \in \{0, 1, 2, \dots\}$ for which

$$d_{\varepsilon_\nu \mu_\nu}(X_k) \geqslant \rho_\nu, \tag{8.2}$$

where $\varepsilon_\nu = \varepsilon_0/2^\nu$, $\mu_\nu = \mu_0/2^\nu$, $\rho_\nu = \rho_0/2^\nu$, and denote it by ν_k. ν_k is clearly finite, since $\rho_\nu \to 0$ as $\nu \to \infty$, whereas the left-hand side of (8.2) tends to $d(X_k)>0$ by (8.1).

We introduce the notation

$$\hat{\varepsilon}_k = \varepsilon_{\nu_k}, \quad \hat{\mu}_k = \mu_{\nu_k}, \quad \hat{\rho}_k = \rho_{\nu_k},$$
$$d_k = d_{\hat{\varepsilon}_k \hat{\mu}_k}(X_k), \quad g_k = g_{\hat{\varepsilon}_k \hat{\mu}_k}(X_k),$$

where $g_{\hat{\varepsilon}_k \hat{\mu}_k}(X_k)$ is the direction of $(\hat{\varepsilon}_k, \hat{\mu}_k)$-steepest descent of $\varphi(X)$ at X_k.

We claim that

$$X_k + \alpha g_k \in \Omega \quad \text{for all} \quad \alpha \in [0, \hat{\mu}_k]. \tag{8.3}$$

Indeed, for $j \in Q_{\hat{\mu}_k}(X_k)$ and all $\alpha \geqslant 0$ we have, by Lemma 8.1,

$$h_j(X_k + \alpha g_k) = h_j(X_k) + \alpha (A_j, g_k) \leqslant 0.$$

Let $j \notin Q_{\hat{\mu}_k}(X_k)$. Since $\| A_j \| = 1$, we get

$$h_j(X_k + \alpha g_k) = h_j(X_k) + \alpha (A_j, g_k) \leqslant -\hat{\mu}_k + \alpha.$$

Thus, for $\alpha \in [0, \hat{\mu}_k]$ and all $j \in [0 : N_1]$,

$$h_j(X_k + \alpha g_k) \leqslant 0,$$

which is equivalent to (8.3). In particular, we see that g_k is a feasible direction: $g_k \in \Gamma(X_k)$.

Now consider the ray

$$X = X_k(\alpha) \overset{\text{df}}{=} X_k + \alpha g_k, \quad \alpha \geqslant 0,$$

and find $\alpha_k \in [0, \infty)$ for which

$$\varphi(X_k(\alpha_k)) = \min_{\substack{\alpha \geqslant 0 \\ X_k(\alpha) \in \Omega}} \varphi(X_k(\alpha)).$$

We now define the $(k+1)$-th approximation thus:

$$X_{k+1} = X_k(\alpha_k).$$

It is clear that

$$X_{k+1} \in M(X_0), \quad \varphi(X_{k+1}) < \varphi(X_k).$$

The sequel is analogous. As a result, we obtain a sequence $\{X_k\}$. If this sequence is finite, we are done: the last term is a stationary point of $\varphi(X)$ on Ω.

If the sequence is infinite:

Lemma 8.2. As $k\to\infty$,

$$\hat{\varepsilon}_k\to 0,\qquad \hat{\mu}_k\to 0,\qquad \hat{\rho}_k\to 0. \tag{8.4}$$

Proof. Suppose the contrary: there exist numbers $\varepsilon^*>0$, $\mu^*>0$, $\rho^*>0$ and a sequence of indices $\{k_s\}$, $k_s\xrightarrow[s\to\infty]{}\infty$ such that

$$\hat{\varepsilon}_{k_s}\geqslant\varepsilon^*,\qquad \hat{\mu}_{k_s}\geqslant\mu^*,\qquad \hat{\rho}_{k_s}\geqslant\rho^*.$$

By (8.3), if $\alpha\in[0,\mu^*]$ then

$$X_{k_s}+\alpha g_{k_s}\in\Omega.$$

In view of Lemma 8.1, it is easy to show (see the proof of Lemma 6.2) that there exists α', $0<\alpha'\leqslant\mu^*$, such that

$$\varphi(X_{k_s}+\alpha' g_{k_s})\leqslant\varphi(X_{k_s})-\beta, \tag{8.5}$$

uniformly in k_s, where $\beta>0$. Inequality (8.5) clearly implies a contradiction.

Theorem 8.1. Any limit point of the sequence $\{X_k\}$ is a stationary point of $\varphi(X)$ on Ω.

Proof. That the sequence $\{X_k\}$ has at least one limit point follows from the fact that it is bounded. Let $X_{k_s}\xrightarrow[k_s\to\infty]{}X^*$. Clearly, $X^*\in\Omega$. We wish to show that $d(X^*)=0$.

If this is false,

$$d(X^*)=\min_{\substack{Z\in L(X^*)\\ V\in\Gamma^+(X^*)}}\|Z-V\|=d^*>0.$$

By Lemma 6.1, there exist numbers $\gamma^*>0$, $\delta^*>0$, $K_1>0$ such that for $k_s\geqslant K_1$, $\varepsilon\in[0,\gamma^*]$, $\mu\in[0,\delta^*]$

$$R_\varepsilon(X_{k_s})\subset R(X^*),\qquad Q_\mu(X_{k_s})\subset Q(X^*).$$

Therefore,

$$d_{\varepsilon\mu}(X_{k_s})\geqslant\min_{\substack{Z\in L'(X_{k_s})\\ V\in\Gamma'^+(X_{k_s})}}\|Z-V\|,$$

where

$$L'(X_{k_s})=\operatorname{co}H'(X_{k_s}),\qquad H'(X_{k_s})=\left\{\frac{\partial f_i(X_{k_s})}{\partial X}\,\middle|\,i\in R(X^*)\right\},$$

$$\Gamma'^+(X_{k_s})=\left\{V=-\sum_{j\in Q(X^*)}\alpha_jA_j\,|\,\alpha_j\geqslant 0\right\}.$$

Obviously,

$$\Gamma'^{+}(X_{k_s}) = \Gamma^{+}(X^*).$$

It is readily shown that for sufficiently large $k_s \geqslant K \geqslant K_1$

$$\min_{\substack{Z \in L'(X_{k_s}) \\ V \in \Gamma'^{+}(X_{k_s})}} \| Z - V \| \geqslant \frac{1}{2} d^*.$$

Thus

$$d_{\varepsilon\mu}(X_{k_s}) \geqslant \frac{1}{2} d^* \tag{8.6}$$

for $\varepsilon \in [0, \gamma^*]$, $\mu \in [0, \delta^*]$ and $k_s \geqslant K$.

Let ν_1 denote the smallest nonnegative integer for which

$$\frac{\varepsilon_0}{2^{\nu_1}} \leqslant \gamma^*, \quad \frac{\mu_0}{2^{\nu_1}} \leqslant \delta^*, \quad \frac{\rho_0}{2^{\nu_1}} \leqslant \frac{d^*}{2}.$$

By (8.6), we have for $k_s \geqslant K$

$$d_{\varepsilon_{\nu_1}\mu_{\nu_1}}(X_{k_s}) \geqslant \frac{d^*}{2} \geqslant \rho_{\nu_1}.$$

Consequently,

$$\hat{\varepsilon}_{k_s} \geqslant \varepsilon_{\nu_1}, \ \hat{\mu}_{k_s} \geqslant \mu_{\nu_1}, \ \hat{\rho}_{k_s} \geqslant \rho_{\nu_1},$$

contradicting (8.4). This completes the proof.

3. Proceeding as in §7, one can devise a method of successive approximations for finding (ε, μ)- ationary points and base upon it a new method for stationary points of $\varphi(X)$ on Ω. Recall that

$$\Omega = \{X \in E_n \mid (A_j, X) + b_j \leqslant 0 \quad \text{for all} \quad j \in [0 : N_1]\}.$$

We leave the details of these algorithms and the convergence proofs to the reader.

§9. NONLINEAR CONSTRAINTS. CORRECTION OF DIRECTIONS

1. Let

$$\Omega = \{X \in E_n \mid h_j(X) \leqslant 0 \quad \text{for all} \quad j \in [0 : N_1]\},$$

where $h_j(X)$ are convex and continuously differentiable functions, and suppose that the Slater condition is fulfilled, i. e., there exists a point $\overline{X} \in \Omega$ such that

$$\max_{j \in [0:N_1]} h_j(\overline{X}) < 0. \tag{9.1}$$

For any point $X_0 \in \Omega$, we set

$$d'_{\varepsilon\mu}(X_0) = \min_{\substack{Z \in L_\varepsilon(X_0) \\ V \in \Gamma'^+_\mu(X_0)}} \| Z - V \|,$$

where (see §2.2)

$$\Gamma'^+_\mu(X_0) = \left\{ V = - \sum_{j \in Q'_\mu(X_0)} \alpha_j \frac{\partial h_j(X_0)}{\partial X} \,\middle|\, \alpha_j \geqslant 0 \right\},$$

$$Q'_\mu(X_0) = \{ j \in [0:N_1] \mid -\mu \leqslant t_j(X_0) h_j(X_0) \leqslant 0 \},$$

$$t_j(X_0) = \left\| \frac{\partial h_j(X_0)}{\partial X} \right\|^{-1}.$$

Clearly, $d'_{00}(X_0) = 0$ if and only if $d(X_0) = 0$. Thus the equation $d'_{00}(X_0) = 0$ defines stationary points of $\varphi(X)$ on Ω. If $d'_{\varepsilon\mu}(X_0) = \| Z_{\varepsilon\mu}(X_0) - V_{\varepsilon\mu}(X_0) \| > 0$, we set

$$g'_{\varepsilon\mu}(X_0) = \frac{V_{\varepsilon\mu}(X_0) - Z_{\varepsilon\mu}(X_0)}{\| V_{\varepsilon\mu}(X_0) - Z_{\varepsilon\mu}(X_0) \|}.$$

The direction $g'_{\varepsilon\mu}(X_0)$ possesses the properties:

$$\left(\frac{\partial f_i(X_0)}{\partial X}, \ g'_{\varepsilon\mu}(X_0) \right) \leqslant - d'_{\varepsilon\mu}(X_0) \quad \text{for} \quad i \in R_\varepsilon(X_0), \tag{9.2}$$

$$\left(\frac{\partial h_j(X_0)}{\partial X}, \ g'_{\varepsilon\mu}(X_0) \right) \leqslant 0 \qquad \text{for} \quad j \in Q'_\mu(X_0). \tag{9.3}$$

Lemma 9.1. Let $X_0 \in \Omega$; assume that the set $M(X_0) = \{ X \in \Omega \mid \varphi(X) < \varphi(X_0) \}$ is bounded and let the Slater condition (9.1) hold. Then there exist $a' > 0$ and $\mu' > 0$ such that for all $\mu \in [0, \mu']$ and all $X \in (\partial'_\mu \Omega) \cap M(X_0)$

$$\max_{j \in Q'_\mu(X)} \left(\frac{\partial h_j(X)}{\partial X}, \ q(X) \right) \leqslant - a',$$

where $q(X) = \dfrac{\overline{X} - X}{\| \overline{X} - X \|}$ and $\overline{X}$ is the point in (9.1).

Proof. If this is false, there exist two sequences $\{a_k\}$ and $\{\mu_k\}$ decreasing monotonically to zero, a sequence of vectors $\{X_k\}$, $X_k \in (\partial'_{\mu_k}\Omega) \cap M(X_0)$, $k = 1, 2, \ldots$, and an index $j_0 \in Q'_{\mu_k}(X_k)$, such that

$$\left(\frac{\partial h_{j_0}(X_k)}{\partial X}, \ q(X_k) \right) > - a_k.$$

We may assume without loss of generality that $X_k \xrightarrow[k\to\infty]{} X^*$, $X^* \in \Omega$. Since $j_0 \in Q'_{\mu_k}(X_k)$ for all k, we have

$$-\mu_k \leqslant t_{j_0}(X_k)\, h_{j_0}(X_k) \leqslant 0.$$

Hence it follows that $h_{j_0}(X^*) = 0$, i. e., $X^* \in \partial\Omega$. In particular, $X^* \neq \overline{X}$. We now have

$$-a_k < \left(\frac{\partial h_{j_0}(X_k)}{\partial X},\ q(X_k)\right) \leqslant \frac{h_{j_0}(\overline{X}) - h_{j_0}(X_k)}{\|\overline{X} - X_k\|}.$$

Letting $k \to \infty$ in this inequality, we get $h_{j_0}(\overline{X}) \geqslant 0$, contradicting (9.1).

2. We now describe a successive approximation procedure for finding stationary points of $\varphi(X)$ on Ω. Fix $\varepsilon_0 > 0$, $\mu_0 > 0$, $\rho_0 > 0$ and select an initial approximation $X_0 \in \Omega$. We shall assume that $M(X_0)$ is bounded and that $\mu_0 \leqslant \mu'$, where μ' is the number figuring in Lemma 9.1.

Suppose we have already found $X_k \in M(X_0)$. If $d'_{00}(X_k) = 0$, then X_k is a stationary point and we are done.

Suppose, then, that $d'_{00}(X_k) > 0$. Find the smallest $\nu \in \{0, 1, 2, \ldots\}$ for which

$$d'_{\varepsilon_\nu \mu_\nu}(X_k) \geqslant \rho_\nu, \tag{9.4}$$

where $\varepsilon_\nu = \varepsilon_0/2^\nu$, $\mu_\nu = \mu_0/2^\nu$, $\rho_\nu = \rho_0/2^\nu$, and denote it by ν_k. Now set

$$\hat{\varepsilon}_k = \varepsilon_{\nu_k}, \quad \hat{\mu}_k = \mu_{\nu_k}, \quad \hat{\rho}_k = \rho_{\nu_k},$$
$$d'_k = d'_{\varepsilon_{\nu_k}\mu_{\nu_k}}(X_k), \quad g'_k = g'_{\hat{\varepsilon}_k\hat{\mu}_k}(X_k).$$

If $Q'_{\hat{\mu}_k}(X_k) = \varnothing$, the direction g'_k is feasible. We then define X_{k+1} as the minimum point of $\varphi(X)$ on the set

$$\{X = X_k + \alpha g'_k \mid \alpha \geqslant 0,\ X \in \Omega\}.$$

Now let $Q'_{\hat{\mu}_k}(X_k) \neq \varnothing$. Since now the direction g'_k need not be feasible, it must be corrected. We set

$$c_k = \max_{i \in [0:\ N]} \left\| \frac{\partial f_i(X_k)}{\partial X} \right\|,$$
$$\beta_k = \min\left\{\frac{1}{3},\ \frac{d'_k}{3c_k}\right\},$$
$$p_k = \frac{g'_k + \beta_k q(X_k)}{\|g'_k + \beta_k q(X_k)\|}.$$

Clearly, $\beta_k > 0$ and $2/3 \leqslant \|g'_k + \beta_k g(X_k)\| \leqslant 4/3$. In addition, it follows from (9.2), (9.3) and Lemma 9.1 that

$$\left(\frac{\partial f_i(X_k)}{\partial X},\ p_k\right) \leqslant -d'_k/2 \quad \text{for} \quad i \in R_{\hat{\varepsilon}_k}(X_k), \tag{9.5}$$

$$\left(\frac{\partial h_j(X_k)}{\partial X},\ p_k\right) \leqslant -3/4\, a'\beta_k \quad \text{for} \quad j \in Q'_{\hat{\mu}_k}(X_k). \tag{9.6}$$

The last inequality implies the existence of $\gamma_k > 0$ such that for $\alpha \in [0, \gamma_k]$

$$X'_k(\alpha) = X_k + \alpha p_k \in \Omega.$$

We now determine $\alpha_k \geqslant 0$ such that

$$\varphi(X'_k(\alpha_k)) = \min_{\substack{\alpha \geqslant 0 \\ X'_k(\alpha) \in \Omega}} \varphi(X'_k(\alpha)),$$

and set $X_{k+1} = X'_k(\alpha_k)$. Obviously,

$$X_{k+1} \in M(X_0) \quad \text{and} \quad \varphi(X_{k+1}) < \varphi(X_k).$$

Continuing in this way, we obtain a sequence $\{X_k\}$. If it is finite, its last term is by construction a stationary point of $\varphi(X)$ on Ω. Otherwise:

Theorem 9.1. Any limit point of the sequence $\{X_k\}$ is a stationary point of $\varphi(X)$ on Ω.

The proof is analogous to that of Theorem 8.1, except that one must also use the relations (9.5) and (9.6), the definition of β_k and Lemma 2.6.

Remark. The method is also applicable when the generalized Slater condition (2.23) is satisfied. One must then assume that the linear constraints are written in normal form:

$$h_j(X) \overset{\text{df}}{=} (A_j, X) + b_j \leqslant 0, \quad \|A_j\| = 1, \quad j \in [N'+1 : N_1].$$

The point $\bar{X}$ satisfies the inequalities

$$h_j(\bar{X}) < 0, \quad j \in [0 : N'],$$
$$h_j(\bar{X}) \leqslant 0, \quad j \in [N'+1 : N_1].$$

3*. The technique used above to correct g'_k involved a direction $q(X_k)$ pointing strictly into the interior of the set Ω. An alternative approach is as follows.

Suppose that the Slater condition is fulfilled. Since the set $M(X_0)$ is bounded, it follows that for some $a'' > 0$

$$\max_{j \in Q'_\mu(X)} t_j(X)\left(\frac{\partial h_j(X)}{\partial X}, q(X)\right) \leqslant -a'' \tag{9.7}$$

for all $\mu \in [0, \mu']$ and all $X \in (\partial'_\mu \Omega) \cap M(X_0)$. This follows directly from Lemma 9.1.

We introduce the sets

$$l_{j\xi}(X) = \left\{ Z \,\middle|\, \left\| Z + t_j(X) \frac{\partial h_j(X)}{\partial X} \right\| \leqslant \xi \right\},$$
$$H_{\mu\xi}(X) = \operatorname{co} \bigcup_{j \in Q'_\mu(X)} l_{j\xi}(X).$$

It is readily checked, using (9.7), that for $\xi \in [0, a'')$, $\mu \in [0, \mu']$ and $X \in (\partial'_\mu \Omega) \cap$ $\cap M(X_0)$ we have

$$0 \notin H_{\mu\xi}(X).$$

By Lemma 2.5 of Appendix II, the cone $\Gamma^+_{\mu\xi}(X)$ spanned by the set $\bigcup_{j\in Q'_\mu(X)} l_{j\xi}(X)$ is closed. If $Q'_\mu(X)=\varnothing$, we define $\Gamma^+_{\mu\xi}(X)=\{0\}$ for all $\xi\geqslant 0$.

Note that if there is a vector $p\in E_n$, $\|p\|=1$, such that

$$(p,\ g)\geqslant 0 \quad \text{for all} \quad g\in\Gamma^+_{\mu\xi}(X), \tag{9.8}$$

then for any $j\in Q'_\mu(X)$

$$\left(p,\ t_j(X)\frac{\partial h_j(X)}{\partial X}\right)\leqslant -\xi. \tag{9.9}$$

Indeed, we need only set

$$g=-\xi p-t_j(X)\frac{\partial h_j(X)}{\partial X}.$$

We now fix $\xi_0\in(0,\ a'')$. Suppose we have already determined $X_k\in\Omega$. If $d'_{00}(X_k)=0$, the point X_k is stationary. Otherwise, we define $\hat{\varepsilon}_k$, $\hat{\mu}_k$, $\hat{\rho}_k$, d'_k as before. Suppose that $Q'_{\hat{\mu}_k}(X_k)\neq\varnothing$.

For each $\xi_s=(1/2)^s\xi_0$, $s=0,\ 1,\ 2,\ \ldots$, we set

$$\bar{d}_{ks}=\min_{\substack{Z\in L_{\hat{\varepsilon}_k}(X_k)\\ V\in\Gamma^+_{\hat{\mu}_k\xi_s}X_k}}\|Z-V\|=\|Z_{ks}-V_{ks}\|.$$

Let s_k denote the first number s for which

$$\bar{d}_{ks}\geqslant\frac{1}{2}d'_k.$$

s_k is clearly finite. Let us denote

$$\bar{d}_k=\bar{d}_{ks_k},\quad \bar{\xi}_k=\xi_{s_k},\quad \bar{g}_k=\bar{d}_k^{-1}\left(V_{ks_k}-Z_{ks_k}\right).$$

We have $\bar{d}_k\geqslant\frac{1}{2}d'_k$, and by (9.8) and (9.9)

$$\left(\bar{g}_k,\ t_j(X_k)\frac{\partial h_j(X_k)}{\partial X}\right)\leqslant-\bar{\xi}_k$$

for $j\in Q'_{\hat{\mu}_k}(X_k)$.

Now consider the ray $X=X_k(\alpha)\overset{\mathrm{df}}{=}X_k+\alpha\bar{g}_k$, $\alpha\geqslant 0$, and determine α_k such that

$$\varphi(X_k(\alpha_k))=\min_{\substack{\alpha\geqslant 0\\ X_k(\alpha)\in\Omega}}\varphi(X_k(\alpha)),$$

finally setting $X_{k+1}=X_k(\alpha_k)$.

The resulting sequence $\{X_k\}$ will yield a stationary point (the proof is analogous to that of Theorem 9.1).

§10. PENALTY FUNCTIONS

1. In this section we shall show how to reduce minimization of the maximum function $\varphi(X)$ on a set Ω defined by a system of inequalities to the solution of a sequence of minimax problems on the entire space E_n.

Let

$$\varphi(X) = \max_{i \in [0:N]} f_i(X),$$
$$\Omega = \{X \in E_n \mid h_j(X) \leqslant 0 \quad \text{for all} \quad j \in [0:N_1]\}.$$

The functions $f_i(X)$ and $h_j(X)$ are assumed continuous on E_n. We also assume that there exists a point $\bar{X} \in \Omega$ such that

$$\varphi(\bar{X}) = \min_{X \in \Omega} \varphi(X),$$

and that the set

$$M_0(\bar{X}) = \{X \in E_n \mid \varphi(X) \leqslant \varphi(\bar{X})\}$$

is bounded. By the continuity of $\varphi(X)$, the set $M_0(\bar{X})$ is also closed.

Consider the function

$$\varphi_\lambda(X) = \varphi(X) + \sum_{j=0}^{N_1} \lambda_j h_j^2(X) \delta_j(X),$$

where

$$\lambda = (\lambda_0, \ldots, \lambda_{N_1})$$

and

$$\delta_j(X) = \begin{cases} 0, & \text{if} \quad h_j(X) \leqslant 0, \\ 1, & \text{if} \quad h_j(X) > 0. \end{cases}$$

Obviously, $\varphi_\lambda(X) = \varphi(X)$, provided $X \in \Omega$. Noting that

$$h_j(X)\delta_j(X) = \max\{h_j(X), 0\},$$

we conclude that for every fixed λ the function $\varphi_\lambda(X)$ is continuous in X on E_n.

Let $m(\lambda) = \min_{j \in [0:N_1]} \lambda_j$. We shall say that $\lambda > \mathbf{0}$ if $m(\lambda) > 0$. Let X_λ denote a point for which

$$\varphi_\lambda(X_\lambda) = \min_{X \in E_n} \varphi_\lambda(X).$$

For any $\lambda > \mathbf{0}$ there exists a point with this property. Indeed, set

$$M_\lambda(\bar{X}) = \{X \mid \varphi_\lambda(X) \leqslant \varphi_\lambda(\bar{X})\}.$$

Then $M_\lambda(\bar{X}) \subset M_0(\bar{X})$, since for every $X \in M_\lambda(\bar{X})$ we have

$$\varphi(X) \leqslant \varphi_\lambda(X) \leqslant \varphi_\lambda(\bar{X}) = \varphi(\bar{X}).$$

Thus the set $M_\lambda(\bar{X})$ is bounded. By the continuity of $\varphi_\lambda(X)$ as a function of X, the set is also closed. Hence there exists a point X_λ for which

$$\varphi_\lambda(X_\lambda) = \min_{X \in M_\lambda(\bar{X})} \varphi_\lambda(X) = \min_{X \in E_n} \varphi_\lambda(X),$$

and $X_\lambda \in M_0(\bar{X})$ for all $\lambda > 0$.

In addition, if $X_\lambda \in \Omega$ for some $\lambda > 0$, then X_λ is a minimum point for $\varphi(X)$ on Ω.

Lemma 10.1. If the functions $f_i(X)$ and are continuous on E_n and the set $M_0(\bar{X})$ is bounded, then

$$\lim_{m(\lambda) \to \infty} \varphi(X_\lambda) = \varphi(\bar{X}).$$

Proof. Suppose the contrary: there exist a sequence of vectors $\lambda_k > 0$, $m(\lambda_k) \xrightarrow[k \to \infty]{} \infty$, and a number $\varepsilon > 0$ such that

$$\varphi(X_{\lambda_k}) \leqslant \varphi(\bar{X}) - \varepsilon. \tag{10.1}$$

Without loss of generality, we may assume that

$$X_{\lambda_k} \xrightarrow[k \to \infty]{} X^* \in M_0(\bar{X}).$$

From (10.1) we get

$$\varphi(X^*) \leqslant \varphi(\bar{X}) - \varepsilon.$$

By the definition of $\bar{X}$, we have $X^* \notin \Omega$. Thus,

$$\sum_{j=0}^{N_1} \lambda_j^{(k)} h_j^2(X_{\lambda_k}) \delta_j(X_{\lambda_k}) \xrightarrow[m(\lambda_k) \to \infty]{} \infty.$$

But the sequence $\{\varphi(X_{\lambda_k})\}$ is bounded, and so

$$\varphi_{\lambda_k}(X_{\lambda_k}) \xrightarrow[k \to \infty]{} \infty. \tag{10.2}$$

On the other hand, recalling the definition of X_{λ_k}, we see that

$$\varphi_{\lambda_k}(X_{\lambda_k}) \leqslant \varphi_{\lambda_k}(\bar{X}) = \varphi(\bar{X}),$$

which contradicts (10.2).

Corollary. Since

$$\varphi(X_\lambda) \leqslant \varphi_\lambda(X_\lambda) \leqslant \varphi(\bar{X})$$

for $\lambda > 0$, it follows that

$$\varphi_\lambda(X_\lambda) \xrightarrow[m(\lambda) \to \infty]{} \varphi(\bar{X}).$$

Theorem 10.1. Under the assumptions of Lemma 10.1, let $\{\lambda_k\}$ *be an arbitrary sequence of vectors such that* $m(\lambda_k)\xrightarrow[k\to\infty]{}\infty$. *Then any limit point* X^* *of the sequence* $\{X_{\lambda_k}\}$, *where* X_{λ_k} *is a minimum point of the function* $\varphi_{\lambda_k}(X)$ *on* E_n, *is a minimum point of* $\varphi(X)$ *on* Ω.

Proof. By Lemma 10.1,

$$\varphi(X^*) = \min_{X\in\Omega} \varphi(X).$$

It remains only to show that $X^*\in\Omega$; this is verified indirectly, as in the proof of Lemma 10.1.

2. Set

$$f_{\lambda i}(X) = f_i(X) + \sum_{j=0}^{N_1} \lambda_j h_j^2(X)\,\delta_j(X).$$

Then

$$\varphi_\lambda(X) = \max_{i\in[0:N]} f_{\lambda i}(X).$$

Theorem 10.1 reduces the minimization of $\varphi(X)$ on Ω to minimization of the maximum functions $\varphi_{\lambda_k}(X)$ on E_n for a sequence of vectors λ_k such that $m(\lambda_k)\xrightarrow[k\to\infty]{}\infty$.

The function $\sum_{j=0}^{N_1} \lambda_j h_j^2(X)\,\delta_j(X)$ is known as a penalty function, and methods for minimizing $\varphi(X)$ on Ω using $\varphi_\lambda(X)$ are methods of penalty functions.

The following remark may be useful. If the functions $f_i(X)$ and $h_j(X)$ are continuously differentiable on E_n, then the functions $f_{\lambda i}(X)$, $i\in[0:N]$ are continuously differentiable on E_n for any fixed λ:

$$\frac{\partial f_{\lambda i}(X)}{\partial X} = \frac{\partial f_i(X)}{\partial X} + \sum_{j=0}^{N_1} \lambda_j(|h_j(X)| + h_j(X))\frac{\partial h_j(X)}{\partial X}. \tag{10.3}$$

It suffices to show that the functions

$$\Phi_j(X) = h_j^2(X)\,\delta_j(X),\quad j\in[0:N_1],$$

are continuously differentiable. Indeed, we note that

$$\Phi_j(X) = \frac{1}{2}h_j(X)(|h_j(X)| + h_j(X)) = \frac{1}{2}(\xi(h_j(X)) + h_j^2(X)),$$

where $\xi(t) = t|t|$, $-\infty < t < \infty$. Now $\xi(t)$ is continuously differentiable on the whole real line, and $\xi'(t) = 2|t|$. Hence $\xi(h_j(X))$ is continuously differentiable on E_n, and

$$\frac{\partial\xi(h_j(X))}{\partial X} = \xi'(h_j(X))\frac{\partial h_j(X)}{\partial X} = 2|h_j(X)|\frac{\partial h_j(X)}{\partial X}.$$

Thus the functions $\Phi_j(X)$ are also continuously differentiable on E_n, and

$$\frac{\partial \Phi_j(X)}{\partial X} = (|h_j(X)| + h_j(X)) \frac{\partial h_j(X)}{\partial X}. \tag{10.4}$$

Formula (10.3) now follows from (10.4) and the definition of $f_{\lambda i}(X)$.

§11. CONCLUDING REMARKS

1. A few remarks are in order concerning successive approximation procedures for finding stationary points of $\varphi(X)$ on Ω. If Ω is defined by linear inequalities, the method based on determination of (ε, μ)-stationary points is highly effective. The idea of this method was described in §8. One must bear in mind, however, that in this case the search for a direction of descent involves a quadratic programming problem:

$$\left\| \sum_i \alpha_i \frac{\partial f_i(X_0)}{\partial X} + \sum_j \beta_j \frac{\partial h_j(X_0)}{\partial X} \right\|^2 \longrightarrow \min,$$

where $\sum_i \alpha_i = 1$, $\alpha_i \geqslant 0$, $\beta_j \geqslant 0$.

The methods of §§6 and 9 are mainly of theoretical value. The method of penalty functions reduces a minimax problem with constraints to a minimax problem without constraints. One can then use the methods of Chapter III. When employing the method of §7 one has to solve the same auxiliary problems as in the absence of constraints (see Chap. III, §9). The resulting direction of descent, however, is not the "steepest" in any reasonable sense of the word.

2. All our methods involve an arbitrary choice of the initial approximation X_0 in Ω. It is not always easy to decide whether a given point lies in Ω or not. In order to tackle this problem, we recommend using one of the methods of Chapter III to solve the minimax problem

$$\max_{j \in [0:N_1]} h_j(X) \to \min_{X \in E_n}.$$

Since the functions $h_j(X)$ are convex on E_n, the result will be a sequence $\{X'_k\}$ any of whose limit points is a minimum point for $\varphi_1(X)$ on E_n. One may then define X_0 to be a point of the sequence $\{X'_k\}$ such that

$$\max_{j \in [0:N_1]} h_j(X'_k) \leqslant 0.$$

Similar arguments obtain with regard to verification of the Slater condition.

3. The original problem

$$\max_{i \in [0:N]} f_i(X) \to \min_{X \in \Omega},$$

where $\Omega=\{X \mid h_j(X)\leqslant 0,\ j\in[0:N_1]\}$, is equivalent to the following nonlinear programming problem in the space of vectors $W=(x_1, \ldots, x_n, u)$: determine $\min u$ under the constraints

$$f_i(x_1, \ldots, x_n)-u\leqslant 0, \quad i\in[0:N];$$
$$h_j(x_1, \ldots, x_n)\leqslant 0, \quad j\in[0:N_1].$$

Hence it follows that our problem reduces to a linear programming problem if the functions $f_i(X)$ and $h_j(X)$ are linear.

4. In Chap. IV, §6, we proved the following assertion: if there exist a point $X_0\in\Omega$ and an index set $Q\subset[0:N]$ such that

$$\min_{Z\in\Omega}\max_{i\in Q}\left(\frac{\partial f_i(X_0)}{\partial X}, Z-X_0\right)=-a\leqslant 0,$$

where the functions $f_i(X)$ are convex on Ω for $i\in Q$, then

$$\min_{i\in Q} f_i(X_0)-a\leqslant \inf_{X\in\Omega}\varphi(X)\leqslant\varphi(X_0). \tag{11.1}$$

Let us assume that the set Ω is defined as $\Omega=\{X \mid h_j(X)\leqslant 0,\ j\in[0:N_1]\}$ and is bounded. In other words, there exist constants K_l, $l\in[1:n]$, such that $|x_l|\leqslant K_l$ for all $X=(x_1, \ldots, x_n)\in\Omega$. Set

$$h_{N_1+l}(X)=x_l-K_l, \qquad l\in[1:n];$$
$$h_{N_1+n+l}(X)=-x_l-K_l, \quad l\in[1:n].$$

Then

$$\Omega=\{X \mid h_j(X)\leqslant 0,\ j\in[0:N_1+2n]\}.$$

Consider the set

$$\Omega(X_0)=\Big\{Z \mid h_j(X_0)+ \\ +\left(\frac{\partial h_j(X_0)}{\partial X}, Z-X_0\right)\leqslant 0,\ j\in[0:N_1+2n]\Big\}.$$

This set is clearly bounded and $\Omega\subset\Omega(X_0)$.

Suppose now that the functions $f_i(X)$ are continuously differentiable on an open set Ω'' containing $\Omega(X_0)$.

Since

$$\min_{Z\in\Omega(X_0)}\max_{i\in Q}\left(\frac{\partial f_i(X_0)}{\partial X}, Z-X_0\right)\stackrel{df}{=}-a'\leqslant -a,$$

it follows from (11.1) that

$$\min_{i\in Q} f_i(X_0)-a'\leqslant \inf_{X\in\Omega}\varphi(X)\leqslant\varphi(X_0). \tag{11.2}$$

The left-hand inequality in (11.2) is less sharp than the corresponding inequality in (11.1) but a' may be evaluated by linear programming methods, whereas the evaluation of a may present serious difficulties.

Chapter VI

THE CONTINUOUS MINIMAX PROBLEM

§1. STATEMENT OF THE PROBLEM

Let Ω' be an open subset of E_n, G a bounded closed subset of E_m, $F(X, Y)$ a function which is continuous and continuously differentiable with respect to X on $\Omega' \times G$. We wish to minimize the function

$$\varphi(X) = \max_{Y \in G} F(X, Y)$$

on a convex closed subset Ω of Ω'.

In §§2 and 3 we shall generalize the main results of the previous chapters to the continuous case (directional differentiability of $\varphi(X)$, necessary conditions for a minimax and their geometric interpretation, some estimates).

§4 is devoted to grid methods for finding stationary points of $\varphi(X)$ on Ω; special consideration is given to the case $\Omega = E_n$.

In §5 we prove a special case of the minimax theorem, i. e., we determine conditions under which

$$\min_{X \in \Omega} \max_{Y \in G} F(X, Y) = \max_{Y \in G} \min_{X \in \Omega} F(X, Y).$$

§6 discusses the determination of saddle points on polyhedra.

In the last two sections, as a corollary of the general theorems, we shall derive some classical results of Chebyshev approximation theory.

§2. FUNDAMENTAL THEOREMS

1. Consider a function $F(X, Y)$, $X \in \Omega'$, $Y \in G$, where Ω' is an open subset of E_n and G a bounded closed subset of E_m. We shall assume that $F(X, Y)$ is continuous together with $\partial F(X, Y)/\partial X$ on the set $\Omega' \times G$.

Set

$$\varphi(X) = \max_{Y \in G} F(X, Y).$$

We first list a few simple properties of the function $\varphi(X)$.

I. $\varphi(X)$ is continuous on Ω'.

II. Suppose that for each $Y \in G$ the function $F(X, Y)$ is convex in X on a convex set $\Omega \subset \Omega'$; then $\varphi(X)$ is convex on Ω.

III. If Ω is a closed subset of Ω' and for some $X_0 \in \Omega$ the set

$$M(X_0) = \{X \in \Omega \mid \varphi(X) \leqslant \varphi(X_0)\}$$

is bounded, then $\varphi(X)$ attains its infimum on Ω; in other words, there exists a point $X^* \in \Omega$ at which

$$\varphi(X^*) = \inf_{X \in \Omega} \varphi(X),$$

or, equivalently,

$$\max_{Y \in G} F(X^*, Y) = \inf_{X \in \Omega} \max_{Y \in G} F(X, Y).$$

The proofs of these simple propositions are left to the reader (see Chap. III, §2).

For fixed $X \in \Omega'$, we define a set

$$R(X) = \{Y \in G \mid F(X, Y) = \varphi(X)\}.$$

It is clear that $R(X) \subset G$ is a bounded closed subset of E_m.

Theorem 2.1. The function $\varphi(X)$ is differentiable at each point $X \in \Omega'$ in any direction $g \in E_n, \|g\| = 1$, and

$$\frac{\partial \varphi(X)}{\partial g} = \max_{Y \in R(X)} \left(\frac{\partial F(X, Y)}{\partial X}, g \right). \tag{2.1}$$

For the proof we require two lemmas.

Lemma 2.1.

$$\lim_{\alpha \to +0} \max_{Y \in R(X+\alpha g)} \min_{V \in R(X)} \|Y - V\| = 0.$$

Proof. Suppose the contrary. Then there exist a sequence of positive numbers $\{\alpha_k\}$ converging to zero and a number $a > 0$ such that

$$\max_{Y \in R(X+\alpha_k g)} \min_{V \in R(X)} \|Y - V\| \geqslant a.$$

Let $Y_k \in R(X + \alpha_k g)$ be points such that

$$\min_{V \in R(X)} \|Y_k - V\| \geqslant a. \tag{2.2}$$

Since all the points Y_k are in G, we may assume without loss of generality that the sequence $\{Y_k\}$ is convergent. Letting $k \to \infty$ in the equality

$$F(X + \alpha_k g, Y_k) = \varphi(X + \alpha_k g),$$

we get

$$F(X, Y^*) = \varphi(X).$$

Thus $Y^* \in R(X)$.

We now write

$$0 \leqslant \min_{V \in R(X)} \| Y_k - V \| \leqslant \| Y_k - Y^* \|.$$

This implies that $\min_{V \in R(X)} \| Y_k - V \| \xrightarrow[k \to \infty]{} 0$, contradicting (2.2) and proving the lemma.

R e m a r k. In more intuitive terms, this lemma may be phrased as follows:

For any $\varepsilon > 0$, there exists $\alpha_0 > 0$ such that for every $\alpha \in (0, \alpha_0)$ and every $Y \in R(X + \alpha g)$ there is a vector $V \in R(X)$ satisfying

$$\| Y - V \| < \varepsilon.$$

We now set $Q(X, \alpha) = R(X + \alpha g) \cup R(X)$.

L e m m a 2.2. Uniformly in $g \in E_n$, $\| g \| = 1$,

$$\lim_{\alpha \to +0} \left\{ \max_{Y \in Q(X, \alpha)} \left(\frac{\partial F(X, Y)}{\partial X}, g \right) - \max_{Y \in R(X)} \left(\frac{\partial F(X, Y)}{\partial X}, g \right) \right\} = 0. \tag{2.3}$$

P r o o f. For all $\alpha > 0$, we have

$$Q(X, \alpha) \supset R(X),$$

and thus the expression in braces in (2.3) is nonnegative. Let $\varepsilon > 0$. Then there exists $\delta > 0$ such that, whenever $\| Y' - Y \| < \delta$ $(Y, Y' \in G)$,

$$\left| \left(\frac{\partial F(X, Y')}{\partial X}, g \right) - \left(\frac{\partial F(X, Y)}{\partial X}, g \right) \right| < \varepsilon$$

uniformly in g, $\| g \| = 1$. By Lemma 2.1, there exists $\alpha_0 > 0$ such that for any $\alpha \in (0, \alpha_0)$ and any $Y \in Q(X, \alpha)$ there is a vector $V \in R(X)$ for which

$$\| Y - V \| < \delta.$$

Fix $\alpha \in (0, \alpha_0)$. Let $Y_\alpha \in Q(X, \alpha)$ be a point such that

$$\max_{Y \in Q(X, \alpha)} \left(\frac{\partial F(X, Y)}{\partial X}, g \right) = \left(\frac{\partial F(X, Y_\alpha)}{\partial X}, g \right).$$

Let $V_\alpha \in R(X)$ be a point for which

$$\| Y_\alpha - V_\alpha \| < \delta.$$

We have

$$0 \leqslant \max_{Y \in Q(X, \alpha)} \left(\frac{\partial F(X, Y)}{\partial X}, g \right) - \max_{Y \in R(X)} \left(\frac{\partial F(X, Y)}{\partial X}, g \right) \leqslant$$
$$\leqslant \left(\frac{\partial F(X, Y_\alpha)}{\partial X}, g \right) - \left(\frac{\partial F(X, V_\alpha)}{\partial X}, g \right) < \varepsilon$$

uniformly in g, $\| g \| = 1$. This proves the lemma.

Proof of Theorem 2.1. We first observe that

$$F(X+\alpha g, Y)=F(X, Y)+\alpha\left(\frac{\partial F(X, Y)}{\partial X}, g\right)+o(Y, g; \alpha),$$

where

$$\frac{o(Y, g; \alpha)}{\alpha} \xrightarrow[\alpha \to 0]{} 0$$

uniformly in $Y \in G$ and g, $\|g\|=1$.

Hence, by the definition of $Q(X, \alpha)$,

$$\begin{aligned}\varphi(X+\alpha g) \stackrel{df}{=} \max_{Y \in G} F(X+\alpha g, Y) = \max_{Y \in R(X+\alpha g)} F(X+\alpha g, Y) \leqslant \\ \leqslant \max_{Y \in Q(X, \alpha)} F(X+\alpha g, Y) \leqslant \max_{Y \in Q(X, \alpha)} F(X, Y) + \\ +\alpha \max_{Y \in Q(X, \alpha)}\left(\frac{\partial F(X, Y)}{\partial X}, g\right)+\max_{Y \in G} o(Y, g; \alpha).\end{aligned} \tag{2.4}$$

Let us denote

$$\max_{\|g\|=1} \max_{Y \in G} |o(Y, g; \alpha)| = o(\alpha).$$

It is clear that $\frac{o(\alpha)}{\alpha} \xrightarrow[\alpha \to +0]{} 0$. Since $\max\limits_{Y \in Q(X, \alpha)} F(X, Y) = \varphi(X)$, it follows in view of (2.4) that

$$\varphi(X+\alpha g)-\varphi(X) \leqslant \alpha \max_{Y \in Q(X, \alpha)}\left(\frac{\partial F(X, Y)}{\partial X}, g\right)+o(\alpha).$$

Using Lemma 2.2, we see that

$$\varphi(X+\alpha g)-\varphi(X) \leqslant \alpha \max_{Y \in R(X)}\left(\frac{\partial F(X, Y)}{\partial X}, g\right)+\bar{o}(\alpha), \tag{2.5}$$

where, as before,

$$\frac{\bar{o}(\alpha)}{\alpha} \xrightarrow[\alpha \to +0]{} 0.$$

On the other hand,

$$\begin{aligned}&\varphi(X+\alpha g) \geqslant \\ &\geqslant \max_{Y \in G}\left[F(X, Y)+\alpha\left(\frac{\partial F(X, Y)}{\partial X}, g\right)\right]-\max_{Y \in G}|o(Y, g; \alpha)| \geqslant \\ &\geqslant \max_{Y \in R(X)}\left[F(X, Y)+\alpha\left(\frac{\partial F(X, Y)}{\partial X}, g\right)\right]-o(\alpha)= \\ &\qquad = \varphi(X)+\alpha \max_{Y \in R(X)}\left(\frac{\partial F(X, Y)}{\partial X}, g\right)-o(\alpha).\end{aligned}$$

Hence

$$\varphi(X+\alpha g)-\varphi(X)\geqslant\alpha\max_{Y\in R(X)}\left(\frac{\partial F(X,Y)}{\partial X},g\right)-o(\alpha). \tag{2.6}$$

Combining (2.5) and (2.6), we obtain

$$\alpha\max_{Y\in R(X)}\left(\frac{\partial F(X,Y)}{\partial X},g\right)-o(\alpha)\leqslant\varphi(X+\alpha g)-\varphi(X)\leqslant$$
$$\leqslant\alpha\max_{Y\in R(X)}\left(\frac{\partial F(X,Y)}{\partial X},g\right)+\bar{o}(\alpha). \tag{2.7}$$

Thus,

$$\frac{\partial\varphi(X)}{\partial g}\overset{\text{df}}{=}\lim_{\alpha\to+0}\frac{1}{\alpha}[\varphi(X+\alpha g)-\varphi(X)]=\max_{Y\in R(X)}\left(\frac{\partial F(X,Y)}{\partial X},g\right).$$

This completes the proof.

In fact, by virtue of (2.7) we have proved even more:

$$\varphi(X+\alpha g)=\varphi(X)+\alpha\frac{\partial\varphi(X)}{\partial g}+o(g;\alpha), \tag{2.8}$$

where $\frac{o(g;\alpha)}{\alpha}\xrightarrow[\alpha\to+0]{}0$ uniformly in g, $\|g\|=1$.

2. We shall now consider the problem of minimizing $\varphi(X)$ on a closed convex set $\Omega\subset\Omega'$.

Theorem 2.2. A necessary condition for the function $\varphi(X)$ to achieve its minimum on Ω at a point $X^*\in\Omega$ is that

$$\inf_{Z\in\Omega}\max_{Y\in R(X^*)}\left(\frac{\partial F(X^*,Y)}{\partial X},\ Z-X^*\right)=0. \tag{2.9}$$

If $\varphi(X)$ is convex on Ω, this condition is also sufficient.

If one uses the concept of directional derivative and formula (2.8), the proof of this theorem follows that of Theorem 2.1, Chap. IV, almost word for word. We nevertheless present a different proof.

Necessity. Let $X^*\in\Omega$ be a point at which $\varphi(X)$ assumes its minimum on Ω:

$$\varphi(X^*)=\min_{X\in\Omega}\varphi(X), \tag{2.10}$$

and suppose that (2.9) fails to hold. Then there is a point $Z_1\in\Omega$ such that

$$\max_{Y\in R(X^*)}\left(\frac{\partial F(X^*,Y)}{\partial X},\ Z_1-X^*\right)\overset{\text{df}}{=}-\varepsilon<0. \tag{2.11}$$

Let $\tilde{R}$ denote the set of all $Y\in G$ such that

$$\left(\frac{\partial F(X^*,Y)}{\partial X},\ Z_1-X^*\right)\geqslant-\frac{\varepsilon}{2}. \tag{2.12}$$

It is clear that $\tilde{R}$ is either empty or compact (=bounded and closed).

Let us assume first that $\tilde{R}$ is nonempty. By (2.11) and (2.12), $\tilde{R}$ is disjoint from $R(X^*)$, and so

$$\rho \overset{df}{=} \max_{Y \in \tilde{R}} F(X^*, Y) < \varphi(X^*). \tag{2.13}$$

Set

$$X(\lambda) = X^* + \lambda (Z_1 - X^*).$$

Since Ω is convex, it follows that for $0 \leqslant \lambda \leqslant 1$

$$X(\lambda) \in \Omega.$$

Denote

$$K = \max_{0 \leqslant \lambda \leqslant 1} \max_{Y \in G} \left| \left(\frac{\partial F(X(\lambda), Y)}{\partial X}, Z_1 - X^* \right) \right|.$$

K is clearly a finite number. We have already selected $\varepsilon > 0$; let $\delta > 0$ be such that, whenever $\| X(\lambda) - X^* \| \leqslant \delta$,

$$\left| \left(\frac{\partial F(X(\lambda), Y)}{\partial X}, Z_1 - X^* \right) - \left(\frac{\partial F(X^*, Y)}{\partial X}, Z_1 - X^* \right) \right| \leqslant \frac{\varepsilon}{4}$$

uniformly in $Y \in G$.

Finally, we set $h = \varphi(X^*) - \rho$ and

$$\lambda_0 = \min \left\{ \frac{\delta}{\| Z_1 - X^* \|}, \frac{h}{2K}, 1 \right\}, \quad 0 < \lambda_0 \leqslant 1.$$

We claim that $\varphi(X(\lambda_0)) < \varphi(X^*)$. Since this inequality contradicts (2.10), this will complete the proof that (2.9) is a necessary condition, provided $\tilde{R}$ is a nonempty set.

By elementary calculus, for arbitrary $Y \in G$,

$$F(X(\lambda_0), Y) = F(X^*, Y) + \int_0^{\lambda_0} \left(\frac{\partial F(X(t), Y)}{\partial X}, Z_1 - X^* \right) dt. \tag{2.14}$$

Let $Y \in \tilde{R}$; then by (2.13), (2.14) and the definition of λ_0 we obtain

$$F(X(\lambda_0), Y) \leqslant \varphi(X^*) - h + K\lambda_0 \leqslant \varphi(X^*) - \frac{h}{2}.$$

Let $Y \in G \setminus \tilde{R}$. Then

$$F(X(\lambda_0), Y) = F(X^*, Y) + \lambda_0 \left(\frac{\partial F(X^*, Y)}{\partial X}, Z_1 - X^* \right) +$$
$$+ \int_0^{\lambda_0} \left[\left(\frac{\partial F(X(t), Y)}{\partial X}, Z_1 - X^* \right) - \left(\frac{\partial F(X^*, Y)}{\partial X}, Z_1 - X^* \right) \right] dt.$$

By (2.12) and the definition of λ_0,

$$F(X(\lambda_0), Y) \leqslant \varphi(X^*) - \lambda_0 \frac{\varepsilon}{2} + \lambda_0 \frac{\varepsilon}{4} = \varphi(X^*) - \frac{\lambda_0 \varepsilon}{4}. \qquad (2.15)$$

Thus

$$\max_{Y \in G} F(X(\lambda_0), Y) < \varphi(X^*), \qquad (2.16)$$

which contradicts (2.10).

If $\tilde{R}$ is empty, then, for all $Y \in G$,

$$\left(\frac{\partial F(X^*, Y)}{\partial X}, Z_1 - X^*\right) < -\frac{\varepsilon}{2}.$$

We then set

$$\lambda_0 = \min\left\{\frac{\delta}{\|Z_1 - X^*\|}, 1\right\}, \qquad 0 < \lambda_0 \leqslant 1.$$

Inequality (2.15) will now hold for all $Y \in G$, and this implies (2.16).

Sufficiency. Let $X^* \in \Omega$ satisfy (2.9) and suppose that $\varphi(X)$ is convex on Ω; in other words, for $X_1, X_2 \in \Omega$ and $0 \leqslant \lambda \leqslant 1$,

$$\varphi(X_1 + \lambda(X_2 - X_1)) \leqslant \varphi(X_1) + \lambda[\varphi(X_2) - \varphi(X_1)]. \qquad (2.17)$$

We wish to prove that

$$\varphi(X^*) = \min_{X \in \Omega} \varphi(X).$$

Suppose the contrary. Then there exists a vector $Z_1 \in \Omega$ such that $\varphi(Z_1) < \varphi(X^*)$.

Set $X(\lambda) = X^* + \lambda(Z_1 - X^*)$, $0 \leqslant \lambda \leqslant 1$, and

$$\varepsilon = \frac{1}{2}[\varphi(X^*) - \varphi(Z_1)] > 0.$$

There exists $\delta > 0$ such that, whenever $\|X(\lambda) - X^*\| \leqslant \delta$,

$$\left|\left(\frac{\partial F(X(\lambda), Y)}{\partial X}, Z_1 - X^*\right) - \left(\frac{\partial F(X^*, Y)}{\partial X}, Z_1 - X^*\right)\right| \leqslant \varepsilon$$

uniformly in $Y \in G$.

Define λ_0 by

$$\lambda_0 = \min\left\{\frac{\delta}{\|Z_1 - X^*\|}, 1\right\}, \qquad 0 < \lambda_0 \leqslant 1,$$

and rewrite (2.14) as

$$\lambda_0\left(\frac{\partial F(X^*, Y)}{\partial X}, Z_1 - X^*\right) = F(X(\lambda_0), Y) - F(X^*, Y) -$$
$$-\int_0^{\lambda_0}\left[\left(\frac{\partial F(X(t), Y)}{\partial X}, Z_1 - X^*\right) - \left(\frac{\partial F(X^*, Y)}{\partial X}, Z_1 - X^*\right)\right] dt.$$

By (2.17) and the choice of ε, we see that for $Y \in R(X^*)$

$$\lambda_0\left(\frac{\partial F(X^*, Y)}{\partial X}, Z_1 - X^*\right) \leqslant \varphi(X(\lambda_0)) - \varphi(X^*) + \varepsilon\lambda_0 \leqslant$$
$$\leqslant \lambda_0[\varphi(Z_1) - \varphi(X^*)] + \varepsilon\lambda_0 = -\varepsilon\lambda_0.$$

Thus

$$\max_{Y \in R(X^*)}\left(\frac{\partial F(X^*, Y)}{\partial X}, Z_1 - X^*\right) \leqslant -\varepsilon < 0,$$

which contradicts (2.9), completing the proof.

Remark. If $\Omega = E_n$, condition (2.9) is equivalent to the inequality

$$\min_{\|g\|=1} \max_{Y \in R(X^*)}\left(\frac{\partial F(X^*, Y)}{\partial X}, g\right) \geqslant 0.$$

3. *Theorem 2.3. Suppose that for some point $X_0 \in \Omega$ and a closed set $Q \subset G$*

$$\inf_{Z \in \Omega} \max_{Y \in Q}\left(\frac{\partial F(X_0, Y)}{\partial X}, Z - X_0\right) \overset{df}{=} -a \leqslant 0, \tag{2.18}$$

and let $F(X, Y)$ be convex in X on Ω for all $Y \in Q$. Then

$$\min_{Y \in Q} F(X_0, Y) - a \leqslant \inf_{X \in \Omega} \varphi(X) \leqslant \varphi(X_0).$$

The proof is analogous in all respects to that of Theorem 5.1, Chap. IV.

Corollary 1. Under the assumptions of the theorem, if $Q \subset R(X_0)$ and $a = 0$, then

$$\inf_{X \in \Omega} \varphi(X) = \varphi(X_0),$$

i.e., X_0 is a minimum point of $\varphi(X)$ on the closed convex set Ω.

Corollary 2. If $\Omega = E_n$ and $a = 0$, then (2.18) is equivalent to the condition

$$\min_{\|g\|=1} \max_{Y \in Q}\left(\frac{\partial F(X_0, Y)}{\partial X}, g\right) \geqslant 0.$$

When this condition is satisfied, or if $F(X, Y)$ is convex in X for all $Y \in Q$,

$$\min_{Y \in Q} F(X_0, Y) \leqslant \inf_{X \in E_n} \varphi(X) \leqslant \varphi(X_0).$$

§3. GEOMETRIC INTERPRETATION OF THE NECESSARY CONDITION FOR A MINIMAX. SOME COROLLARIES

1. Fix some $X \in \Omega$. As before, let

$$\Gamma(X) = \{V = \lambda(Z - X) \mid \lambda > 0,\ Z \in \Omega\}.$$

Let $\overline{\Gamma}(X)$ denote the cone of feasible directions for Ω at X (i. e., the closure of the cone $\Gamma(X)$) and $\Gamma^+(X)$ the dual cone of $\overline{\Gamma}(X)$.

Set

$$H(X) = \left\{ Z = \frac{\partial F(X, Y)}{\partial X} \,\middle|\, Y \in R(X) \right\}.$$

It is readily seen that $H(X)$ is a bounded closed subset of E_n. Let $L(X)$ denote the convex hull of the set $H(X)$:

$$L(X) = \left\{ Z = \sum_{k=1}^{r} \alpha_k Z_k \,\middle|\, Z_k \in H(X), \right.$$

$$\left. \alpha_k \geqslant 0,\ \sum_{k=1}^{r} \alpha_k = 1,\ r = 1, 2 \ldots . \right\}.$$

By Lemma 1.2 of Appendix II, $L(X)$ is a bounded closed convex set.

Theorem 3.1. Condition (2.9) is equivalent to

$$L(X^*) \cap \Gamma^+(X^*) \neq \varnothing. \tag{3.1}$$

Proof. We first show that (2.9) implies (3.1). Suppose on the contrary that (2.9) holds but

$$L(X^*) \cap \Gamma^+(X^*) = \varnothing.$$

By the separation theorem (Appendix II, Theorem 2.1), there exists a vector $V_0 \in \Gamma^{++}(X^*) = \overline{\Gamma}(X^*)$ such that

$$\max_{V \in L(X^*)} (V_0, Z) \stackrel{df}{=} -a < 0.$$

The function (V, Z) is continuous in V, uniformly in $Z \in L(X^*)$. Therefore, for $V_0 \in \overline{\Gamma}(X^*)$ we can find a vector $V_1 = \lambda_1(X_1 - X^*) \in \Gamma(X^*)$ such that

$$\max_{Z \in L(X^*)} (V_1, Z) \leqslant -\frac{a}{2}.$$

Hence it follows that

$$\max_{Z \in L(X^*)} (Z, X_1 - X^*) \leqslant -\frac{a}{2\lambda_1}.$$

In particular,

$$\max_{Y \in R(X^*)} \left(\frac{\partial F(X^*, Y)}{\partial X}, X_1 - X^* \right) \leqslant -\frac{a}{2\lambda_1},$$

contradicting (2.9).

We now show that (3.1) implies (2.9). If this is not true, there is a vector $V_0 \in \Gamma(X^*) \subset \bar{\Gamma}(X^*)$ such that

$$\max_{Y \in R(X^*)} \left(\frac{\partial F(X^*, Y)}{\partial X}, V_0 \right) < 0.$$

This may be rewritten as

$$\max_{Z \in H(X^*)} (V_0, Z) < 0.$$

It is easy to see (compare the proof of Lemma 3.1 in Chap. III) that

$$\max_{Z \in L(X^*)} (V_0, Z) = \max_{Z \in H(X^*)} (V_0, Z) < 0. \tag{3.2}$$

Since $V_0 \in \bar{\Gamma}(X^*) = \Gamma^{++}(X^*)$, we have

$$(V_0, Z) \geqslant 0 \quad \text{for all} \quad Z \in \Gamma^+(X^*). \tag{3.3}$$

Comparing (3.2) and (3.3), we conclude that

$$L(X^*) \cap \Gamma^+(X^*) = \varnothing.$$

This contradicts (3.1).

Corollary. If $\Gamma(X^*) = E_n$, condition (3.1) is equivalent to

$$0 \in L(X^*). \tag{3.4}$$

Indeed, in this case $\Gamma^+(X^*) = \{0\}$.

2. *Theorem 3.2. A necessary condition for $\varphi(X)$ to achieve its minimum on Ω at a point $X^* \in \Omega$ is that there exist r points $Y_1, \ldots, Y_r$ in $R(X^*), 1 \leqslant r \leqslant n+1$, and r nonnegative numbers $\alpha_k \geqslant 0$, $\sum_{k=1}^{r} \alpha_k = 1$, such that*

$$\inf_{Z \in \Omega} \sum_{k=1}^{r} \alpha_k \left(\frac{\partial F(X^*, Y_k)}{\partial X}, Z - X^* \right) = 0. \tag{3.5}$$

If $\varphi(X)$ is convex on Ω, this condition is also sufficient.

Proof. Necessity. Let $X^* \in \Omega$ be a minimum point of $\varphi(X)$ on Ω. Then condition (3.1) is fulfilled, so that there exists a point Z_0 in the intersection of $L(X^*)$ and $\Gamma^+(X^*)$. Since any point of the convex hull may be expressed as a convex combination of at most $n+1$ points of the original set (see Appendix II, Lemma 1.1), there exist points $Y_1, \ldots, Y_r$ in $R(X^*)$ and numbers $\alpha_k \geqslant 0$, $\sum_{k=1}^{r} \alpha_k = 1$, $1 \leqslant r \leqslant n+1$, such that

$$Z_0 = \sum_{k=1}^{r} \alpha_k \frac{\partial F(X^*, Y_k)}{\partial X}.$$

Since $Z_0 \in \Gamma^+(X^*)$, it follows that for any $V \in \bar{\Gamma}(X^*)$

$$(Z_0, V) \geqslant 0.$$

In particular, for all $Z \in \Omega$,

$$\sum_{k=1}^{r} \alpha_k \left(\frac{\partial F(X^*, Y_k)}{\partial X}, Z - X^* \right) \geqslant 0,$$

whence follows (3.5).

S u f f i c i e n c y. Let (3.5) hold. Then for all $V \in \Gamma(X^*)$

$$\left(\sum_{k=1}^{r} \alpha_k \frac{\partial F(X^*, Y_k)}{\partial X}, V \right) \geqslant 0.$$

By the continuity of the scalar product, this is true for all $V \in \bar{\Gamma}(X^*)$. Thus

$$\sum_{k=1}^{r} \alpha_k \frac{\partial F(X^*, Y_k)}{\partial X} \in \Gamma^+(X^*).$$

On the other hand, by the definition of $L(X^*)$,

$$\sum_{k=1}^{r} \alpha_k \frac{\partial F(X^*, Y_k)}{\partial X} \in L(X^*)$$

(recall that $Y_k \in R(X^*)$, $k \in [1:r]$). Consequently, condition (3.1) is satisfied and it remains to refer to Theorems 3.1 and 2.2.

R e m a r k. If $\Gamma(X^*) = E_n$, condition (3.5) is equivalent to

$$\sum_{k=1}^{r} \alpha_k \frac{\partial F(X^*, Y_k)}{\partial X} = 0. \tag{3.6}$$

Indeed, in this case (3.1) is equivalent to (3.4), and the latter in turn (by Lemma 1.1 in Appendix II) to (3.6).

T h e o r e m 3.3. If $F(X, Y)$ is a convex function of X on Ω for every fixed $Y \in G$, and there exists a point $X^* \in \Omega$ for which

$$\inf_{X \in \Omega} \max_{Y \in G} F(X, Y) = \max_{Y \in G} F(X^*, Y), \tag{3.7}$$

then there exists a finite set $G_r = \{Y_k \in G \mid k \in [1:r]\}$, $1 \leqslant r \leqslant n+1$, such that

$$\inf_{X \in \Omega} \max_{Y \in G} F(X, Y) = \inf_{X \in \Omega} \max_{Y \in G_r} F(X, Y),$$

and moreover

$$\inf_{X\in\Omega}\max_{Y\in G_r} F(X, Y)=\max_{Y\in G_r} F(X^*, Y).$$

P r o o f. By (3.7) and Theorem 3.2, there exist points $Y_1, \ldots, Y_r$ in $R(X^*)$ such that (3.5) is satisfied. We claim that the set $G_r \overset{df}{=} \{Y_k \mid k\in[1:r]\}$ meets the requirements.

Indeed, since the function

$$\max_{Y\in G_r} F(X, Y)$$

is convex on Ω and condition (3.5) is fulfilled, it follows by Theorem 3.2 that

$$\inf_{X\in\Omega}\max_{Y\in G_r} F(X, Y)=\max_{Y\in G_r} F(X^*, Y). \tag{3.8}$$

It now remains to observe that, since G_r is a subset of $R(X^*)$,

$$\max_{Y\in G} F(X^*, Y)=\max_{Y\in G_r} F(X^*, Y). \tag{3.9}$$

Combining (3.7), (3.8) and (3.9), we complete the proof.

3*. We now consider the case that the set Ω is defined in a special manner.

Let

$$\Omega=\{X\in E_n \mid h(X, \beta)\leqslant 0 \quad \text{for all} \quad \beta\in\omega\},$$

where $\omega\subset E_s$ is a bounded closed set and the function $h(X, \beta)$ is continuous together with $\partial h(X, \beta)/\partial X$ in all variables on $E_n\times\omega$. Assume moreover that $h(X, \beta)$ is convex in X for each fixed $\beta\in\omega$ and that there is a point $\bar{X}\in E_n$ such that

$$\max_{\beta\in\omega} h(\bar{X}, \beta)<0. \tag{3.10}$$

We shall refer to (3.10) as the Slater condition. Under the above assumptions, Ω is a convex closed set. By (3.10), it contains interior points.

Let $X_0\in\Omega$. Consider the set

$$\Gamma(X_0)=\{V\in E_n \mid \max_{\beta\in\omega} h(X_0+\alpha V, \beta)\leqslant 0 \quad \text{for all} \quad \alpha\in[0, \alpha_0(V)],\ \alpha_0(V)>0\}.$$

Let $\bar{\Gamma}(X_0)$ denote the closure of $\Gamma(X_0)$.

It is clear that $\bar{\Gamma}(X_0)$ is the cone of feasible directions for Ω at X_0 (see subsection 1).

We introduce the notation

$$B(X_0)=\left\{V\in E_n \middle| \left(V, \frac{\partial h(X_0, \beta)}{\partial X}\right)\leqslant 0 \quad \text{for all} \quad \beta\in Q(X_0)\right\},$$

where $Q(X_0)=\{\beta\in\omega \mid h(X_0, \beta)=0\}$. If $Q(X_0)=\varnothing$, we set $B(X_0)=E_n$.

Lemma 3.1.

$$\bar{\Gamma}(X_0)=B(X_0).$$

Proof. If $Q(X_0)=\varnothing$, the lemma is trivially true. Indeed, in that case X_0 is an interior point of Ω, so that

$$\Gamma(X_0)=\bar{\Gamma}(X_0)=E_n.$$

On the other hand, by definition, $B(X_0)=E_n$. Thus, $\bar{\Gamma}(X_0)=B(X_0)$.

Now assume that $Q(X_0)\neq\varnothing$. In that case,

$$\max_{\beta\in\omega} h(X_0, \beta)=0.$$

We first prove that

$$\bar{\Gamma}(X_0)\subset B(X_0).$$

Let $V_0\in\bar{\Gamma}(X_0)$; in other words, there exists a sequence of vectors $\{V_i\}$ such that

$$V_0=\lim_{i\to\infty} V_i;\quad V_i\in\Gamma(X_0),\quad i=1, 2, \ldots$$

Fix i arbitrarily. For all $\beta\in\omega$ and $\alpha\in(0, \alpha_0(V_i))$, it follows from the definition of $\Gamma(X_0)$ that

$$h(X_0+\alpha V_i, \beta)\leqslant 0.$$

Hence,

$$h(X_0+\alpha V_i, \beta)=h(X_0, \beta)+\alpha\left(\frac{\partial h(X_0+\tau_i V_i, \beta)}{\partial X}, V_i\right)\leqslant 0,$$

where $\tau_i\overset{df}{=}\tau_i(\alpha)\in(0, \alpha)$.

If $\beta\in Q(X_0)$, then $h(X_0, \beta)=0$, and so

$$\left(\frac{\partial h(X_0+\tau_i(\alpha)V_i, \beta)}{\partial X}, V_i\right)\leqslant 0. \tag{3.11}$$

Inequality (3.11) is valid for all $\alpha\in(0, \alpha_0(V_i))$. Letting α tend to zero, we get

$$\left(\frac{\partial h(X_0, \beta)}{\partial X}, V_i\right)\leqslant 0. \tag{3.12}$$

Now let $i\to\infty$ in (3.12). It follows that for all $\beta\in Q(X_0)$

$$\left(\frac{\partial h(X_0, \beta)}{\partial X}, V_0\right)\leqslant 0.$$

Thus $V_0 \in B(X_0)$, and the inclusion $\bar{\Gamma}(X_0) \subset B(X_0)$ is proved.

We now prove the inverse inclusion

$$B(X_0) \subset \bar{\Gamma}(X_0).$$

Let $V_0 \in B(X_0)$. This means that

$$\left(V_0, \frac{\partial h(X_0, \beta)}{\partial X}\right) \leqslant 0 \quad \text{for all} \quad \beta \in Q(X_0). \tag{3.13}$$

We shall first show that for any $a > 0$ there exists $\bar{\varepsilon} > 0$ such that, for all $\varepsilon \in (0, \bar{\varepsilon}]$,

$$\max_{\beta \in Q_\varepsilon(X_0)} \left(V_0, \frac{\partial h(X_0, \beta)}{\partial X}\right) \leqslant a, \tag{3.14}$$

where $Q_\varepsilon(X_0) = \{\beta \in \omega \mid -\varepsilon \leqslant h(X_0, \beta) \leqslant 0\}$.

Indeed, suppose this is false. Then for some $\bar{a} > 0$ and some sequence of positive numbers $\{\varepsilon_\nu\}$ tending to zero we have

$$\max_{\beta \in Q_{\varepsilon_\nu}(X_0)} \left(V_0, \frac{\partial h(X_0, \beta)}{\partial X}\right) \geqslant \bar{a} > 0, \quad \nu = 1, 2, \ldots \tag{3.15}$$

The maximum in (3.15) is obviously achieved. Thus, for every ν there is a point $\beta_\nu \in Q_{\varepsilon_\nu}(X_0)$ such that

$$\left(V_0, \frac{\partial h(X_0, \beta_\nu)}{\partial X}\right) \geqslant \bar{a}. \tag{3.16}$$

Moreover,

$$-\varepsilon_\nu \leqslant h(X_0, \beta_\nu) \leqslant 0. \tag{3.17}$$

Since all the points β_ν, $\nu = 1, 2, \ldots$, are in ω, the sequence $\{\beta_\nu\}$ contains a convergent subsequence, and we may assume without loss of generality that the entire sequence $\{\beta_\nu\}$ converges to a point $\beta^* \in \omega$. Letting $\nu \to \infty$ in (3.16) and (3.17), we obtain

$$\left(V_0, \frac{\partial h(X_0, \beta^*)}{\partial X}\right) \geqslant \bar{a},$$

where $h(X_0, \beta^*) = 0$, i. e., $\beta^* \in Q(X_0)$.

A fortiori, then,

$$\max_{\beta \in Q(X_0)} \left(V_0, \frac{\partial h(X_0, \beta)}{\partial X}\right) \geqslant \bar{a},$$

contradicting (3.13). We have thus proved that for any $a > 0$ there exists $\bar{\varepsilon} > 0$ such that inequality (3.14) holds for all $\varepsilon \in (0, \bar{\varepsilon}]$.

Now set $W_0=\bar{X}-X_0$, where $\bar{X}$ is the point figuring in (3.10); we claim that for any $\gamma>0$ the point $V_0(\gamma)\overset{df}{=}V_0+\gamma W_0$ lies in $\Gamma(X_0)$; i. e., we must show that for all $\alpha\in[0, \alpha_0(\gamma)]$, $\alpha_0(\gamma)>0$,

$$\max_{\beta\in\omega} h(X_0+\alpha V_0(\gamma), \beta)\leqslant 0.$$

Fix $\gamma>0$ and denote

$$\varphi_1(X)=\max_{\beta\in\omega} h(X, \beta),$$
$$a=-\frac{1}{8}\gamma\varphi_1(\bar{X})>0.$$

As shown above, given $a>0$ we can find $\bar{\varepsilon}>0$ such that for $\varepsilon\in(0, \bar{\varepsilon}]$

$$\max_{\beta\in Q_\varepsilon(X_0)}\left(\frac{\partial h(X_0, \beta)}{\partial X}, V_0\right)\leqslant a.$$

Further, we note (see Appendix II, §3) that since $h(X, \beta)$ is convex with respect to X

$$\left(\frac{\partial h(X_0, \beta)}{\partial X}, \bar{X}-X_0\right)\leqslant h(\bar{X}, \beta)-h(X_0, \beta). \tag{3.18}$$

If $\varepsilon\in\left(0, -\frac{1}{2}\varphi_1(\bar{X})\right]$, it follows from (3.18) that for $\beta\in Q_\varepsilon(X_0)$

$$\left(\frac{\partial h(X_0, \beta)}{\partial X}, W_0\right)\leqslant\varphi_1(\bar{X})+\varepsilon\leqslant\frac{1}{2}\varphi_1(\bar{X})=-\frac{4a}{\gamma}.$$

Fix some ε, $0<\varepsilon<\min\left\{\bar{\varepsilon}, -\frac{1}{2}\varphi_1(\bar{X})\right\}$. Then

$$\max_{\beta\in Q_\varepsilon(X_0)}\left(\frac{\partial h(X_0, \beta)}{\partial X}, V_0\right)\leqslant a,$$
$$\gamma\max_{\beta\in Q_\varepsilon(X_0)}\left(\frac{\partial h(X_0, \beta)}{\partial X}, W_0\right)\leqslant -4a,$$

whence it follows that

$$\max_{\beta\in Q_\varepsilon(X_0)}\left(\frac{\partial h(X_0, \beta)}{\partial X}, V_0(\gamma)\right)\leqslant -3a. \tag{3.19}$$

Next,

$$\begin{aligned} h(X_0+\alpha V_0(\gamma), \beta)&=\\ &=h(X_0, \beta)+\alpha\left(\frac{\partial h(X_0+\tau V_0(\gamma), \beta)}{\partial X}, V_0(\gamma)\right)\leqslant\\ &\leqslant\alpha\left(\frac{\partial h(X_0+\tau V_0(\gamma), \beta)}{\partial X}, V_0(\gamma)\right), \end{aligned} \tag{3.20}$$

where $\tau \in (0, \alpha)$. By (3.19), (3.20) and the continuity of $\partial h(X, \beta)/\partial X$ on $E_n \times \omega$, we infer that for sufficiently small $\alpha \in [0, \alpha_0]$, $\alpha_0 > 0$, and all $\beta \in Q_\varepsilon(X_0)$

$$h(X_0 + \alpha V_0(\gamma), \beta) \leqslant -2\alpha a \leqslant 0. \tag{3.21}$$

Now let $\beta \notin Q_\varepsilon(X_0)$. Then $h(X_0, \beta) < -\varepsilon$. It is easy to see that there exists $\alpha_0(\gamma)$, $0 < \alpha_0(\gamma) \leqslant \alpha_0$, such that for $\alpha \in [0, \alpha_0(\gamma)]$ and $\beta \notin Q_\varepsilon(X_0)$

$$h(X_0 + \alpha V_0(\gamma), \beta) \leqslant -\frac{1}{2}\varepsilon. \tag{3.22}$$

Combining (3.21) and (3.22), we see that

$$\max_{\beta \in \omega} h(X_0 + \alpha V_0(\gamma), \beta) \leqslant 0$$

for all $\alpha \in [0, \alpha_0(\gamma)]$. But this means that $V_0(\gamma) \in \Gamma(X_0)$. Since this is true for any $\gamma > 0$, we conclude, letting γ tend to zero, that $V_0 \in \bar{\Gamma}(X_0)$.

We have thus shown that

$$B(X_0) \subset \bar{\Gamma}(X_0).$$

This completes the proof of the lemma.

Remark. The Slater condition (3.10) is essential. This is evident from the following example.

Let

$$\Omega = \{X = (x_1, x_2) \mid (x_1 - 1)^2 + x_2^2 - 1 \leqslant 0,$$
$$(x_1 - 4)^2 + x_2^2 - 4 \leqslant 0\}$$

(see Figure 33). Then $\omega = \{1, 2\}$,

$$h(X, 1) = (x_1 - 1)^2 + x_2^2 - 1,$$
$$h(X, 2) = (x_1 - 4)^2 + x_2^2 - 4.$$

The set Ω is the singleton containing $X_0 = (2, 0)$. Therefore, $\Gamma(X_0) = \bar{\Gamma}(X_0) = \{0\}$. Let us determine the set $B(X_0)$. Since

$$Q(X_0) = \{1, 2\} \text{ and } \frac{\partial h(X_0, 1)}{\partial X} = (2, 0), \quad \frac{\partial h(X_0, 2)}{\partial X} = (-4, 0),$$

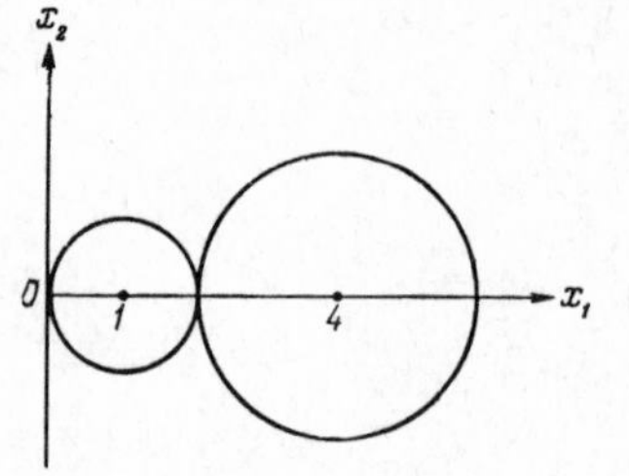

FIGURE 33.

It follows that

$$B(X_0) = \{V = (v_1, v_2) \mid v_1 = 0\}.$$

Thus $\bar{\Gamma}(X_0) \neq B(X_0)$. The reason is that the set Ω contains no interior points.

4. We have just established that if the Slater condition is fulfilled, then

$$\bar{\Gamma}(X_0) = \begin{cases} \left\{V \,\middle|\, \left(V, \frac{\partial h(X_0, \beta)}{\partial X}\right) \leqslant 0 \quad \text{for all} \quad \beta \in Q(X_0),\right\}, \\ \qquad\qquad\qquad\qquad \text{if} \quad Q(X_0) \neq \varnothing, \\ E_n, \qquad\qquad\qquad \text{if} \quad Q(X_0) = \varnothing. \end{cases}$$

Let $K(X_0)$ denote the convex conical hull of the set $H'(X_0) = \left\{- \frac{\partial h(X_0, \beta)}{\partial X} \,\middle|\, \beta \in Q(X_0)\right\}$:

$$K(X_0) = \left\{V = -\sum_{i=1}^{r} \lambda_i \frac{\partial h(X_0, \beta)}{\partial X} \,\middle|\, \lambda_i \geqslant 0,\ \beta_i \in Q(X_0),\ r = 1, 2 \ldots \right\}.$$

If $Q(X_0) = \varnothing$, we set $K(X_0) = \{0\}$. It is readily checked (see Appendix II, Lemma 2.5) that $K(X_0)$ is a closed convex cone.

Lemma 3.2. If the Slater condition is fulfilled, then

$$\Gamma^+(X_0) = K(X_0).$$

Proof. If $Q(X_0) = \varnothing$, there is nothing to prove. We may therefore assume that $Q(X_0) \neq \varnothing$. We claim that

$$B(X_0) = K^+(X_0). \tag{3.23}$$

Let $V_0 \in B(X_0)$. Then

$$\left(-\frac{\partial h(X_0, \beta)}{\partial X}, V_0\right) \geqslant 0 \quad \text{for all} \quad \beta \in Q(X_0).$$

Hence it clearly follows that for all $Z \in K(X_0)$

$$(Z, V_0) \geqslant 0,$$

i. e., $V_0 \in K^+(X_0)$. Thus,

$$B(X_0) \subset K^+(X_0). \tag{3.24}$$

Conversely, let $V_0 \in K^+(X_0)$. Then for all $Z \in K(X_0)$

$$(Z, V_0) \geqslant 0.$$

In particular, for all $\beta \in Q(X_0)$

$$\left(-\frac{\partial h(X_0, \beta)}{\partial X}, V_0\right) \geqslant 0,$$

or, equivalently,

$$\left(\frac{\partial h(X_0, \beta)}{\partial X}, V_0\right) \leqslant 0.$$

But this means that $V_0 \in B(X_0)$. Thus,

$$K^+(X_0) \subset B(X_0). \tag{3.25}$$

Combining (3.24) and (3.25), we obtain (3.23). Dualizing both sides of (3.23), we get

$$\Gamma^+(X_0) = K(X_0),$$

since $B(X_0) = \bar{\Gamma}(X_0)$ and $K^{++}(X_0) = K(X_0)$.

Theorem 3.4. Let

$$\Omega = \{X \in E_n \mid h(X, \beta) \leqslant 0 \quad \textit{for all} \quad \beta \in \omega\},$$

and suppose that the Slater condition (3.10) is satisfied. A necessary condition for X^* to be a minimum point of $\varphi(X)$ on Ω is that there exist points $Y_1, \ldots, Y_r$ in $R(X^*)$, $\beta_1, \ldots, \beta_{r_1}$ in $Q(X^*)$, and nonnegative numbers

$$\lambda_{11}, \ldots, \lambda_{1r}; \lambda_{21}, \ldots, \lambda_{2r_1}; \quad \sum_{i=1}^{r} \lambda_{1i} = 1,$$

where $1 \leqslant r \leqslant n+1$, $1 \leqslant r_1 \leqslant n$, such that

$$\sum_{i=1}^{r} \lambda_{1i} \frac{\partial F(X^*, Y_i)}{\partial X} + \sum_{i=1}^{r_1} \lambda_{2i} \frac{\partial h(X^*, \beta_i)}{\partial X} = 0. \tag{3.26}$$

If $\varphi(X)$ is convex on Ω, this condition is also sufficient.

Proof. By Theorems 2.2 and 3.1, it will suffice to show that (3.26) is equivalent to (3.1). We first show that (3.26) follows from (3.1). Let $Z^* \in L(X^*) \cap \Gamma^+(X^*)$. We then obtain (3.26) by using Lemmas 1.1 and 2.4 of Appendix II and Lemma 3.2. That (3.1) follows from (3.26) is obvious.

§4. CONVERGENCE OF THE GRID METHOD

1. Definition. A point $X^* \in \Omega$ for which

$$\inf_{Z \in \Omega} \max_{Y \in R(X^*)} \left(\frac{\partial F(X^*, Y)}{\partial X}, Z - X^*\right) = 0$$

is called a stationary point of $\varphi(X)$ on Ω.

Stationary points may be found by a grid method, the gist of which is as follows.

For each natural number N, we introduce a finite grid $G_N = \{Y_{Ni} \in G \mid i \in [0:N]\}$ on the set G. We shall make the following assumption concerning the behavior of the grids G_N as $N \to \infty$:

For any $\varepsilon > 0$, there exists N_0 such that for $N > N_0$ the distance between any point $Y \in G$ and the point of G_N nearest Y is less than ε.

Any sequence of grids $\{G_N\}$ satisfying this condition is said to be dense in G.

Let

$$f_{Ni}(X) = F(X, Y_{Ni}), \quad i \in [0:N];$$
$$\varphi_N(X) \stackrel{df}{=} \max_{i \in [0:N]} f_{Ni}(X) = \max_{Y \in G_N} F(X, Y).$$

We shall assume that for sufficiently large N the function $\varphi_N(X)$ has at least one stationary point $X_N \in \Omega$ on the set Ω.

Thus, for sufficiently large N, we have

$$\inf_{Z \in \Omega} \max_{i \in \tilde{R}(X_N)} \left(\frac{\partial f_{Ni}(X_N)}{\partial X}, Z - X_N \right) = 0, \tag{4.1}$$

where

$$\tilde{R}(X) = \{i \in [0:N] \mid f_{Ni}(X) = \varphi_N(X)\}.$$

Under these assumptions:

Theorem 4.1. Every limit point of the sequence $\{X_N\}$ is a stationary point of $\varphi(X)$ on Ω.

Proof. To fix ideas, let us assume that the entire sequence $\{X_N\}$ converges to a point $X^* \in \Omega$, but X^* is not a stationary point of $\varphi(X)$ on Ω. Then there exists a point $\bar{Z} \in \Omega$ such that

$$\max_{Y \in R(X^*)} \left(\frac{\partial F(X^*, Y)}{\partial X}, \bar{Z} - X^* \right) < 0. \tag{4.2}$$

We shall prove that there is a number $\bar{\varepsilon} > 0$ for which

$$\max_{Y \in R_{\bar{\varepsilon}}(X^*)} \left(\frac{\partial F(X^*, Y)}{\partial X}, \bar{Z} - X^* \right) \stackrel{df}{=} -r_{\bar{\varepsilon}} < 0, \tag{4.3}$$

where

$$R_\varepsilon(X) = \{Y \in G \mid \varphi(X) - F(X, Y) \leqslant \varepsilon\}.$$

If this were not true, any sequence $\{\varepsilon_\nu\}$, $\varepsilon_\nu > 0$, converging monotonically to zero would be such that

$$\max_{Y \in R_{\varepsilon_\nu}(X^*)} \left(\frac{\partial F(X^*, Y)}{\partial X}, \bar{Z} - X^* \right) \geqslant 0. \tag{4.4}$$

Since $R_{\varepsilon_\nu}(X^*)$ are bounded closed subsets of E_m and so the maxima in (4.4) are all achieved, we can find $Y_\nu \in R_{\varepsilon_\nu}(X^*)$ such that

$$\left(\frac{\partial F(X^*, Y_\nu)}{\partial X}, \ \bar{Z} - X^*\right) \geqslant 0. \tag{4.5}$$

We have, moreover,

$$\varphi(X^*) - F(X^*, Y_\nu) \leqslant \varepsilon_\nu. \tag{4.6}$$

Now, since $Y_\nu \in G$, the sequence $\{Y_\nu\}$ contains a convergent subsequence and we may assume without loss of generality that the entire sequence converges to some $Y^* \in G$. Then, on the one hand, it follows from (4.6) that $Y^* \in R(X^*)$; on the other, by (4.5)

$$\left(\frac{\partial F(X^*, Y^*)}{\partial X}, \ \bar{Z} - X^*\right) \geqslant 0.$$

A fortiori,

$$\max_{Y \in R(X^*)} \left(\frac{\partial F(X^*, Y)}{\partial X}, \ \bar{Z} - X^*\right) \geqslant 0,$$

which contradicts (4.2).

Thus there exists $\bar{\varepsilon} > 0$ satisfying (4.3).

Now let $\delta > 0$ be such that, whenever $\| Y' - Y'' \| < \delta$,

$$| F(X^*, Y') - F(X^*, Y'') | < \frac{\bar{\varepsilon}}{3}.$$

Take N_0 so large that for $N > N_0$ the distance between any $Y \in G$ and the nearest point of the grid G_N is less than δ and moreover

$$\max_{Y \in G} | F(X_N, Y) - F(X^*, Y) | < \frac{\bar{\varepsilon}}{3}.$$

We claim that then, for all $i \in \tilde{R}(X_N)$,

$$Y_{Ni} \in R_{\bar{\varepsilon}}(X^*). \tag{4.7}$$

Indeed, let $i \in \tilde{R}(X_N)$. Then

$$F(X_N, Y_{Ni}) \stackrel{df}{=} f_{Ni}(X_N) = \max_{k \in [0:N]} f_{Nk}(X_N) =$$
$$= \max_{k \in [0:N]} F(X_N, Y_{Nk}) = \max_{Y \in G_N} F(X_N, Y).$$

Next,

$$\max_{Y\in G} F(X^*, Y) = F(X^*, \bar{Y}) \leqslant$$
$$\leqslant \max_{Y\in G_N} F(X^*, Y) + \frac{\bar{\varepsilon}}{3} \leqslant \max_{Y\in G_N} F(X_N, Y) + \frac{2\bar{\varepsilon}}{3} =$$
$$= F(X_N, Y_{Ni}) + \frac{2\bar{\varepsilon}}{3} \leqslant F(X^*, Y_{Ni}) + \bar{\varepsilon}.$$

Thus $Y_{Ni} \in R_{\bar{\varepsilon}}(X^*)$, so that (4.7) is indeed true for $N > N_0$.

Now let $\bar{N} > N_0$ be so large that

$$\max_{Y\in G}\left|\left(\frac{\partial F(X^*, Y)}{\partial X},\ X_{\bar{N}} - X^*\right)\right| < \frac{r_{\bar{\varepsilon}}}{3},$$
$$\max_{Y\in G}\left|\left(\frac{\partial F(X^*, Y)}{\partial X} - \frac{\partial F(X_{\bar{N}}, Y)}{\partial X},\ \bar{Z} - X_{\bar{N}}\right)\right| < \frac{r_{\bar{\varepsilon}}}{3}.$$

By (4.7) and (4.3)

$$\max_{i\in \tilde{R}(X_{\bar{N}})} \left(\frac{\partial f_{\bar{N}i}(X_{\bar{N}})}{\partial X},\ \bar{Z} - X_{\bar{N}}\right) \leqslant$$
$$\leqslant \max_{Y\in R_{\bar{\varepsilon}}(X^*)} \left(\frac{\partial F(X_{\bar{N}}, Y)}{\partial X},\ \bar{Z} - X_{\bar{N}}\right) \leqslant$$
$$\leqslant \max_{Y\in R_{\bar{\varepsilon}}(X^*)} \left(\frac{\partial F(X^*, Y)}{\partial X},\ \bar{Z} - X_{\bar{N}}\right) + \frac{r_{\bar{\varepsilon}}}{3} \leqslant$$
$$\leqslant \max_{Y\in R_{\bar{\varepsilon}}(X^*)} \left(\frac{\partial F(X^*, Y)}{\partial X},\ \bar{Z} - X^*\right) + \frac{2r_{\bar{\varepsilon}}}{3} = -\frac{r_{\bar{\varepsilon}}}{3} < 0,$$

which contradicts (4.1), proving the theorem.

Remark. The points X_N are stationary points for discrete minimax problems. They may be determined by the methods set forth in Chaps. III–V.

2* We now consider the case $\Omega = E_n$ separately. As before, let $\{G_N\}$ be a sequence of grids dense in G as $N \to \infty$ (see subsection 1). We introduce the following notation:

$$R_\varepsilon(X) = \{Y \in G \mid \varphi(X) - F(X, Y) \leqslant \varepsilon\},$$
$$R_{N\varepsilon}(X) = \{Y \in G_N \mid \varphi_N(X) - F(X, Y) \leqslant \varepsilon\},$$
$$\chi_\varepsilon(X, g) = \max_{Y\in R_\varepsilon(X)} \left(\frac{\partial F(X, Y)}{\partial X},\ g\right),$$
$$\chi_{N\varepsilon}(X, g) = \max_{Y\in R_{N\varepsilon}(X)} \left(\frac{\partial F(X, Y)}{\partial X},\ g\right),$$
$$\psi_\varepsilon(X) = \min_{\|g\|=1} \chi_\varepsilon(X, g),$$
$$\psi_{N\varepsilon}(X) = \min_{\|g\|=1} \chi_{N\varepsilon}(X, g),$$
$$D(X) = \inf_{\varepsilon\in[0,\ \bar{\varepsilon}]} \varepsilon\psi_\varepsilon(X),$$
$$D_N(X) = \inf_{\varepsilon\in[0,\ \bar{\varepsilon}]} \varepsilon\psi_{N\varepsilon}(X),$$

where $\bar{\varepsilon} > 0$ is arbitrary but fixed.

Our goal is to prove the following

Theorem 4.2. As $N\to\infty$, we have

$$D_N(X)\to D(X),$$

and the convergence is uniform in $X\in\bar{S}$, where $\bar{S}\subset E_n$ is an arbitrary bounded closed set.

We first prove a lemma.

Lemma 4.1. Fix $\varepsilon_0>0$. For any $\rho>0$, $\delta>0\,(\delta\leqslant\varepsilon_0)$ and $\varepsilon\geqslant\varepsilon_0$, there exists N_0 such that for $N>N_0$ and all $X\in\bar{S}$

$$\psi_{N,\varepsilon-\delta}(X)\leqslant\psi_\varepsilon(X)\leqslant\psi_{N,\varepsilon+\delta}(X)+\rho. \tag{4.8}$$

Proof. We first note that if $\delta\leqslant\varepsilon_0$ and $\varepsilon\geqslant\varepsilon_0$, then $\varepsilon-\delta\geqslant 0$.

It is readily shown that

$$\varphi_N(X)\xrightarrow[N\to\infty]{}\varphi(X)$$

uniformly in $X\in\bar{S}$. Thus there exists N_1 so large that for $N>N_1$ and all $X\in\bar{S}$

$$\varphi(X)-\varphi_N(X)\leqslant\delta. \tag{4.9}$$

Now fix $X\in\bar{S}$ and $g\in E_n$, $\|g\|=1$, and let

$$\chi_{N,\varepsilon-\delta}(X,g)=\left(\frac{\partial F(X,\bar{Y}_N)}{\partial X},g\right),$$

where $\bar{Y}_N\stackrel{\mathrm{df}}{=}\bar{Y}_N(X,g)\in R_{N,\varepsilon-\delta}(X)$. Then

$$\varphi_N(X)-F(X,\bar{Y}_N)\leqslant\varepsilon-\delta. \tag{4.10}$$

In view of (4.9) and (4.10), we see that for $N>N_1$

$$\varphi(X)-F(X,\bar{Y}_N)=(\varphi(X)-\varphi_N(X))+(\varphi_N(X)-F(X,\bar{Y}_N))\leqslant\varepsilon.$$

Thus $\bar{Y}_N\in R_\varepsilon(X)$. We now have

$$\chi_{N,\varepsilon-\delta}(X,g)=\left(\frac{\partial F(X,\bar{Y}_N)}{\partial X},g\right)\leqslant\max_{Y\in R_\varepsilon(X)}\left(\frac{\partial F(X,Y)}{\partial X},g\right)=\chi_\varepsilon(X,g).$$

Since this is true for all g, $\|g\|=1$, it follows that

$$\psi_{N,\varepsilon-\delta}(X)\leqslant\psi_\varepsilon(X).$$

We have thus shown that the left-hand inequality of (4.8) is valid for $N>N_1$ and $X\in\bar{S}$.

We now prove the right-hand inequality. Since G is closed and bounded, given $\rho>0$ and $\delta>0$ we can find $\gamma>0$ such that whenever $\|Y'-Y''\|\leqslant\gamma$, Y', $Y''\in G$, the following inequalities hold uniformly in $X\in\bar{S}$:

$$|F(X,Y')-F(X,Y'')|\leqslant\delta, \tag{4.11}$$

$$\left\| \frac{\partial F(X, Y')}{\partial X} - \frac{\partial F(X, Y'')}{\partial X} \right\| \leqslant \rho. \tag{4.12}$$

Now choose N_2 so large that for $N > N_2$ the distance between any $Y \in G$ and the nearest point of the grid G_N is less than γ. Fix $X \in \bar{S}$ and $g \in E_n$, $\|g\| = 1$, and let

$$\chi_\varepsilon(X, g) = \left(\frac{\partial F(X, \tilde{Y})}{\partial X}, g \right),$$

where $\tilde{Y} \stackrel{df}{=} \tilde{Y}(X, g) \in R_\varepsilon(X)$. Then

$$\varphi(X) - F(X, \tilde{Y}) \leqslant \varepsilon. \tag{4.13}$$

Let $Y'_N \in G_N$ be the point nearest $\tilde{Y}$ in the grid G_N. By (4.11) and (4.13), for any $N > N_2$ we have

$$\varphi_N(X) - F(X, Y'_N) \leqslant \varphi(X) - F(X, Y'_N) =$$
$$= (\varphi(X) - F(X, \tilde{Y})) + (F(X, \tilde{Y}) - F(X, Y'_N)) \leqslant \varepsilon + \delta.$$

Thus $Y'_N \in R_{N,\, \varepsilon+\delta}(X)$. Hence, via (4.12), we obtain

$$\chi_\varepsilon(X, g) = \left(\frac{\partial F(X, \tilde{Y})}{\partial X}, g \right) =$$
$$= \left(\frac{\partial F(X, Y'_N)}{\partial X}, g \right) + \left(\frac{\partial F(X, \tilde{Y})}{\partial X} - \frac{\partial F(X, Y'_N)}{\partial X}, g \right) \leqslant$$
$$\leqslant \max_{Y \in R_{N,\, \varepsilon+\delta}(X)} \left(\frac{\partial F(X, Y)}{\partial X}, g \right) + \rho = \chi_{N,\, \varepsilon+\delta}(X, g) + \rho.$$

This is true for all g, $\|g\| = 1$, and so

$$\psi_\varepsilon(X) \leqslant \psi_{N,\, \varepsilon+\delta}(X) + \rho.$$

We have thus shown that the right-hand inequality of (4.8) holds for $N > N_2$, uniformly in $X \in \bar{S}$. Both inequalities will now hold for $N > N_0 \stackrel{df}{=} \max\{N_1, N_2\}$.

Proof of Theorem 4.2. We set

$$c = \max_{X \in \bar{S}} \max_{Y \in G} \left\| \frac{\partial F(X, Y)}{\partial X} \right\|.$$

Clearly,

$$|\psi_{N\varepsilon}(X)| \leqslant c, \quad |\psi_\varepsilon(X)| \leqslant c. \tag{4.14}$$

Let $\alpha_0 > 0$ be arbitrary, and choose numbers $\varepsilon_0 > 0$, $\rho > 0$ and $\delta > 0$ so that

$$4\varepsilon_0 < \bar{\varepsilon}, \quad 4\varepsilon_0 c \leqslant \alpha_0, \quad \delta < \varepsilon_0, \quad \delta c + \bar{\varepsilon}\rho < \varepsilon_0 c. \tag{4.15}$$

By Lemma 4.1, there exists N_0 such that for all $N > N_0$ and $\varepsilon \geqslant \varepsilon_0$

$$\psi_{N,\, \varepsilon-\delta}(X) \leqslant \psi_\varepsilon(X) \leqslant \psi_{N,\, \varepsilon+\delta}(X) + \rho$$

uniformly in $X \in \bar{S}$. Hence, for $\varepsilon \geqslant \varepsilon_0$,

$$\varepsilon\psi_{N,\,\varepsilon-\delta}(X) \leqslant \varepsilon\psi_\varepsilon(X) \leqslant \varepsilon\psi_{N,\,\varepsilon+\delta}(X) + \varepsilon\rho.$$

We can thus write

$$(\varepsilon-\delta)\,\psi_{N,\,\varepsilon-\delta}(X) + \delta\psi_{N,\,\varepsilon-\delta}(X) \leqslant \varepsilon\psi_\varepsilon(X) \leqslant$$
$$\leqslant (\varepsilon+\delta)\,\psi_{N,\,\varepsilon+\delta}(X) - \delta\psi_{N,\,\varepsilon+\delta}(X) + \varepsilon\rho.$$

In view of (4.14), we see that for $\varepsilon_0 \leqslant \varepsilon \leqslant \bar{\varepsilon}$

$$(\varepsilon-\delta)\,\psi_{N,\,\varepsilon-\delta}(X) - \delta c \leqslant \varepsilon\psi_\varepsilon(X) \leqslant$$
$$\leqslant (\varepsilon+\delta)\,\psi_{N,\,\varepsilon+\delta}(X) + \delta c + \bar{\varepsilon}\rho.$$

Thus

$$\inf_{\varepsilon \in [\varepsilon_0,\,\bar{\varepsilon}]} (\varepsilon-\delta)\,\psi_{N,\,\varepsilon-\delta}(X) - \delta c \leqslant \inf_{\varepsilon \in [\varepsilon_0,\,\bar{\varepsilon}]} \varepsilon\psi_\varepsilon(X) \leqslant$$
$$\leqslant \inf_{\varepsilon \in [\varepsilon_0,\,\bar{\varepsilon}]} (\varepsilon+\delta)\,\psi_{N,\,\varepsilon+\delta}(X) + \delta c + \bar{\varepsilon}\rho. \tag{4.16}$$

Since

$$D_N(X) \overset{df}{=} \inf_{\varepsilon \in [0,\,\bar{\varepsilon}]} \varepsilon\psi_{N\varepsilon}(X) \leqslant \inf_{\varepsilon \in [\varepsilon_0-\delta,\,\bar{\varepsilon}-\delta]} \varepsilon\psi_{N\varepsilon}(X),$$

$$\inf_{\varepsilon \in [\varepsilon_0,\,\bar{\varepsilon}]} (\varepsilon-\delta)\,\psi_{N,\,\varepsilon-\delta}(X) = \inf_{\varepsilon \in [\varepsilon_0-\delta,\,\bar{\varepsilon}-\delta]} \varepsilon\psi_{N\varepsilon}(X),$$

$$\inf_{\varepsilon \in [\varepsilon_0,\,\bar{\varepsilon}]} (\varepsilon+\delta)\,\psi_{N,\,\varepsilon+\delta}(X) =$$
$$= \inf_{\varepsilon \in [\varepsilon_0+\delta,\,\bar{\varepsilon}+\delta]} \varepsilon\psi_{N\varepsilon}(X) \leqslant \inf_{\varepsilon \in [\varepsilon_0+\delta,\,\bar{\varepsilon}]} \varepsilon\psi_{N\varepsilon}(X),$$

it follows from (4.16) and (4.15) that

$$D_N(X) - \varepsilon_0 c \leqslant \inf_{\varepsilon \in [\varepsilon_0,\,\bar{\varepsilon}]} \varepsilon\psi_\varepsilon(X) \leqslant$$
$$\leqslant \inf_{\varepsilon \in [\varepsilon_0+\delta,\,\bar{\varepsilon}]} \varepsilon\psi_{N\varepsilon}(X) + \varepsilon_0 c. \tag{4.17}$$

We now note that (see Appendix III, §2)

$$D(X) \overset{df}{=} \inf_{\varepsilon \in [0,\,\bar{\varepsilon}]} \varepsilon\psi_\varepsilon(X) = \min\{\inf_{\varepsilon \in [0,\,\varepsilon_0]} \varepsilon\psi_\varepsilon(X),\ \inf_{\varepsilon \in [\varepsilon_0,\,\bar{\varepsilon}]} \varepsilon\psi_\varepsilon(X)\}, \tag{4.18}$$

$$D_N(X) = \min\{\inf_{\varepsilon \in [0,\,2\varepsilon_0]} \varepsilon\psi_{N\varepsilon}(X),\ \inf_{\varepsilon \in [\varepsilon_0+\delta,\,\bar{\varepsilon}]} \varepsilon\psi_{N\varepsilon}(X)\}. \tag{4.19}$$

We now distinguish two cases: $D(X) \leqslant -3\varepsilon_0 c$ and $D(X) > -3\varepsilon_0 c$.
I. $D(X) \leqslant -3\varepsilon_0 c$.
Using (4.18) and (4.14), we write

$$D(X) = \inf_{\varepsilon \in [\varepsilon_0,\,\bar{\varepsilon}]} \varepsilon\psi_\varepsilon(X). \tag{4.20}$$

It now follows from (4.17) and I that

$$D_N(X) \leqslant D(X) + \varepsilon_0 c \leqslant -2\varepsilon_0 c,$$

so that, by (4.19),

$$D_N(X) = \inf_{\varepsilon \in [\varepsilon_0+\delta,\ \bar{\varepsilon}]} \varepsilon \psi_{N\varepsilon}(X). \tag{4.21}$$

Combining (4.17), (4.20) and (4.21) and using (4.15), we obtain for all $N > N_0$,

$$|D(X) - D_N(X)| \leqslant \varepsilon_0 c \leqslant \alpha_0. \tag{4.22}$$

II. $D(X) > -3\varepsilon_0 c$.

In this case, we claim that for $N > N_0$

$$D_N(X) \geqslant -4\varepsilon_0 c. \tag{4.23}$$

Suppose the contrary: $D_N(X) < -4\varepsilon_0 c$. Then there exists $\varepsilon' \in [0, \bar{\varepsilon}]$ such that

$$\varepsilon' \psi_{N\varepsilon'}(X) < -4\varepsilon_0 c.$$

Hence, by (4.14),

$$\varepsilon' > 4\varepsilon_0.$$

Since in particular $\varepsilon' - \delta \geqslant \varepsilon_0$, we have

$$(\varepsilon' - \delta)\psi_{\varepsilon'-\delta}(X) \leqslant (\varepsilon' - \delta)\psi_{N\varepsilon'}(X) + (\varepsilon' - \delta)\rho.$$

We may strengthen this inequality to

$$(\varepsilon' - \delta)\psi_{\varepsilon'-\delta}(X) \leqslant \varepsilon' \psi_{N\varepsilon'}(X) + \delta c + \bar{\varepsilon}\rho \leqslant -3\varepsilon_0 c.$$

But then

$$D(X) \overset{df}{=} \inf_{\varepsilon \in [0,\ \bar{\varepsilon}]} \varepsilon \psi_\varepsilon(X) \leqslant (\varepsilon' - \delta)\psi_{\varepsilon'-\delta}(X) \leqslant -3\varepsilon_0 c,$$

which contradicts inequality II. This proves (4.23).

We now have

$$-3\varepsilon_0 c < D(x) \leqslant 0.$$
$$-4\varepsilon_0 c \leqslant D_N(X) \leqslant 0.$$

Hence, by (4.15), we see that for all $N > N_0$

$$|D(X) - D_N(X)| \leqslant 4\varepsilon_0 c \leqslant \alpha_0. \tag{4.24}$$

This completes the proof, because inequalities (4.22) and (4.24) are true for all $N > N_0$ and $X \in \bar{S}$.

Corollary. *The function* $D(X)$ *is continuous on* E_n.

This follows from the continuity of the functions $D_N(X)$ on E_n (see Chap. III, 8) and Theorem 4.2.

3*. *Theorem 4.3. A point X^* is a stationary point of $\varphi(X)$ on E_n if and only if*

$$D(X^*)=0.$$

We shall need the following

Lemma 4.2. For fixed $X\in E_n$, the function $\psi_\varepsilon(X)$, as a function of ε, $0\leqslant\varepsilon<\infty$, is right continuous at $\varepsilon=0$.

Proof. Let

$$\lambda(\varepsilon)=\psi_\varepsilon(X), \quad 0\leqslant\varepsilon<\infty.$$

It is clear that the function $\lambda(\varepsilon)$ is nondecreasing and bounded:

$$|\lambda(\varepsilon)|\leqslant\max_{Y\in G}\left\|\frac{\partial F(X, Y)}{\partial X}\right\|.$$

Suppose that $\lambda(\varepsilon)$ is discontinuous on the right at $\varepsilon=0$. Then there exists a sequence of positive numbers $\{\varepsilon_\nu\}$, tending monotonically to zero, such that

$$\lim_{\nu\to\infty}\lambda(\varepsilon_\nu)=\bar{\lambda}>\lambda(0). \tag{4.25}$$

In particular, for all ν,

$$\min_{\|g\|=1}\max_{Y\in R_{\varepsilon_\nu}(X)}\left(\frac{\partial F(X, Y)}{\partial X}, g\right)\geqslant\bar{\lambda}. \tag{4.26}$$

Fix g, $\|g\|=1$. By (4.26),

$$\max_{Y\in R_{\varepsilon_\nu}(X)}\left(\frac{\partial F(X, Y)}{\partial X}, g\right)\geqslant\bar{\lambda}. \tag{4.27}$$

Since $R_{\varepsilon_\nu}(X)\subset G$ is bounded and closed, the maximum in (4.27) is attained. Thus, for all ν there exists a point $Y_\nu\in R_{\varepsilon_\nu}(X)$ such that

$$\left(\frac{\partial F(X, Y_\nu)}{\partial X}, g\right)\geqslant\bar{\lambda}. \tag{4.28}$$

Moreover,

$$0\leqslant\varphi(X)-F(X, Y_\nu)\leqslant\varepsilon_\nu. \tag{4.29}$$

We may assume without loss of generality that the sequence $\{Y_\nu\}$ converges to some $Y^*\in G$. Then, by (4.29), $Y^*\in R(X)$. Letting $\nu\to\infty$ in (4.28), we find that

$$\left(\frac{\partial F(X, Y^*)}{\partial X}, g\right)\geqslant\bar{\lambda}.$$

A fortiori,

$$\max_{Y \in R(X)} \left(\frac{\partial F(X, Y)}{\partial X}, g \right) \geqslant \bar{\lambda}. \tag{4.30}$$

Inequality (4.30) is true for arbitrary $g, \| g \| = 1$, and so

$$\min_{\|g\|=1} \max_{Y \in R(X)} \left(\frac{\partial F(X, Y)}{\partial X}, g \right) \geqslant \bar{\lambda}.$$

We have thus shown that $\lambda(0) \geqslant \bar{\lambda}$, but this contradicts (4.25), proving the lemma.

Proof of Theorem 4.3. Necessity. Let X^* be a stationary point of $\varphi(X)$ on E_n. Then $\psi_0(X^*) \geqslant 0$, so that certainly

$$\psi_\varepsilon(X^*) \geqslant 0, \quad 0 \leqslant \varepsilon \leqslant \bar{\varepsilon}.$$

Hence

$$D(X^*) \geqslant 0.$$

The inverse inequality follows from the definition of $D(X)$, so that

$$D(X^*) = 0.$$

Sufficiency. Let $D(X^*) = 0$. Then, for all $\varepsilon \in (0, \bar{\varepsilon}]$,

$$\psi_\varepsilon(X^*) \geqslant 0.$$

In view of Lemma 4.2, we conclude that $\psi_0(X^*) \geqslant 0$. But this means that X^* is indeed a stationary point as required.

4*. We are now in a position to describe another grid method for finding stationary points of $\varphi(X)$ on E_n. We shall assume that the grids G_N are so constructed that $G_N \subset G_{N'}$ whenever $N' > N$.

Let $X_0 \in E_n$ be some initial approximation. Suppose that for some N_0 the set

$$\bar{M}(X_0) = \{X \in E_n \mid \varphi_{N_0}(X) \leqslant \varphi(X_0)\}$$

is bounded. Since

$$M(X_0) \stackrel{df}{=} \{X \mid \varphi(X) \leqslant \varphi(X_0)\} \subset \bar{M}(X_0)$$

and for $N \geqslant N_0$

$$M_N(X_0) \stackrel{df}{=} \{X \mid \varphi_N(X) \leqslant \varphi_N(X_0)\} \subset \bar{M}(X_0),$$

it follows, in particular, that there are points at which the functions $\varphi(X)$ and $\varphi_N(X)$, $N \geqslant N_0$, assume their minima.

We now fix a parameter $\varepsilon_0>0$ arbitrarily and consider the function $\varphi_{N_0}(X)$. Starting from the initial approximation X_0 and performing a finite number of iterations of the D-method (see Chap. III, §8), we obtain a point X_1 such that

$$-D_{N_0}(X_1)\leqslant\varepsilon_0.$$

Since $\varphi_{N_0}(X_1)\leqslant\varphi_{N_0}(X_0)$, we have $X_1\in\bar{M}(X_0)$. Now set $N_1=2N_0$, $\varepsilon_1=\varepsilon_0/2$, and consider the function $\varphi_{N_1}(X)$. Let

$$X_1'=\begin{cases} X_1, & \text{if} \quad \varphi_{N_1}(X_1)\leqslant\varphi_{N_1}(X_0), \\ X_0 - \text{otherwise}. \end{cases}$$

It is clear that $X_1'\in M_{N_1}(X_0)$. Letting X_1' play the role of initial approximation and using finitely many iterations of the D-method, we find a point X_2 for which

$$-D_{N_1}(X_2)\leqslant\varepsilon_1.$$

Since $\varphi_{N_1}(X_2)\leqslant\varphi_{N_1}(X_1')\leqslant\varphi_{N_1}(X_0)$, we have

$$X_2\in M_{N_1}(X_0)$$

and, a fortiori,

$$X_2\in\bar{M}(X_0).$$

Suppose we have found the k-th approximation $X_k\in\bar{M}(X_0)$. Set $N_k=2N_{k-1}$, $\varepsilon_k=\varepsilon_{k-1}/2$ and consider the function $\varphi_{N_k}(X)$. Let

$$X_k'=\begin{cases} X_k, & \text{if} \quad \varphi_{N_k}(X_k)\leqslant\varphi_{N_k}(X_0), \\ X_0 - \text{otherwise}. \end{cases}$$

Clearly, $X_k'\in M_{N_k}(X_0)$. Taking X_k' as initial approximation for finitely many iterations of the D-method, we find a point X_{k+1} such that

$$-D_{N_k}(X_{k+1})\leqslant\varepsilon_k.$$

Since $\varphi_{N_k}(X_{k+1})\leqslant\varphi_{N_k}(X_k')\leqslant\varphi_{N_k}(X_0)$, we have $X_k\in M_{N_k}(X_0)$, and a fortiori $X_k\in\bar{M}(X_0)$.

Proceeding in this way, we obtain a sequence $\{X_k\}$ such that $X_k\in\bar{M}(X_0)$ and

$$-D_{N_{k-1}}(X_k)\leqslant\varepsilon_{k-1}, \tag{4.31}$$

where $N_{k-1}=2^{k-1}N_0$, $\varepsilon_{k-1}=\varepsilon_0/2^{k-1}$, $k=1, 2, \ldots$. Since $\bar{M}(X_0)$ is bounded and closed, this sequence has at least one limit point.

Theorem 4.4. Any limit point of the sequence $\{X_k\}$ is a stationary point of $\varphi(X)$ on E_n.

Proof. Let

$$X_{k_j} \xrightarrow[k_j\to\infty]{} X^*.$$

We have

$$|D(X^*)| \leqslant |D(X^*) - D(X_{k_j})| + \\ + |D(X_{k_j}) - D_{N_{k_j-1}}(X_{k_j})| + |D_{N_{k_j-1}}(X_{k_j})|. \tag{4.32}$$

The first term on the right tends to zero as $k_j \to \infty$, in view of the continuity of $D(X)$; the second does so because the sequence $\{X_{k_j}\}$ is bounded and by Theorem 4.2; the third tends to zero by construction (see (4.31)). Letting $k_j \to \infty$ in (4.32), we obtain

$$D(X^*) = 0.$$

It now remains to refer to Theorem 4.3.

5. We have already seen that for every $X \in E_n$

$$\varphi_N(X) \xrightarrow[N\to\infty]{} \varphi(X).$$

Despite this, it is not necessarily true that

$$\frac{\partial \varphi_N(X)}{\partial g} \xrightarrow[N\to\infty]{} \frac{\partial \varphi(X)}{\partial g}$$

for all $X \in E_n$ and $g \in E_n$, $\| g \| = 1$.

Example. Let

$$F(x, y) = \cos\left(x\left(y + \frac{\pi}{4}\right)\right), \quad G = \left[0, \frac{3\pi}{2}\right];$$
$$\varphi(x) = \max_{y \in \left[0, \frac{3\pi}{2}\right]} F(x, y).$$

Consider the point $x = 1$. Then (Figure 34)

$$R(x) = \left\{0, \frac{3\pi}{2}\right\}, \quad \varphi(x) = \cos\frac{\pi}{4} = \frac{\sqrt{2}}{2}.$$

By (2.1),

$$\frac{\partial \varphi(x)}{\partial g} = \max\left\{-g\frac{\pi}{4}\sin\frac{\pi}{4}, -g\frac{7\pi}{4}\sin\frac{7\pi}{4}\right\} = \\ = \max\left\{-g\frac{\pi\sqrt{2}}{8}, g\frac{7\pi\sqrt{2}}{8}\right\}.$$

For $g_1 = (+1)$, we have

$$\frac{\partial \varphi(x)}{\partial g_1} = \frac{7\pi\sqrt{2}}{8}, \tag{4.33}$$

and for $g_2=(-1)$

$$\frac{\partial\varphi(x)}{\partial g_2}=\frac{\pi\sqrt{2}}{8}.$$

It is clear that $\psi(x)=\min\limits_{\|g\|=1}\frac{\partial\varphi(x)}{\partial g}=\frac{\pi\sqrt{2}}{8}>0$, so that $x=1$ is a stationary point.

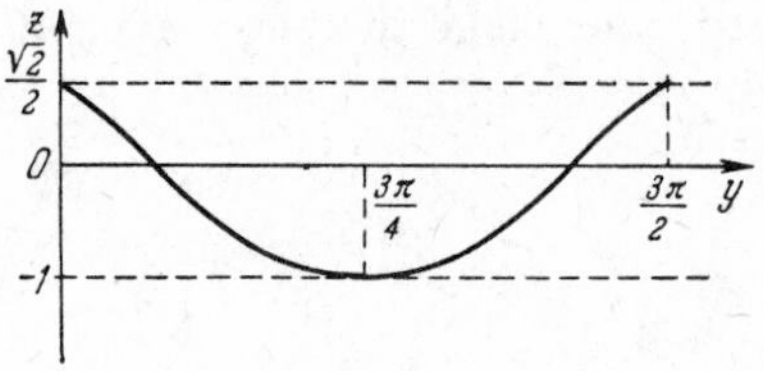

FIGURE 34.

We now introduce finite grids

$$G_N=\left\{\frac{k}{N}\,\middle|\,k\in\left[0:E\left(\frac{3\pi}{2}N\right)\right]\right\}.$$

For any natural number N, the number k/N is rational, so that $3\pi/2\notin G_N$ for all N. At the same time, $0\in G_N$.

Now the point $x=1$ is such that $R_N(x)=\{0\}$. Therefore,

$$\frac{\partial\varphi_N(x)}{\partial g}=-g\frac{\pi\sqrt{2}}{8}.$$

In particular, for $x=1$, $g_1=(+1)$ and all natural N,

$$\frac{\partial\varphi_N(x)}{\partial g_1}=-\frac{\pi\sqrt{2}}{8}.$$

Comparing this with (4.33), we see that indeed

$$\frac{\partial\varphi_N(x)}{\partial g_1}\not\to\frac{\partial\varphi(x)}{\partial g_1}.$$

§5. SPECIAL CASE OF THE MINIMAX THEOREM

1. In this section we shall deduce certain sufficient conditions for the equality

$$\min_{X\in\Omega}\max_{Y\in G}F(X,Y)=\max_{Y\in G}\min_{X\in\Omega}F(X,Y).$$

We first prove an auxiliary proposition, which is also of independent interest.

Let $G \subset E_m$ be a bounded closed convex set,

$$\varphi(X) = \max_{Y \in G} F(X, Y),$$
$$R(X) = \{Y \in G \mid F(X, Y) = \varphi(X)\},$$
$$H(X) = \left\{ \frac{\partial F(X, Y)}{\partial X} \,\middle|\, Y \in R(X) \right\}.$$

Let $L(X)$ denote the convex hull of $H(X)$:

$$L(X) = \operatorname{co} H(X).$$

Theorem 5.1. Suppose that, for every fixed X in a neighborhood $S_\delta(X_0)$ of X_0, the function $F(X, Y)$ is concave in Y on G. Then

$$L(X_0) = H(X_0).$$

For the proof, we need the following:

Lemma 5.1. Under the assumptions of Theorem 5.1, for any $Y_1, Y_2 \in R(X_0)$ and $\alpha \in [0, 1]$ the point $Y_\alpha \overset{df}{=} \alpha Y_1 + (1 - \alpha) Y_2$ is in $R(X_0)$ and

$$\frac{\partial F(X_0, Y_\alpha)}{\partial X} = \alpha \frac{\partial F(X_0, Y_1)}{\partial X} + (1 - \alpha) \frac{\partial F(X_0, Y_2)}{\partial X}. \tag{5.1}$$

Proof. Since $F(X_0, Y)$ is concave in Y, we have

$$F(X_0, Y_\alpha) \geqslant \alpha F(X_0, Y_1) + (1 - \alpha) F(X_0, Y_2). \tag{5.2}$$

On the other hand,

$$F(X_0, Y_1) = F(X_0, Y_2) = \varphi(X_0),$$

and so

$$F(X_0, Y_\alpha) \leqslant \varphi(X_0) = \alpha F(X_0, Y_1) + (1 - \alpha) F(X_0, Y_2). \tag{5.3}$$

Combining (5.2) and (5.3), we get

$$F(X_0, Y_\alpha) = \alpha F(X_0, Y_1) + (1 - \alpha) F(X_0, Y_2), \tag{5.4}$$

or, equivalently,

$$F(X_0, Y_\alpha) = \varphi(X_0).$$

Thus $Y_\alpha \in R(X_0)$.

Now, since $F(X, Y)$ is concave in Y for any $X \in S_\delta(X_0)$, it follows that

$$F(X, Y_\alpha) \geqslant \alpha F(X, Y_1) + (1 - \alpha) F(X, Y_2). \tag{5.5}$$

Fix $\alpha \in [0, 1]$. We consider an arbitrary vector $g \in E_n$, $\|g\| = 1$, and evaluate the derivative of the function $\Phi(X) \overset{df}{=} F(X, Y_\alpha)$ at X_0 in the direction g:

$$\frac{\partial \Phi(X_0)}{\partial g} = \lim_{\beta \to +0} \frac{F(X_0 + \beta g, Y_\alpha) - F(X_0, Y_\alpha)}{\beta} = \left(\frac{\partial F(X_0, Y_\alpha)}{\partial X}, g\right). \tag{5.6}$$

It follows from (5.4) and (5.5) that for small $\beta > 0$

$$F(X_0 + \beta g, Y_\alpha) - F(X_0, Y_\alpha) \geqslant \alpha [F(X_0 + \beta g, Y_1) - F(X_0, Y_1)] + (1 - \alpha)[F(X_0 + \beta g, Y_2) - F(X_0, Y_2)].$$

Hence

$$\frac{\partial \Phi(X_0)}{\partial g} \geqslant \alpha \left(\frac{\partial F(X_0, Y_1)}{\partial X}, g\right) + (1 - \alpha)\left(\frac{\partial F(X_0, Y_2)}{\partial X}, g\right). \tag{5.7}$$

From (5.6) and (5.7) we obtain

$$\left(\frac{\partial F(X_0, Y_\alpha)}{\partial X} - \left[\alpha \frac{\partial F(X_0, Y_1)}{\partial X} + (1 - \alpha) \frac{\partial F(X_0, Y_2)}{\partial X}\right], g\right) \geqslant 0.$$

Since this inequality is true for any $g \in E_n$, $\|g\| = 1$, we have

$$\frac{\partial F(X_0, Y_\alpha)}{\partial X} - \left[\alpha \frac{\partial F(X_0, Y_1)}{\partial X} + (1 - \alpha) \frac{\partial F(X_0, Y_2)}{\partial X}\right] = 0,$$

which is equivalent to (5.1).

Proof of Theorem 5.1. The gist of Lemma 5.1 is that the set $H(X_0)$ is convex. On the other hand, the convex hull of a convex set is the set itself (Appendix II, §1), and so $L(X_0) = H(X_0)$. This completes the proof of the theorem.

2. *Theorem 5.2. Let $F(X, Y)$ be continuous together with $\frac{\partial F(X, Y)}{\partial X}$ on $\Omega' \times G$, where $\Omega' \subset E_n$, $G \subset E_m$. Assume that Ω' is open and let $\Omega \subset \Omega'$ and G be bounded closed convex sets. Suppose moreover that for every fixed $X_0 \in \Omega'$ the function $F(X_0, Y)$ is concave in Y on G, and for every fixed $Y_0 \in G$ the function $F(X, Y_0)$ is concave in X on Ω. Then*

$$\min_{X \in \Omega} \max_{Y \in G} F(X, Y) = \max_{Y \in G} \min_{X \in \Omega} F(X, Y).$$

Proof. Consider the function

$$\varphi(X) = \max_{Y \in G} F(X, Y).$$

Let $X^* \in \Omega$ be a minimum point of $\varphi(X)$ on Ω. By Theorems 2.2 and 3.1,

$$L(X^*) \cap \Gamma^+(X^*) \neq \varnothing.$$

Let $Z^* \in L(X^*) \cap \Gamma^+(X^*)$. By Theorem 5.1, $Z^* \in H(X^*)$, i. e., there exists a point $Y^* \in R(X^*)$ such that

$$Z^* = \frac{\partial F(X^*, Y^*)}{\partial X}.$$

Since $Z^* \in \Gamma^+(X^*)$, it follows that

$$\frac{\partial F(X^*, Y^*)}{\partial X} \in \Gamma^+(X^*). \tag{5.8}$$

By Remark 2 to Theorem 3.1 in Chap. IV and (5.8),

$$F(X^*, Y^*) \geqslant \min_{X \in \Omega} F(X, Y^*). \tag{5.9}$$

On the other hand, $Y^* \in R(X^*)$, so that

$$F(X^*, Y^*) = \max_{Y \in G} F(X^*, Y). \tag{5.10}$$

In view of (5.9) and (5.10), we see that for any $X \in \Omega$ and $Y \in G$

$$F(X^*, Y) \leqslant F(X^*, Y^*) \leqslant F(X, Y^*). \tag{5.11}$$

It follows from (5.11) that

$$\max_{Y \in G} F(X^*, Y) \leqslant \min_{X \in \Omega} F(X, Y^*).$$

A fortiori

$$\min_{X \in \Omega} \max_{Y \in G} F(X, Y) \leqslant \max_{Y \in G} \min_{X \in \Omega} F(X, Y). \tag{5.12}$$

The inverse inequality is trivial: for any $X \in \Omega$ and $Y \in G$,

$$F(X, Y) \leqslant \max_{Y' \in G} F(X, Y').$$

Hence

$$\min_{X \in \Omega} F(X, Y) \leqslant \min_{X \in \Omega} \max_{Y' \in G} F(X, Y').$$

Since this inequality is valid for all $Y \in G$, we have

$$\max_{Y \in G} \min_{X \in \Omega} F(X, Y) \leqslant \min_{X \in \Omega} \max_{Y \in G} F(X, Y). \tag{5.13}$$

Combining (5.12) and (5.13), we complete the proof.

Remark. Our proof of Theorem 5.2 uses a superfluous assumption, namely, that the function $F(X, Y)$ is continuously differentiable with respect to Y. However, the proof is constructive in the sense that, given X^*, one can actually exhibit Y^* as a point for which

$$\frac{\partial F(X^*, Y^*)}{\partial X} \in H(X^*) \cap \Gamma^+(X^*).$$

A pair $[X^*, Y^*]$ satisfying inequality (5.11) is known as a saddle point of $F(X, Y)$.

3. We now consider the following minimax problem (see Remark to Theorem 6.1, Chap. IV):

$$\max_{V \in L} (V, Z - X) \to \min_{Z \in \Omega}. \tag{5.14}$$

We shall assume that Ω and L are bounded closed subsets of E_n, L is convex and Ω strictly convex. By Theorem 5.2 (the minimax theorem),

$$\min_{Z \in \Omega} \max_{V \in L} (V, Z - X) = \max_{V \in L} \min_{Z \in \Omega} (V, Z - X). \tag{5.15}$$

Consider the function

$$\theta(V) = \min_{Z \in \Omega} (V, Z - X).$$

We claim that $\theta(V)$ is continuously differentiable everywhere on E_n, with the possible exception of the point $V = \mathbf{0}$. Set

$$\hat{R}(V) = \{Z \in \Omega \mid (V, Z - X) = \theta(V)\}.$$

Lemma 5.2. If Ω is strictly convex and $V \neq \mathbf{0}$, then $\hat{R}(V)$ contains a single point $Z(V)$, and $Z(V)$ is a continuous function of V.

Proof. Let $V \neq \mathbf{0}$ and suppose that $\hat{R}(V)$ contains two distinct points Z_1 and Z_2:

$$(V, Z_1 - X) = (V, Z_2 - X) = \min_{Z \in \Omega} (V, Z - X).$$

Since Ω is strictly convex, $Z' \overset{df}{=} \frac{1}{2}(Z_1 + Z_2)$ is an interior point of Ω. Thus there exists $\rho > 0$ such that

$$(Z' + Z) \in \Omega \quad \text{for all} \quad Z, \|Z\| < \rho.$$

Set $Z_0 = \frac{-\rho V}{2\|V\|}$. Then

$$\min_{Z \in \Omega} (V, Z - X) \leqslant (V, Z' + Z_0 - X) =$$
$$= \frac{1}{2}(V, Z_1 - X) + \frac{1}{2}(V, Z_2 - X) - \frac{\rho}{2\|V\|}(V, V) =$$
$$= \min_{Z \in \Omega} (V, Z - X) - \frac{1}{2}\rho \|V\|,$$

which is impossible. Thus $\hat{R}(V)$ consists of a single point, which we denote by $Z(V)$.

We must show that $Z(V)$ is continuous at any point $V_0 \neq \mathbf{0}$. Otherwise, there exists a sequence $\{V'_k\}$ converging to V_0 such that $\{Z(V'_k)\}$ does not converge to $Z(V_0)$. Since $Z(V'_k) \in \Omega$ for all k, there exists a subsequence $\{Z(V'_{k_j})\}$ that converges to some point Z^* other than $Z(V_0)$. Clearly, $Z^* \in \Omega$.
We note that

$$(V'_{k_j}, Z(V'_{k_j}) - X) = \min_{Z \in \Omega}(V'_{k_j}, Z - X). \tag{5.16}$$

Letting $k_j \to \infty$ in (5.16), we obtain

$$(V_0, Z^* - X) = \min_{Z \in \Omega}(V_0, Z - X).$$

Thus $Z^* \in \hat{R}(V_0)$, which is impossible since $\hat{R}(V_0)$ contains the single point $Z(V_0)$. This completes the proof of the lemma.

We can now prove that the function $\theta(V)$, $V \neq \mathbf{0}$, is continuously differentiable.

Since

$$-\theta(V) = \max_{Z \in \Omega}(V, X - Z),$$

it follows from Theorem 2.1 that $\theta(V)$ is differentiable at any point $V \in E_n$ in any direction $g \in E_n$, $\|g\| = 1$. In fact,

$$\frac{\partial \theta(V)}{\partial g} = -\frac{\partial(-\theta(V))}{\partial g} = \min_{Z \in \hat{R}(V)}(Z - X, g).$$

If $V \neq \mathbf{0}$, then $\hat{R}(V)$ contains a single point $Z(V)$. Thus, for all $g \in E_n$, $\|g\| = 1$, and $V \neq \mathbf{0}$, we have

$$\frac{\partial \theta(V)}{\partial g} = (Z(V) - X, g).$$

Since $Z(V)$ is continuous in V, we conclude (see Appendix III, §3), that the function $\theta(V)$ is continuously differentiable for $V \neq \mathbf{0}$ and

$$\frac{\partial \theta(V)}{\partial V} = Z(V) - X. \tag{5.17}$$

Thus, if $\mathbf{0} \notin L$, the minimax problem (5.14) is reduced by (5.15) to a simpler problem — maximization on L of the continuously differentiable function $\theta(V)$, whose gradient is given by formula (5.17).

§6*. DETERMINATION OF SADDLE POINTS ON POLYHEDRA

1. Let $\Omega \subset E_n$, $G \subset E_m$ be bounded closed sets, and $F(X, Y)$ a continuous function defined on $\Omega \times G$. For the reader's convenience, we recall the definition of a saddle point.

Definition. A point $[X^*, Y^*] \in \Omega \times G$ is called a saddle point of $F(X, Y)$ on $\Omega \times G$ if

$$F(X^*, Y) \leqslant F(X^*, Y^*) \leqslant F(X, Y^*) \tag{6.1}$$

for all $X \in \Omega$, $Y \in G$.

Lemma 6.1. The function $F(X, Y)$ has a saddle point on $\Omega \times G$ if and only if

$$\min_{X \in \Omega} \max_{Y \in G} F(X, Y) = \max_{Y \in G} \min_{X \in \Omega} F(X, Y). \tag{6.2}$$

Proof. We first observe that all the maxima and minima involved are actually achieved since $F(X, Y)$ is continuous and the sets Ω and G bounded and closed.

We first prove that the existence of a saddle point implies (6.2). By (6.1), we have

$$\max_{Y \in G} F(X^*, Y) \leqslant \min_{X \in \Omega} F(X, Y^*).$$

Hence

$$\min_{X \in \Omega} \max_{Y \in G} F(X, Y) \leqslant \max_{Y \in G} \min_{X \in \Omega} F(X, Y).$$

Comparing this with the obvious inequality

$$\max_{Y \in G} \min_{X \in \Omega} F(X, Y) \leqslant \min_{X \in \Omega} \max_{Y \in G} F(X, Y),$$

we obtain (6.2).

Now suppose that (6.2) is true. We set

$$\varphi(X) = \max_{Y \in G} F(X, Y), \qquad \Phi(Y) = \min_{X \in \Omega} F(X, Y).$$

It follows from (6.2) and the continuity of $\varphi(X)$ and $\Phi(Y)$ that there exist points $X^* \in \Omega$ and $Y^* \in G$ for which

$$\min_{X \in \Omega} \varphi(X) = \varphi(X^*) = \max_{Y \in G} \Phi(Y) = \Phi(Y^*). \tag{6.3}$$

We claim that $[X^*, Y^*]$ is a saddle point of $F(X, Y)$ on $\Omega \times G$.

Indeed, it follows from (6.3) that for any $[X, Y] \in \Omega \times G$

$$F(X^*, Y) \leqslant \max_{Y' \in G} F(X^*, Y') = \varphi(X^*) = \Phi(Y^*) = \\ = \min_{X \in \Omega} F(X, Y^*) \leqslant F(X^*, Y^*). \tag{6.4}$$

Similarly,

$$F(X, Y^*) \geqslant \min_{X' \in \Omega} F(X', Y^*) = \Phi(Y^*) = \varphi(X^*) = \\ = \max_{Y \in G} F(X^*, Y) \geqslant F(X^*, Y^*). \qquad (6.5)$$

Combining (6.4) and (6.5), we get (6.1), proving the lemma.

2. Let $\Omega' \subset E_n$ and $G' \subset E_m$ be open sets, and let $F(X, Y)$ be defined and twice continuously differentiable on $\Omega' \times G'$

We consider the subset of E_n defined by

$$\Omega = \{X \mid (A_i, X) + a_i \leqslant 0, \ i \in [1 : N_1]\},$$

and the subset of E_m

$$G = \{Y \mid (B_j, Y) + b_j \leqslant 0, \ j \in [1 : N_2]\}.$$

Ω and G are clearly convex sets.

We shall assume that Ω and G are nonempty and bounded, and that $\Omega \subset \Omega'$, $G \subset G'$.

We may assume without loss of generality that

$$\| A_i \| = \| B_j \| = 1, \quad i \in [1 : N_1], \quad j \in [1 : N_2]. \qquad (6.6)$$

Suppose that the function $F(X, Y)$ is strictly convex in X and strictly concave in Y on $\Omega \times G$, i. e., there exist $m_1 > 0$ and $m_2 > 0$ such that, for all $[X, Y] \in \Omega \times G$,

$$\left(V, \frac{\partial^2 F(X, Y)}{\partial X^2} V\right) \geqslant m_1 \| V \|^2, \qquad (6.7)$$

$$-\left(W, \frac{\partial^2 F(X, Y)}{\partial Y^2} W\right) \geqslant m_2 \| W \|^2 \qquad (6.8)$$

for any $V \in E_n$, $W \in E_m$.

We wish to find a saddle point of $F(X, Y)$ on $\Omega \times G$, i. e., a point satisfying inequalities (6.1). The existence of a saddle point follows from Theorem 5.2 and Lemma 6.1.

Let $[X, Y] \in \Omega \times G$. Consider the cones

$$\Gamma_1^+(X) = \left\{ g \in E_n \mid g = -\sum_{i \in Q_1(X)} \alpha_i A_i, \ \alpha_i \geqslant 0 \right\},$$
$$\Gamma_2^+(Y) = \left\{ q \in E_m \mid q = -\sum_{j \in Q_2(Y)} \beta_j B_j, \ \beta_j \geqslant 0 \right\},$$

where

$$Q_1(X) = \{i \in [1 : N_1] \mid (A_i, X) + a_i = 0\},$$
$$Q_2(Y) = \{j \in [1 : N_2] \mid (B_j, Y) + b_j = 0\}.$$

If $Q_1(X)=\varnothing$, we set $\Gamma_1^+(X)=\{0\}$; if $Q_2(Y)=\varnothing$, we set $\Gamma_2^+(Y)=\{0\}$.

As shown in Chap. V, §2, $\Gamma_1^+(X)$ is the dual cone of the cone of feasible directions of Ω at X, and $\Gamma_2^+(Y)$ is the dual cone of the cone of feasible directions of G at Y.

We introduce the functions

$$d_1(X, Y)=\min_{Z\in\Gamma_1^+(X)}\left\|Z-\frac{\partial F(X,Y)}{\partial X}\right\|\stackrel{df}{=}\left\|Z_1(X,Y)-\frac{\partial F(X,Y)}{\partial X}\right\|,$$

$$d_2(X, Y)=\min_{Z\in\Gamma_2^+(Y)}\left\|Z+\frac{\partial F(X,Y)}{\partial Y}\right\|\stackrel{df}{=}\left\|Z_2(X,Y)+\frac{\partial F(X,Y)}{\partial Y}\right\|.$$

Note that for any $[X, Y]\in\Omega\times G$ the point $[Z_1(X, Y),\ Z_2(X, Y)]\in E_n\times E_m$ is unique.

If $d_1(X, Y)>0$, then

$$g(X, Y)=\frac{1}{d_1(X,Y)}\left(Z_1(X,Y)-\frac{\partial F(X,Y)}{\partial X}\right)$$

is the direction of steepest descent of the function $f_1(Z)=F(Z, Y)$ on Ω at $Z=X$.

If $d_2(X, Y)>0$, then

$$q(X, Y)=\frac{1}{d_2(X,Y)}\left(Z_2(X,Y)+\frac{\partial F(X,Y)}{\partial Y}\right)$$

is the direction of steepest descent of $f_2(Z)=-F(X, Z)$ on G at $Z=Y$.

The next theorem follows from the definition of a saddle point (see (6.1)) and from Theorem 3.1 of Chap. IV:

Theorem 6.1. A point $[X^*, Y^*]\in\Omega\times G$ is a saddle point of the function $F(X, Y)$ on the set $\Omega\times G$ if and only if

$$d_1(X^*, Y^*)=d_2(X^*, Y^*)=0.$$

The geometric meaning of this condition is

$$\frac{\partial F(X^*,Y^*)}{\partial X}\in\Gamma_1^+(X^*),\qquad -\frac{\partial F(X^*,Y^*)}{\partial Y}\in\Gamma_2^+(Y^*).$$

Let $\varepsilon>0$ be fixed, and consider the sets

$$Q_{1\varepsilon}(X)=\{i\in[1:N_1]\mid -\varepsilon\leqslant(A_i, X)+a_i\leqslant 0\},$$
$$Q_{2\varepsilon}(Y)=\{j\in[1:N_2]\mid -\varepsilon\leqslant(B_j, Y)+b_j\leqslant 0\}$$

and cones

$$\Gamma_{1\varepsilon}^+(X)=\left\{g\in E_n\mid g=-\sum_{i\in Q_{1\varepsilon}(X)}\alpha_i A_i,\ \alpha_i\geqslant 0\right\},$$
$$\Gamma_{2\varepsilon}^+(Y)=\left\{q\in E_m\mid q=-\sum_{j\in Q_{2\varepsilon}(Y)}\beta_j B_j,\ \beta_j\geqslant 0\right\}.$$

Set

$$d_{1\varepsilon}(X, Y)= \min_{Z\in\Gamma_{1\varepsilon}^{+}(X)} \left\| Z-\frac{\partial F(X, Y)}{\partial X}\right\| \stackrel{df}{=} \left\| Z_{1\varepsilon}(X, Y)-\frac{\partial F(X, Y)}{\partial X}\right\| \stackrel{df}{=} \| g_{\varepsilon}(X, Y)\|, \qquad (6.9)$$

$$d_{2\varepsilon}(X, Y)= \min_{Z\in\Gamma_{2\varepsilon}^{+}(Y)} \left\| Z+\frac{\partial F(X, Y)}{\partial Y}\right\| \stackrel{df}{=} \left\| Z_{2\varepsilon}(X, Y)+\frac{\partial F(X, Y)}{\partial Y}\right\| \stackrel{df}{=} \| q_{\varepsilon}(X, Y)\|. \qquad (6.10)$$

It is easily checked that for all $i\in Q_{1\varepsilon}(X)$ and $j\in Q_{2\varepsilon}(Y)$

$$(A_i,\, g_{\varepsilon}(X, Y))\leqslant 0, \quad (B_j,\, g_{\varepsilon}(X, Y))\leqslant 0. \qquad (6.11)$$

Definition. A point $[X^*, Y^*]\in\Omega\times G$ is called an ε-saddle point of $F(X, Y)$ on $\Omega\times G$ if

$$d_{1\varepsilon}(X^*, Y^*)=d_{2\varepsilon}(X^*, Y^*)=0. \qquad (6.12)$$

Lemma 6.2. The point $Z_{1\varepsilon}(X, Y)$ may be expressed as

$$Z_{1\varepsilon}(X, Y)=-\sum_{i\in\tilde{Q}_{1\varepsilon}(X, Y)} \alpha_i(X, Y)\, A_i,$$

where $\alpha_i(X, Y)\geqslant 0$ and

$$\tilde{Q}_{1\varepsilon}(X, Y)=\{i\in Q_{1\varepsilon}(X)\,|\,(A_i,\, g_{\varepsilon}(X, Y))=0\}.$$

Proof. The assertion is trivial if $Z_{1\varepsilon}(X, Y)=\mathbf{0}$. Suppose, then, that $Z_{1\varepsilon}(X, Y)\neq\mathbf{0}$.

If $g_{\varepsilon}(X, Y)=\mathbf{0}$, the assertion is again obvious, for then $\tilde{Q}_{1\varepsilon}(X, Y)=Q_{1\varepsilon}(X)$. Now let $g_{\varepsilon}(X, Y)\neq\mathbf{0}$. Then it follows at once from (6.9) and (6.11) that

$$(Z_{1\varepsilon}(X, Y),\, g_{\varepsilon}(X, Y))=0, \qquad (6.13)$$

$$(A_i,\, g_{\varepsilon}(X, Y))<0 \qquad (6.14)$$

for all $i\in\hat{Q}_{1\varepsilon}(X, Y)\stackrel{df}{=}Q_{1\varepsilon}(X)\setminus\tilde{Q}_{1\varepsilon}(X, Y)$.

Indeed, if (6.13) were false, there would be a point $Z'_{1\varepsilon}\in\Gamma_{1\varepsilon}^{+}(X)$ such that

$$\left\| Z'_{1\varepsilon}-\frac{\partial F(X, Y)}{\partial X}\right\| < d_{1\varepsilon}(X, Y),$$

and this is impossible by (6.9).

Inequality (6.14) follows from (6.11) and the definition of the set $\tilde{Q}_{1\varepsilon}(X, Y)$.

The set $\tilde{Q}_{1\varepsilon}(X, Y)$ is not empty, since $Z_{1\varepsilon}(X, Y)\neq\mathbf{0}$.

Let $Z_{1\varepsilon}(X, Y)=-\sum_{i\in Q_{1\varepsilon}(X)} \alpha_i(X, Y)\, A_i$, $\alpha_i(X, Y)\geqslant 0$. We may write $Z_{1\varepsilon}(X, Y)$ as

$$Z_{1\varepsilon}(X, Y)=C_1+C_2,$$

where

$$C_1 = -\sum_{i \in \tilde{Q}_{1\varepsilon}(X, Y)} \alpha_i(X, Y) A_i, \quad C_2 = -\sum_{i \in \hat{Q}_{1\varepsilon}(X, Y)} \alpha_i(X, Y) A_i.$$

We claim that $C_2 = \mathbf{0}$. Indeed, by (6.13),

$$\begin{aligned}(Z_{1\varepsilon}(X, Y), g_\varepsilon(X, Y)) &= (C_1, g_\varepsilon(X, Y)) + (C_2, g_\varepsilon(X, Y)) = \\ &= (C_2, g_\varepsilon(X, Y)) = -\sum_{i \in \hat{Q}_{1\varepsilon}(X, Y)} \alpha_i(X, Y)(A_i, g_\varepsilon(X, Y)) = 0. \qquad (6.15)\end{aligned}$$

Since it follows from (6.14) that $(A_i, g_\varepsilon(X, Y)) < 0$ and $\alpha_i(X, Y) \geqslant 0$, we see that (6.15) can hold only if $\alpha_i(X, Y) = 0$, for all $i \in \hat{Q}_{1\varepsilon}(X, Y)$, and the lemma is proved.

Similarly, one proves:

Lemma 6.3. The point $Z_{2\varepsilon}(X, Y)$ may be expressed as

$$Z_{2\varepsilon}(X, Y) = -\sum_{j \in \tilde{Q}_{2\varepsilon}(X, Y)} \beta_j(X, Y) B_j,$$

where $\beta_j(X, Y) \geqslant 0$ and

$$\tilde{Q}_{2\varepsilon}(X, Y) = \{j \in Q_{2\varepsilon}(Y) \mid (q_\varepsilon(X, Y), B_j) = 0\}.$$

3. Fix some $\varepsilon > 0$. We shall now describe a successive approximation procedure which determines ε-saddle points of $F(X, Y)$ on $\Omega \times G$. We define a function

$$d_\varepsilon(X, Y) = \frac{1}{2}[d_{1\varepsilon}^2(X, Y) + d_{2\varepsilon}^2(X, Y)].$$

Condition (6.12) is clearly equivalent to

$$d_\varepsilon(X^*, Y^*) = 0.$$

As initial approximation we select any point $[X_0, Y_0] \in \Omega \times G$. Suppose we have already found the point $[X_k, Y_k] \in \Omega \times G$.

If $d_\varepsilon(X_k, Y_k) = 0$, the point $[X_k, Y_k]$ is an ε-saddle point and we are done. If $d_\varepsilon(X_k, Y_k) > 0$, we consider the rays

$$X = X_k(\alpha) \stackrel{df}{=} X_k + \alpha g_k, \quad Y = Y_k(\alpha) \stackrel{df}{=} Y_k + \alpha q_k, \quad \alpha \geqslant 0,$$

where $g_k = g_\varepsilon(X_k, Y_k)$, $q_k = q_\varepsilon(X_k, Y_k)$.

Let $\alpha_k \geqslant 0$ be such that

$$d_\varepsilon(X_k(\alpha_k), Y_k(\alpha_k)) = \min_{\substack{\alpha \geqslant 0 \\ X_k(\alpha) \in \Omega,\ Y_k(\alpha) \in G}} d_\varepsilon(X_k(\alpha), Y_k(\alpha)).$$

Now set

$$X_{k+1} = X_k(\alpha_k), \quad Y_{k+1} = Y_k(\alpha_k).$$

It is clear that

$$[X_{k+1}, Y_{k+1}] \in \Omega \times G, \quad d_\varepsilon(X_{k+1}, Y_{k+1}) \leqslant d_\varepsilon(X_k, Y_k). \tag{6.16}$$

Continuing in this way, we construct a sequence

$$\{[X_k, Y_k]\}, \quad [X_k, Y_k] \in \Omega \times G, \quad k = 0, 1, 2, \ldots$$

If this sequence is finite, its last term is by construction an ε-saddle point of $F(X, Y)$ on $\Omega \times G$. Otherwise:

Theorem 6.2. Any limit point of the sequence $\{[X_k, Y_k]\}$ is an ε-saddle point of $F(X, Y)$ on $\Omega \times G$.

Proof. That there are indeed limit points follows from the fact that $\Omega \times G$ is bounded and closed. Let

$$d^* = \lim_{k \to \infty} d_\varepsilon(X_k, Y_k).$$

Clearly, for all k, by (6.16),

$$d_\varepsilon(X_k, Y_k) \geqslant d^*. \tag{6.17}$$

Let

$$[X_{k_s}, Y_{k_s}] \xrightarrow[k_s \to \infty]{} [X^*, Y^*].$$

We claim that $d_\varepsilon(X_{k_s}, Y_{k_s}) \xrightarrow[k_s \to \infty]{} 0$; this will show, by (6.16), that

$$d_\varepsilon(X_k, Y_k) \xrightarrow[k \to \infty]{} 0. \tag{6.18}$$

Suppose the contrary:

$$d_\varepsilon(X_{k_s}, Y_{k_s}) \geqslant 2a^2 > 0, \quad a > 0.$$

Then either $d_{1\varepsilon}(X_{k_s}, Y_{k_s}) \geqslant a$ or $d_{2\varepsilon}(X_{k_s}, Y_{k_s}) \geqslant a$. Suppose, say, that for all k_s

$$d_{1\varepsilon}(X_{k_s}, Y_{k_s}) \geqslant a$$

(if the inequality $d_{1\varepsilon}(X_{k_s}, Y_{k_s}) \geqslant a$ occurs infinitely many times, one can assure, if necessary by extracting a suitable subsequence of $\{[X_{k_s}, Y_{k_s}]\}$, that the inequality is true for all k_s).

We may also assume without loss of generality that

$$Q_{1\varepsilon}(X_{k_s}, Y_{k_s}) = Q' \subset [1 : N_1],$$
$$Q_{2\varepsilon}(X_{k_s}, Y_{k_s}) = Q'' \subset [1 : N_2].$$

Note that

$$Q_{1\varepsilon}(X^*, Y^*) \supset Q', \quad Q_{2\varepsilon}(X^*, Y^*) \supset Q''.$$

In view of Lemmas 6.2, 6.3 and the definition of $\tilde{Q}_{1\varepsilon}$ and $\tilde{Q}_{2\varepsilon}$, we see that for all $\alpha > 0$

$$Z_{1\varepsilon}(X_{k_s}, Y_{k_s}) \in \Gamma^+_{1\varepsilon}(X_{k_s}(\alpha)),$$
$$Z_{2\varepsilon}(X_{k_s}, Y_{k_s}) \in \Gamma^+_{2\varepsilon}(Y_{k_s}(\alpha)),$$

and so

$$d_{1\varepsilon}(X_{k_s}(\alpha), Y_{k_s}(\alpha)) \leqslant \left\| Z_{1\varepsilon}(X_{k_s}, Y_{k_s}) - \frac{\partial F(X_{k_s}(\alpha), Y_{k_s}(\alpha))}{\partial X} \right\|, \tag{6.19}$$

$$d_{2\varepsilon}(X_{k_s}(\alpha), Y_{k_s}(\alpha)) \leqslant \left\| Z_{2\varepsilon}(X_{k_s}, Y_{k_s}) + \frac{\partial F(X_{k_s}(\alpha), Y_{k_s}(\alpha))}{\partial Y} \right\|. \tag{6.20}$$

For $\alpha > 0$ such that

$$\alpha d_{1\varepsilon}(X_k, Y_k) \leqslant \varepsilon, \quad \alpha d_{2\varepsilon}(X_k, Y_k) \leqslant \varepsilon,$$

we have $X_k(\alpha) \in \Omega$, $Y_k(\alpha) \in G$.

This is easily proved using (6.11), (6.6) and the equalities

$$\alpha d_{1\varepsilon}(X_k, Y_k) = \| \alpha g_k \|, \quad \alpha d_{2\varepsilon}(X_k, Y_k) = \| \alpha q_k \|.$$

Moreover, since $F(X, Y)$ is continuously differentiable and $\Omega \times G$ bounded, there exists $\alpha_0 > 0$ such that for all k and $X_k(\alpha) \in \Omega$ we have $Y_k(\alpha) \in G$, $\alpha \in [0, \alpha_0]$

We now use the definition of $d_\varepsilon(X, Y)$, the equality $\| X \|^2 = (X, X)$ and formulas (6.9), (6.10), (6.19), (6.20), (6.7) and (6.8), to obtain

$$\begin{aligned} d_\varepsilon(X_{k_s}(\alpha), Y_{k_s}(\alpha)) \leqslant \\ \leqslant d_\varepsilon(X_{k_s}, Y_{k_s}) + \alpha\Big[-\Big(g_{k_s}, \frac{\partial^2 F(X_{k_s}, Y_{k_s})}{\partial X^2} g_{k_s}\Big) - \\ -\Big(g_{k_s}, \frac{\partial^2 F(X_{k_s}, Y_{k_s})}{\partial X \, \partial Y} q_{k_s}\Big) + \Big(q_{k_s}, \frac{\partial^2 F(X_{k_s}, Y_{k_s})}{\partial Y \, \partial X} g_{k_s}\Big) + \\ + \Big(q_{k_s}, \frac{\partial^2 F(X_{k_s}, Y_{k_s})}{\partial Y^2} q_{k_s}\Big)\Big] + o_{k_s}(\alpha) \leqslant d_\varepsilon(X_{k_s}, Y_{k_s}) - \\ - \alpha\left[m_1 d^2_{1\varepsilon}(X_{k_s}, Y_{k_s}) + m_2 d^2_{2\varepsilon}(X_{k_s}, Y_{k_s})\right] + o_{k_s}(\alpha), \end{aligned}$$

where $\frac{o_{k_s}(\alpha)}{\alpha} \xrightarrow[\alpha \to 0]{} 0$ uniformly in k_s.

Since by assumption $d_{1\varepsilon}(X_{k_s}, Y_{k_s}) \geqslant a$, it follows that there exists a number $\alpha_1 \in (0, \alpha_0]$, independent of k_s, such that for all $\alpha \in [0, \alpha_1]$

$$d_\varepsilon(X_{k_s}(\alpha), Y_{k_s}(\alpha)) \leqslant d_\varepsilon(X_{k_s}, Y_{k_s}) - \frac{1}{2} m_1 a^2 \alpha.$$

Fix any $\alpha' \in (0, \alpha_1]$. For all k_s, we have

$$\begin{aligned} d_\varepsilon(X_{k_s+1}, Y_{k_s+1}) \leqslant d_\varepsilon(X_{k_s}(\alpha'), Y_{k_s}(\alpha')) \leqslant \\ \leqslant d_\varepsilon(X_{k_s}, Y_{k_s}) - \frac{1}{2} m_1 a^2 \alpha', \end{aligned}$$

which contradicts (6.17). We have thus proved that

$$d_{\varepsilon}(X_{k_s}, Y_{k_s}) \xrightarrow[k_s \to \infty]{} 0. \tag{6.21}$$

Now, for sufficiently large k_s, we have

$$Q_{1\varepsilon}(X_{k_s}) \subset Q_{1\varepsilon}(X^*), \quad Q_{2\varepsilon}(Y_{k_s}) \subset Q_{2\varepsilon}(Y^*),$$

and so it follows from (6.21) that

$$d_{\varepsilon}(X^*, Y^*) = 0,$$

i. e., $[X^*, Y^*]$ is an ε-saddle point of $F(X, Y)$ on the set $\Omega \times G$. This completes the proof.

Remark. The fact that $\varepsilon > 0$ was essential for the proof of Theorem 6.2, since only then can one find $\alpha_0 > 0$ (and hence also $\alpha_1 > 0$) which is independent of k_s.

4. We now present a successive approximation procedure for finding saddle points of $F(X, Y)$ on $\Omega \times G$.

We fix $\varepsilon_1 > 0$, $a > 0$, and select an initial approximation $[X_0, Y_0] \in \Omega \times G$. Using finitely many iterations of the method of subsection 3, with $\varepsilon = \varepsilon_1$, we determine a point $[X_1, Y_1] \in \Omega \times G$ such that $d_{\varepsilon_1}(X_1, Y_1) \leqslant a\varepsilon_1$; that this is indeed possible in finitely many steps follows from (6.18). Set $\varepsilon_2 = \frac{1}{2}\varepsilon_1$, and again use the method of subsection 3 with initial approximation $[X_1, Y_1]$ and $\varepsilon = \varepsilon_2$. Finitely many iterations yield a point $[X_2, X_2] \in \Omega \times G$ for which

$$d_{2\varepsilon}(X_2, Y_2) \leqslant a\varepsilon_2.$$

The sequel is analogous.

The sum result is a sequence

$$\{[X_k, Y_k]\}, \quad [X_k, Y_k] \in \Omega \times G,$$
$$d_{\varepsilon_k}(X_k, Y_k) \leqslant a\varepsilon_k, \quad \varepsilon_k = 2^{1-k}\varepsilon_1, \quad k = 1, 2, \ldots \tag{6.22}$$

Theorem 6.3. Any limit point of the sequence $\{[X_k, Y_k]\}$ is a saddle point of $F(X, Y)$ on $\Omega \times G$.

Proof. The existence of at least one limit point is obvious. Let

$$[X_{k_s}, Y_{k_s}] \xrightarrow[k_s \to \infty]{} [X^*, Y^*] \in \Omega \times G.$$

We claim that

$$d(X^*, Y^*) \overset{\mathrm{df}}{=} \frac{1}{2}\left(d_1^2(X^*, Y^*) + d_2^2(X^*, Y^*)\right) = 0.$$

If not, let $d(X^*, Y^*) = a > 0$. For sufficiently large k_s, we have

$$Q_{1\varepsilon_{k_s}}(X_{k_s}) \subset Q_1(X^*), \quad Q_{2\varepsilon_{k_s}}(Y_{k_s}) \subset Q_2(Y^*),$$

and so

$$d_{\varepsilon_{k_s}}(X_{k_s}, Y_{k_s}) \geqslant \frac{a}{2},$$

which contradicts (6.22), proving the theorem.

Remark 1. We have assumed throughout that the function $F(X, Y)$ is strictly convex-concave. If $F(X, Y)$ is only convex-concave, we can replace it by the strictly convex-concave function

$$F_1(X, Y) = F(X, Y) + c(X, X) - d(Y, Y),$$

where $c > 0$ and $d > 0$ are arbitrary numbers. For sufficiently small c and d, the saddle points of $F_1(X, Y)$ lie near those of $F(X, Y)$.

Remark 2. By Lemma 3.2 in Appendix II and the remark following it, any convex-concave function satisfies the inequalities

$$F(X_0, Y_0) + \min_{X \in \Omega} \left(\frac{\partial F(X_0, Y_0)}{\partial X}, X - X_0 \right) \leqslant$$
$$\leqslant \min_{X \in \Omega} \max_{Y \in G} F(X, Y) \leqslant F(X_0, Y_0) + \max_{Y \in G} \left(\frac{\partial F(X_0, Y_0)}{\partial Y}, Y - Y_0 \right).$$

These inequalities enable one to shorten the search for a saddle point (or ε-saddle point) while retaining sufficient accuracy.

Remark 3. It is essential that the functions defining the sets Ω and G be linear. If this is not the case, the sets Ω and G may first be approximated by sets of the proper type, after which the above methods may be applied.

§7. BEST APPROXIMATION OF FUNCTIONS OF SEVERAL VARIABLES BY GENERALIZED POLYNOMIALS

1. Let $G \subset E_m$ be a bounded closed set.

Definition. A system of functions

$$u_0(Y),\ u_1(Y),\ \ldots,\ u_n(Y) \tag{7.1}$$

continuous on G is said to be linearly independent on G when the generalized polynomial

$$\Phi_n(A, Y) = \sum_{k=0}^{n} a_k u_k(Y), \text{ where } A = (a_0, a_1, \ldots, a_n),$$

vanishes identically on G if and only if

$$A = \mathbf{0}.$$

We shall assume henceforth that the system (7.1) is linearly independent on G.

Let $f(Y)$ be continuous on G, and set

$$\rho = \inf_{A \in \Omega} \max_{Y \in G} |f(Y) - \Phi_n(A, Y)|, \tag{7.2}$$

where $\Omega \subset E_{n+1}$ is a closed convex set.

A generalized polynomial $\Phi_n(A^*, Y)$, $A^* \in \Omega$, is called a polynomial of best approximation for $f(Y)$ on G under the constraint Ω (or simply a polynomial of best approximation), if

$$\max_{Y \in G} |f(Y) - \Phi_n(A^*, Y)| = \rho. \tag{7.3}$$

The set of vectors $A^* \in \Omega$ satisfying (7.3) is precisely the set of solutions of the following minimax problem:

$$\max_{Y \in G} (f(Y) - \Phi_n(A, Y))^2 \to \inf_{A \in \Omega}. \tag{7.4}$$

Indeed, let $\Phi_n(A^*, Y)$, $A^* \in \Omega$ be a polynomial of best approximation. Then for all $A \in \Omega$

$$\max_{Y \in G} |\Delta_n(A^*, Y)| \leqslant \max_{Y \in G} |\Delta_n(A, Y)|, \tag{7.5}$$

where $\Delta_n(A, Y) = f(Y) - \Phi_n(A, Y)$. Squaring inequality (7.5) and using the equality (see Appendix III, §2)

$$(\max_{Y \in G} |\Delta_n(A, Y)|)^2 = \max_{Y \in G} (\Delta_n(A, Y))^2,$$

we get

$$\max_{Y \in G} (\Delta_n(A^*, Y))^2 \leqslant \max_{Y \in G} (\Delta_n(A, Y))^2.$$

Thus A^* is a solution of problem (7.4). One proves in analogous fashion that every solution of problem (7.4) yields the coefficient vector of a polynomial of best approximation.

We have thus proved that the minimax problems (7.2) and (7.4) are equivalent in the sense that their solution sets coincide, and that

$$\rho^2 = \inf_{A \in \Omega} \max_{Y \in G} (f(Y) - \Phi_n(A, Y))^2.$$

In the sequel, the term "polynomial of best approximation" will be applied to a generalized polynomial $\Phi_n(A^*, Y)$, $A^* \in \Omega$, for which

$$\max_{Y \in G} (f(Y) - \Phi_n(A^*, Y))^2 = \rho^2.$$

We introduce the notation

$$F(A, Y) = (\Delta_n(A, Y))^2 = (f(Y) - \Phi_n(A, Y))^2,$$
$$\varphi(A) = \max_{Y \in G} F(A, Y).$$

Theorem 7.1. There exists a generalized polynomial of best approximation.

Proof. Let

$$A_k \stackrel{df}{=} (a_0^{(k)}, a_1^{(k)}, \ldots, a_n^{(k)}) \in \Omega, \quad k = 1, 2, \ldots,$$

be a sequence such that

$$\varphi(A_k) \to \rho^2. \tag{7.6}$$

To prove the theorem, it will suffice to show that this sequence $\{A_k\}$ is bounded. Suppose, on the contrary, that the sequence

$$\tau_k = \sum_{i=0}^{n} |a_i^{(k)}|, \quad k = 1, 2, \ldots,$$

contains a subsequence which goes to infinity. We may assume without loss of generality that the entire sequence $\{\tau_k\}$ is divergent:

$$\tau_k \xrightarrow[k \to \infty]{} \infty. \tag{7.7}$$

By (7.6), we have

$$\max_{Y \in G} \left| \sum_{i=0}^{n} a_i^{(k)} u_i(Y) \right| \leqslant c, \tag{7.8}$$

uniformly in k, where c is a constant.

Set

$$\bar{A}_k = (\bar{a}_0^{(k)}, \bar{a}_1^{(k)}, \ldots, \bar{a}_n^{(k)}),$$

where $\bar{a}_i^{(k)} = \dfrac{a_i^{(k)}}{\tau_k}$, $i \in [0:n]$. Then

$$\sum_{i=0}^{n} |\bar{a}_i^{(k)}| = 1, \quad k = 1, 2, \ldots, \tag{7.9}$$

and by (7.8)

$$\max_{Y \in G} \left| \sum_{i=0}^{n} \bar{a}_i^{(k)} u_i(Y) \right| \leqslant \frac{c}{\tau_k}. \tag{7.10}$$

Since the sequence $\{\bar{A}_k\}$ is bounded, it contains a convergent subsequence. We may assume again that the entire sequence $\{\bar{A}_k\}$ converges to a point $\bar{A} = (\bar{a}_0, \bar{a}_1, \ldots, \bar{a}_n)$.

Letting $k \to \infty$ in (7.9) and (7.10) and using (7.7), we obtain

$$\sum_{i=0}^{k} |\bar{a}_i| = 1,$$

$$\sum_{i=0}^{k} \bar{a}_i u_i(Y) = 0 \quad \text{for all} \quad Y \in G.$$

But this contradicts the assumption that the functions $u_0(Y)$, $u_1(Y)$, ..., $u_n(Y)$ are linearly independent on G.

Remark. It is evident from the proof that for any $A_0 \in E_{n+1}$ the set $\{A \mid \varphi(A) \leqslant \varphi(A_0)\}$ is bounded. The same is true, a fortiori, of the set

$$M(A_0) = \{A \in \Omega \mid \varphi(A) \leqslant \varphi(A_0)\}$$

where $A_0 \in \Omega$.

2. The function $F(A, Y)$ is convex in A on E_{n+1} for any $Y \in G$. Indeed, let $\alpha \in [0, 1]$. Then

$$\begin{aligned} F(\alpha A_1 + (1-\alpha) A_2, Y) \leqslant (\alpha \Delta_n(A_1, Y) + (1-\alpha)\Delta_n(A_2, Y))^2 + \\ + \alpha(1-\alpha)(\Delta_n(A_1, Y) - \Delta_n(A_2, Y))^2 = \\ = \alpha(\Delta_n(A_1, Y))^2 + (1-\alpha)(\Delta_n(A_2, Y))^2 = \\ = \alpha F(A_1, Y) + (1-\alpha) F(A_2, Y). \end{aligned}$$

For $i \in [0:n]$,

$$\left(\frac{\partial F(A, Y)}{\partial A}\right)_i = -2\Delta_n(A, Y)\, u_i(Y). \qquad (7.11)$$

We set

$$R(A) = \{Y \in G \mid F(A, Y) = \varphi(A)\}.$$

It is obvious that

$$R(A) = \left\{Y \in G \,\middle|\, |\Delta_n(A, Y)| = \max_{V \in G} |\Delta_n(A, V)|\right\}.$$

Theorem 7.2. $\Phi_n(A^*, Y)$, $A^* \in \Omega$, is a polynomial of best approximation for $f(Y)$ on the set G under the constraints Ω if and only if

$$\inf_{A \in \Omega} \max_{Y \in R(A^*)} \Delta_n(A^*, Y)[\Phi_n(A^*, Y) - \Phi_n(A, Y)] = 0. \qquad (7.12)$$

Proof. The theorem is an obvious corollary of Theorem 2.2. We need only observe that, by (7.11),

$$\left(\frac{\partial F(A^*, Y)}{\partial A}, A - A^*\right) = 2\Delta_n(A^*, Y)[\Phi_n(A^*, Y) - \Phi_n(A, Y)].$$

If $\Omega = E_{n+1}$, formula (7.12) may be replaced by the simpler equality

$$\inf_{A \in E_{n+1}} \max_{Y \in R(A^*)} \Delta_n(A^*, Y)\Phi_n(A, Y) = 0.$$

Theorem 7.3. $\Phi_n(A^*, Y)$, $A^* \in \Omega$, is a polynomial of best approximation for $f(Y)$ on G under the constraints Ω if and only if there exist r distinct *points* $Y_1, \ldots, Y_r$ *in* $R(A^*)$ *and* r *nonnegative numbers* $\alpha_1, \ldots, \alpha_r$, $\sum_{i=1}^{r} \alpha_i = 1$, $1 \leqslant r \leqslant n+2$, *such that*

$$\inf_{A \in \Omega} \sum_{i=1}^{r} \alpha_i \Delta_n (A^*, Y_i) [\Phi_n (A^*, Y_i) - \Phi_n (A, Y_i)] = 0. \tag{7.13}$$

If $\Omega = E_{n+1}$, *formula (7.13) may be replaced by*

$$\sum_{i=1}^{r} \alpha_i \Delta_n (A^*, Y_i) \begin{pmatrix} u_0(Y_i) \\ u_1(Y_i) \\ \cdot \\ \cdot \\ \cdot \\ u_n(Y_i) \end{pmatrix} = \mathbf{0}.$$

This is a corollary of Theorem 3.2.

Theorem 7.4. There exists a finite set

$$G_r = \{Y_i \in G \mid i \in [1:r],\ 1 \leqslant r \leqslant n+2\},$$

such that the problem of best approximation of $f(Y)$ *on* G *under the constraints* Ω *is equivalent to best approximation of* $f(Y)$ *on* G_r *under the constraints* Ω.

This theorem follows from Theorem 3.3.

3. The polynomial of best approximation may be determined by the grid method. For this purpose, we define on G a sequence of grids G_N which is dense in G as $N \to \infty$ (see §4). Let $\Phi_n(A_N^*, Y)$, $A_N^* \in \Omega$, be a polynomial of best approximation for $f(Y)$ on G_N under the constraints Ω.

Theorem 7.5. Every limit point of the sequence $\{A_N^*\}$ *is the coefficient vector of a polynomial of best approximation for* $f(Y)$ *on* G *under the constraints* Ω.

We shall need two lemmas.

Lemma 7.1. There exists N_0 *such that for* $N > N_0$ *the system of functions*

$$u_0(Y),\ u_1(Y),\ \ldots,\ u_n(Y)$$

is linearly independent on G_N.

Proof. Suppose the contrary. Then for some sequence N_s, $s = 1, 2, \ldots$, we have

$$\sum_{k=0}^{n} a_k^{(s)} u_k(Y) = 0; \quad Y \in G_{N_s}; \quad \sum_{k=0}^{n} |a_k^{(s)}| > 0. \tag{7.14}$$

Let $a_{k_s}^{(s)}$ be the coefficient of largest absolute value among $a_k^{(s)}$, $k \in [0:n]$. Setting

$$\bar{a}_k^{(s)} = \frac{a_k^{(s)}}{a_{k_s}^{(s)}},$$

we obtain a system equivalent to (7.14):

$$\sum_{k=0}^{n} \bar{a}_k^{(s)} u_k(Y) = 0, \quad Y \in G_{N_s}. \tag{7.15}$$

We have $|\bar{a}_k^{(s)}| \leqslant 1$ and one of the coefficients $\bar{a}_k^{(s)}$, $k \in [0:n]$, is unity. Extracting a subsequence if necessary, we may assume that $\bar{a}_{k_0}^{(s)} = 1$ for some k_0 and all s, and that the sequence $\{\bar{a}_k^{(s)}\}$ converges as $s \to \infty$:

$$\bar{a}_k^{(s)} \to \bar{a}_k, \quad k \in [0:n].$$

Then, using the fact that the functions $u_k(Y)$ are continuous and the grids G_N dense in G as $N \to \infty$, we see by (7.15) that for all $Y \in G$

$$u_{k_0}(Y) + \sum_{\substack{k=0 \\ k \neq k_0}}^{n} \bar{a}_k u_k(Y) = 0,$$

contrary to our assumption that system (7.1) is linearly independent on G.

Lemma 7.2. If the sequence of vectors $\{A_N\}$ is such that

$$\max_{Y \in G_N} |\Phi_n(A_N, Y)| \leqslant c$$

uniformly in N, then the sequence $\{A_N\}$ is bounded.

Proof. If this is false, then, as in the proof of Theorem 7.1, we can find a sequence of vectors $\bar{A}_N = (\bar{a}_0^{(N)}, \bar{a}_1^{(N)}, \ldots, \bar{a}_n^{(N)})$, $N = 1, 2, \ldots$, with the properties:

I. $$\sum_{i=0}^{n} |\bar{a}_i^{(N)}| = 1, \quad N = 1, 2, \ldots$$

II. $$\max_{Y \in G_N} \left| \sum_{i=0}^{n} \bar{a}_i^{(N)} u_i(Y) \right| \leqslant \frac{c}{\tau_N},$$

where $\tau_N \xrightarrow[N \to \infty]{} \infty$.

III. $$\bar{A}_N \xrightarrow[N \to \infty]{} \bar{A} \overset{df}{=} (\bar{a}_0, \bar{a}_1, \ldots, \bar{a}_n).$$

Hence it follows that for all $Y \in G$

$$\sum_{i=0}^{n} \bar{a}_i u_i(Y) = 0,$$

where $\sum_{i=0}^{n} |\bar{a}_i| = 1$, but this contradicts the assumption that system (7.1) is linearly independent on G.

Proof of Theorem 7.5. In view of Lemma 7.1, it follows from Theorem 7.1 that for sufficiently large $N > N_0$ there exists a polynomial of best approximation $\Phi_n(A_N^*, Y)$, $A_N^* \in \Omega$, for $f(Y)$ on the grid G_N under the constraints Ω. Now it is readily seen that

$$\max_{Y \in G_N} |\Phi_N(A_N^*, Y)| \leqslant \bar{c},$$

uniformly in $N > N_0$, where $\bar{c}$ is a constant.

Indeed,

$$\max_{Y \in G_N} |\Phi_N(A_N^*, Y)| \leqslant$$
$$\leqslant \max_{Y \in G_N} |f(Y) - \Phi_n(A_N^*, Y)| + \max_{Y \in G} |f(Y)| =$$
$$= \min_{\{A\}} \max_{Y \in G_N} |f(Y) - \Phi_n(A, Y)| + \max_{Y \in G} |f(Y)| \leqslant$$
$$\leqslant \rho + \max_{Y \in G} |f(Y)| \stackrel{df}{=} \bar{c}.$$

In view of Lemma 7.2, we conclude that the sequence $\{A_N^*\}$ is bounded, and therefore has at least one limit point.

We now need only refer to Theorem 4.1 and observe that, since $F(A, Y)$ is convex in A for every fixed $Y \in G$, any stationary point of $\varphi(A)$ on Ω yields the coefficient vector of a polynomial of best approximation for $f(Y)$ on G under the constraints Ω.

§8. BEST APPROXIMATION OF FUNCTIONS BY ALGEBRAIC POLYNOMIALS ON AN INTERVAL

1. We return once more to the problem of best approximation for functions defined on a closed interval, using algebraic polynomials; Chebyshev's theorem will follow as a corollary from the other results of this chapter.

Let

$$\rho = \min_{\{A\}} \max_{c \leqslant t \leqslant d} |f(t) - P_n(A, t)|, \tag{8.1}$$

where $f(t)$ is continuous on $[c, d]$,

$$A = (a_0, a_1, \ldots, a_n) \in E_{n+1}$$

and

$$P_n(A, t) = \sum_{i=0}^{n} a_i t^i.$$

We shall assume throughout that $\rho > 0$, so that for any $A \in E_{n+1}$

$$\max_{c \leqslant t \leqslant d} |f(t) - P_n(A, t)| > 0.$$

Set $Y = (\xi, t) \in E_2$,

$$F(A, Y) = \xi(-f(t) + P_n(A, t)),$$
$$G = \{(\xi, t) \mid \xi = \pm 1, \ t \in [c, d]\}.$$

Clearly, for any $A \in E_{n+1}$,

$$\max_{Y \in G} F(A, Y) = \max_{c \leqslant t \leqslant d} |f(t) - P_n(A, t)|. \tag{8.2}$$

Thus,

$$\rho = \min_{\{A\}} \max_{Y \in G} F(A, Y). \tag{8.3}$$

The minimax problem (8.3) is equivalent to (8.1) and more easily handled.

Since

$$\frac{\partial F(A, Y)}{\partial A} = \xi(1, t, \ldots, t^n), \tag{8.4}$$

the function $F(A, Y)$ is continuous together with $\partial F(A, Y)/\partial A$ on the set $E_{n+1} \times G$. In addition, $F(A, Y)$ is convex (even linear) in A for every fixed $Y \in G$.

We set

$$Z(Y) = \xi(1, t, \ldots, t^n),$$
$$R(A) = \{Y \in G \mid F(A, Y) = \max_{W \in G} F(A, W)\}.$$

Let $L(A)$ denote the convex hull of the points $Z(Y)$, $Y \in R(A)$.

Theorem 8.1. *A^* is the coefficient vector of a polynomial of best approximation for $f(t)$ on $[c, d]$ if and only if*

$$0 \in L(A^*).$$

The truth of this theorem follows in an obvious manner from Theorem 2.2, the corollary to Theorem 3.1, and formula (8.4).

2. Lemma 8.1. *Let*

$$Y_i = (\xi_i, t_i) \in G, \quad i \in [0 : n+1],$$

be $n+2$ given points such that $\xi_i = \pm 1$ and

$$t_0 < t_1 < \ldots < t_{n+1}.$$

The following two conditions are equivalent:

I. $\operatorname{sign} \xi_i = -\operatorname{sign} \xi_{i+1}$, $i \in [0 : n]$.

II. *The origin lies in the convex hull of the points* $Z(Y_i)$, $i \in [0 : n+1]$.

Proof. We first show that I implies II. It is known (see Appendix I, §1) that there exist positive numbers $\alpha_i > 0$, $\sum_{i=0}^{n+1} \alpha_i = 1$, such that for any algebraic polynomial $P_n(A, t)$ of degree $\leqslant n$

$$\sum_{i=0}^{n+1} (-1)^i \alpha_i P_n(A, t_i) = 0.$$

Using this result together with I, we obtain

$$\sum_{i=0}^{n+1} \alpha_i Z(Y_i) = (\operatorname{sign} \xi_0) \sum_{i=0}^{n+1} (-1)^i \alpha_i \begin{pmatrix} 1 \\ t_i \\ \cdot \\ \cdot \\ t_i^n \end{pmatrix} = 0,$$

which is equivalent to II.

Conversely, suppose there exist nonnegative numbers $\alpha_i \geqslant 0$, $\sum_{i=0}^{n+1} \alpha_i = 1$, such that

$$\sum_{i=0}^{n+1} \alpha_i Z(Y_i) = 0. \tag{8.5}$$

We wish to prove that I is true.

We first observe that all the numbers α_i, $i \in [0:n+1]$, are strictly positive. Indeed, if, say, were to vanish, it would follow from (8.5) that for any algebraic polynomial $P_n(A, t)$ of degree $\leqslant n$

$$\sum_{i=0}^{n} \alpha_i \xi_i P_n(A, t_i) = 0. \tag{8.6}$$

In particular, this would be true for the interpolating polynomial $P_n(A_0, t)$ such that

$$P_n(A_0, t_i) = \alpha_i \xi_i, \quad i \in [0:n].$$

Substituting $P_n(A_0, t_i)$ into (8.6), we obtain

$$\sum_{i=0}^{n} \alpha_i^2 = 0,$$

contradicting the assumption $\sum_{i=0}^{n} \alpha_i = 1$.

Thus $\alpha_i > 0$, $i \in [0:n+1]$.

We can write (8.5) as

$$\sum_{i=0}^{n} \alpha_i \xi_i \begin{pmatrix} 1 \\ t_i \\ \cdot \\ \cdot \\ t_i^n \end{pmatrix} = -\alpha_{n+1}\xi_{n+1} \begin{pmatrix} 1 \\ t_{n+1} \\ \cdot \\ \cdot \\ t_{n+1}^n \end{pmatrix}.$$

Solving by Cramer's rule for $i \in [0:n]$, we get

$$\alpha_i \xi_i = -\alpha_{n+1}\xi_{n+1} \frac{V(t_0, \ldots, t_{i-1}, t_{n+1}, t_{i+1}, \ldots, t_n)}{V(t_0, \ldots, t_n)},$$

where $V(u_0, \ldots, u_n)$ is the Vandermonde determinant.

Using the elementary identity

$$V(t_0, \ldots, t_{i-1}, t_{n+1}, t_{i+1}, \ldots, t_n) = \\ = (-1)^{n-i} V(t_0, \ldots, t_{i-1}, t_{i+1}, \ldots, t_{n+1}),$$

we see that

$$\alpha_i \xi_i = -(-1)^{n-i} \alpha_{n+1} \xi_{n+1} \frac{V(t_0, \ldots, t_{i-1}, t_{i+1}, \ldots, t_{n+1})}{V(t_0, \ldots, t_n)}. \tag{8.7}$$

Since the numbers t_i are arranged in increasing order, the determinants in (8.7) are positive, as are the numbers α_i, $i \in [0 : n+1]$.

Hence it follows that for all $i \in [0 : n]$

$$\xi_i = -(-1)^{n-i} \xi_{n+1}.$$

But this is equivalent to I, and the proof is complete.

We now set

$$\Delta(A, t) = f(t) - P_n(A, t).$$

Chebyshev's theorem. A polynomial $P_n(A^*, t)$ is a polynomial of best approximation for $f(t)$ on $[c, d]$ if and only if there exist $n+2$ points t_i,

$$c \leqslant t_0 < t_1 < \ldots < t_{n+1} \leqslant d,$$

at which the deviation $\Delta(A^*, t)$ attains its maximum absolute value with alternating signs, i.e.,

$$|\Delta(A^*, t_i)| = \max_{c \leqslant t \leqslant d} |\Delta(A^*, t)| \tag{8.8}$$

and

$$\operatorname{sign} \Delta(A^*, t_i) = -\operatorname{sign} \Delta(A^*, t_{i+1}). \tag{8.9}$$

Proof. We shall show that this is a corollary of Theorem 8.1.

Sufficiency. Suppose there exist $n+2$ points t_i, $i \in [0 : n+1]$, with the stipulated properties. Then, setting

$$\xi_i = -\operatorname{sign} \Delta(A^*, t_i),$$

we conclude that the points $Y_i \stackrel{df}{=} (\xi_i, t_i)$ are in $R(A^*)$. Indeed, by (8.2) and (8.8),

$$F(A^*, Y_i) = -\xi_i \Delta(A^*, t_i) = |\Delta(A^*, t_i)| = \\ = \max_{c \leqslant t \leqslant d} |\Delta(A^*, t)| = \max_{Y \in G} F(A^*, Y).$$

It now follows from (8.9) that

$$\operatorname{sign} \xi_i = -\operatorname{sign} \xi_{i+1}.$$

Using Lemma 8.1, we see that the origin is in the convex hull of the points $Z(Y_i)$, $i \in [0:n+1]$, $Y_i \in R(A^*)$, and so it is surely in $L(A^*)$ (see the definition of $L(A)$).

Thus, by Theorem 8.1, A^* is the coefficient vector of a polynomial of best approximation, and sufficiency is proved.

Necessity. Let

$$\mathbf{0} \in L(A^*) \subset E_{n+1}.$$

By Appendix II, Lemma 1.1, there exist r distinct points

$$Z(Y_i), \quad i \in [0:r-1], \quad 1 \leqslant r \leqslant n+2, \text{ where } Y_i \in R(A^*),$$

and nonnegative numbers $\alpha_i \geqslant 0$, $\sum_{i=0}^{r-1} \alpha_i = 1$, such that

$$\sum_{i=0}^{r-1} \alpha_i Z(Y_i) = \mathbf{0}.$$

Now the points $Y_i = (\xi_i, t_i)$ are distinct, and, since $Y_i \in R(A^*)$, their second coordinates t_i, $i \in [0:r-1]$, are also distinct. One now shows as in the proof of Lemma 8.1 that $r = n+2$.

We now know that there exist $n+2$ points Y_i, $i \in [0:n+1]$, with distinct second coordinates, such that

$$\sum_{i=0}^{n+1} \alpha_i Z(Y_i) = \mathbf{0}; \quad \alpha_i > 0, \quad \sum_{i=0}^{n+1} \alpha_i = 1.$$

We may assume that

$$t_0 < t_1 < \ldots < t_{n+1}.$$

Then it follows from Lemma 8.1 that

$$\operatorname{sign} \xi_i = -\operatorname{sign} \xi_{i+1}. \tag{8.10}$$

Since $Y_i \in R(A^*)$, we see in view of (8.2) that

$$\max_{c \leqslant t \leqslant d} |f(t) - P_n(A^*, t)| = \max_{Y \in G} F(A^*, Y) = \\ = F(A^*, Y_i) = -\xi_i (f(t_i) - P_n(A^*, t_i)).$$

Together with (8.10), this clearly implies the validity of (8.8) and (8.9), completing the proof.

3. *de la Vallée-Poussin's theorem. Let t_i be $n+2$ points in the interval $[c, d]$,*

$$c \leqslant t_0 < t_1 < \ldots < t_{n+1} \leqslant d,$$

at which the deviation $\Delta(A_0, t)$ assumes values with alternating signs. Then

$$\rho \geqslant \min_{i \in [0:n+1]} |\Delta(A_0, t)|.$$

Proof. Set

$$\xi_i = -\operatorname{sign} \Delta(A_0, t)$$

and consider the points $Y_i = (\xi_i, t_i)$, $i \in [0:n+1]$. By Lemma 8.1, the origin of the space E_{n+1} is in the convex hull of the points $Z(Y_i)$, $i \in [0:n+1]$. Hence, by (8.4), it follows that

$$\min_{\|g\|=1} \max_{i \in [0:n+1]} \left(\frac{\partial F(A_0, Y_i)}{\partial A}, g \right) \geqslant 0.$$

Using Corollary 2 to Theorem 2.3, we finally obtain

$$\rho \geqslant \min_{i \in [0:n+1]} F(A_0, Y_i) = \min_{i \in [0:n+1]} [-\xi_i \Delta(A_0, t_i)] =$$
$$= \min_{i \in [0:n+1]} |\Delta(A_0, t_i)|.$$

This completes the proof.

Appendix I

ALGEBRAIC INTERPOLATION

§1. DIVIDED DIFFERENCES

Suppose we are given a table of values (ordinates) of some function:

$$y_k = y(t_k), \quad k \in [0 : N].$$

The values of the argument — the abscissas — are assumed to be distinct; no other restrictions are imposed on them.

The first divided differences are defined as the quotients

$$y[t_s, t_{s+1}] = \frac{y_{s+1} - y_s}{t_{s+1} - t_s}, \quad s \in [0 : N-1],$$

The second divided differences are defined by

$$y[t_s, t_{s+1}, t_{s+2}] = \frac{y[t_{s+1}, t_{s+2}] - y[t_s, t_{s+1}]}{t_{s+2} - t_s}, \quad s \in [0 : N-2].$$

In general, the n-th divided differences ($n \leqslant N$) are defined inductively, in terms of the $(n-1)$-th divided differences:

$$y[t_s, t_{s+1}, \ldots, t_{s+n-1}, t_{s+n}] = \\ = \frac{y[t_{s+1}, \ldots, t_{s+n}] - y[t_s, \ldots, t_{s+n-1}]}{t_{s+n} - t_s}, \quad s \in [0 : N-n].$$

The table presented below is a convenient representation of the values of a function together with its divided differences ($N = 3$):

t	y	Divided differences		
		I	II	III
t_0	y_0			
		$y[t_0, t_1]$		
t_1	y_1		$y[t_0, t_1, t_2]$	
		$y[t_1, t_2]$		$y[t_0, t_1, t_2, t_3]$
t_2	y_2		$y[t_1, t_2, t_3]$	
		$y[t_2, t_3]$		
t_3	y_3			

The following representation of the divided differences in terms of the ordinates is well known:

$$y[t_s, t_{s+1}, \ldots, t_{s+n}] = \sum_{k=0}^{n} \frac{y_{s+k}}{(t_{s+k} - t_s)\ldots(\ \ )\ldots(t_{s+k} - t_{s+n})}, \tag{1.1}$$

where $s \in [0 : N - n]$ and

$$(t_{s+k} - t_s)\ldots(\ \ )\ldots(t_{s+k} - t_{s+n}) = \prod_{i=0,\ i \neq k}^{n} (t_{s+k} - t_{s+i}).$$

We list a few of the basic properties of divided differences.

(1) The divided differences are symmetric functions of their variables:

$$y[t_s, \ldots, t_{s+i}, \ldots, t_{s+j}, \ldots, t_{s+n}] = \\ = y[t_s, \ldots, t_{s+j}, \ldots, t_{s+i}, \ldots, t_{s+n}], \\ i, j \in [0 : n], \quad i \neq j.$$

(2) Suppose we are given two tables

$$y_k = y(t_k), \quad z_k = z(t_k), \quad k \in [0 : N].$$

Based on these, we can set up another table

$$W_k = W(t_k) \overset{df}{=} \alpha y(t_k) + \beta z(t_k),$$

where α and β are arbitrary constants. Then

$$W[t_s, t_{s+1}, \ldots, t_{s+n}] = \alpha y[t_s, t_{s+1}, \ldots, t_{s+n}] + \\ + \beta z[t_s, t_{s+1}, \ldots, t_{s+n}].$$

(3) Let $t_0, t_1, \ldots, t_{n+1}$ be arbitrary abscissas,

$$y(t) = t^{n+1} \quad \text{and} \quad y_k = y(t_k), \quad k \in [0 : n+1].$$

Then

$$y[t_0, t_1, \ldots, t_{n+1}] = 1.$$

(4) Under the assumptions of (3), if $y(t) = Q_n(t)$, where $Q_n(t)$ is an arbitrary algebraic polynomial of degree $\leqslant n$ then

$$y[t_0, t_1, \ldots, t_{n+1}] = 0.$$

In view of (1.1), we may rewrite this result as

$$\sum_{k=0}^{n+1} \frac{Q_n(t_k)}{(t_k - t_0)\ldots(\ \ )\ldots(t_k - t_{n+1})} = 0. \tag{1.2}$$

For $k \in [0:n+1]$, we set

$$\alpha_k = \frac{\dfrac{(-1)^k}{(t_k - t_0)\ldots(\ )\ldots(t_k - t_{n+1})}}{\sum\limits_{i=0}^{n+1} \dfrac{(-1)^i}{(t_i - t_0)\ldots(\ )\ldots(t_i - t_{n+1})}}.$$

Clearly, $\sum\limits_{k=0}^{n+1} \alpha_k = 1$. In terms of the numbers α_k, formula (1.2) becomes

$$\sum_{k=0}^{n+1} (-1)^k \alpha_k Q_n(t_k) = 0.$$

If the abscissas t_k are arranged in increasing order,

$$t_0 < t_1 < \ldots < t_{n+1},$$

it is easily seen that the numbers α_k, $k \in [0:n+1]$, are positive:

$$\alpha_k = \frac{\dfrac{1}{|t_k - t_0|\ldots(\ )\ldots|t_k - t_{n+1}|}}{\sum\limits_{i=0}^{n+1} \dfrac{1}{|t_i - t_0|\ldots(\ )\ldots|t_i - t_{n+1}|}} > 0.$$

§2. INTERPOLATING POLYNOMIALS

1. An algebraic polynomial $L_n(t)$ of degree $\leqslant n$,

$$L_n(t) = b_0 + b_1 t + \ldots + b_n t^n,$$

satisfying the conditions

$$L_n(t_k) = y_k, \quad k \in [0:n],$$

is called an *interpolating polynomial* for the table (t_k, y_k), $k \in [0:n]$.

If the abscissas t_k are pairwise distinct, the interpolating polynomial exists and is unique. Indeed,

$$b_i = \frac{\Delta_i}{\Delta}, \quad i \in [0:n],$$

where Δ is the Vandermonde determinant

$$\Delta \stackrel{df}{=} \begin{vmatrix} 1 & t_0 & \ldots & t_0^n \\ 1 & t_1 & \ldots & t_1^n \\ \cdot & \cdot & \cdot & \cdot \\ 1 & t_n & \ldots & t_n^n \end{vmatrix} = \prod_{n \geqslant k > j \geqslant 0} (t_k - t_j),$$

and Δ_i is the determinant obtained from Δ when the i-th column is replaced by the column $\{y_k\}$, $k \in [0:n]$.

Expanding the determinant Δ_i in terms of the elements of its i-th column, we obtain

$$b_i = \frac{\sum_{k=0}^{n} y_k \Delta_{ki}}{\Delta}, \quad i \in [0:n], \tag{2.1}$$

where Δ_{ki} is the cofactor of the element of Δ at the intersection of the k-th row and i-th column. Note that the numbers Δ and Δ_{ki} depend only on the interpolation abscissas $\{t_k\}$.

Let

$$P_n(t) = a_0 + a_1 t + \ldots + a_n t^n$$

and $y_k = P_n(t_k)$, $k \in [0:n]$. Then, since the interpolating polynomial $L_n(t)$ is unique, it must coincide with $P_n(t)$.

In view of (2.1), we obtain

$$a_i = \frac{\sum_{k=0}^{n} P_n(t_k)\, \Delta_{ki}}{\Delta}, \quad i \in [0:n]. \tag{2.2}$$

Hence it follows, in particular, that if $P_n(t)$ vanishes at $n+1$ points it must vanish identically (i. e., all its coefficients vanish).

2. We now let A denote the coefficient vector of the polynomial $P_n(A, t) = \sum_{i=0}^{n} a_i t^i$,

$$A = (a_0, a_1, \ldots, a_n).$$

We shall say that a sequence of vectors $A_s = (a_0^{(s)}, a_1^{(s)}, \ldots, a_n^{(s)})$, $s = 1, 2 \ldots$, c o n v e r g e s to a vector $A^* = (a_0^*, a_1^*, \ldots, a_n^*)$ as $s \to \infty$, writing

$$A_s \xrightarrow[s\to\infty]{} A^*,$$

if each coordinate sequence converges separately (coordinatewise convergence):

$$a_i^{(s)} \xrightarrow[s\to\infty]{} a_i^*, \quad i \in [0:n].$$

Lemma 2.1. Let $P_n(A_s, t)$, $s = 1, 2, \ldots$ *be a sequence of polynomials of degree* $\leqslant n$ *and* $P_n(A^*, t)$ *a polynomial.*

A sufficient condition for

$$A_s \xrightarrow[s\to\infty]{} A^*,$$

is that for some set of pairwise distinct points t_k, $k \in [0:n]$,

$$P_n(A_s, t_k) \xrightarrow[s\to\infty]{} P_n(A^*, t_k).$$

This follows from (2.2).

Lemma 2.2 (compactness lemma). Given a sequence of polynomials $P_n(A_s, t)$, $s=1, 2, \ldots$, *uniformly bounded on a set of* $n+1$ *points* t_k, $k \in [0:n]$:

$$|P_n(A_s, t_k)| \leqslant M.$$

Then there exist a subsequence $\{A_{s_i}\}$ *and a vector* A^* *such that*

$$A_{s_i} \xrightarrow[i\to\infty]{} A^*.$$

Proof. For any $k \in [0:n]$, the number sequences $P_n(A_s, t_k)$, $s=1, 2, \ldots$, are bounded; hence there exists an index sequence $\{s_i\}$ such that for $k \in [0:n]$,

$$P_n(A_{s_i}, t_k) \xrightarrow[i\to\infty]{} y_k^*.$$

Let $P_n(A^*, t)$ be an interpolating polynomial such that

$$P_n(A^*, t_k) = y_k^*, \quad k \in [0:n].$$

We now have

$$P_n(A_{s_i}, t_k) \xrightarrow[i\to\infty]{} P_n(A^*, t_k), \quad k \in [0:n].$$

By Lemma 2.1, this implies that

$$A_{s_i} \xrightarrow[i\to\infty]{} A^*.$$

This proves our lemma.

3. Returning to the interpolation problem

$$L_n(t_k) = y_k, \quad k \in [0:n],$$

we cite Lagrange's formula for the interpolating polynomial:

$$L_n(t) = \sum_{k=0}^{n} y_k l_k(t),$$

where

$$l_k(t) = \frac{(t-t_0)(t-t_1)\ldots(t-t_{k-1})(t-t_{k+1})\ldots(t-t_n)}{(t_k-t_0)(t_k-t_1)\ldots(t_k-t_{k-1})(t_k-t_{k+1})\ldots(t_k-t_n)}.$$

In particular, any algebraic polynomial $P_n(A, t)$ of degree $\leqslant n$ admits a representation

$$P_n(A, t) = \sum_{k=0}^{n} P_n(A, t_k) l_k(t). \tag{2.3}$$

Lemma 2.3. Given a sequence of algebraic polynomials $P_n(A_s, t)$, $s=1, 2, \ldots$, *uniformly bounded on a set of* $n+1$ *points in* $[c, d]$:

$$|P_n(A_s, t_k)| \leqslant M, \quad k \in [0:n], \quad s=1, 2, \ldots$$

Then for any natural number r *the derivatives* $P_n^{(r)}(A_s, t)$ *are bounded on* $[c, d]$ *uniformly in* $s=1, 2, \ldots,$

$$|P_n^{(r)}(A_s, t)| \leqslant M^{(r)}.$$

Proof. By (2.3),

$$|P_n^{(r)}(A_s, t)| = \left|\sum_{k=0}^{n} P_n(A_s, t_k) l_k^{(r)}(t)\right| \leqslant M \sum_{k=0}^{n} |l_k^{(r)}(t)|.$$

The desired conclusion now follows from the fact that the right-hand side of this inequality is independent of s.

Lemma 2.4 (continuity lemma). Let $P_n(A^*, t)$ *and* $P_n(A_s, t)$, $s=1, 2, \ldots,$ *be interpolating polynomials defined by*

$$P_n(A^*, t_k^*) = y_k^*,$$
$$P_n(A_s, t_k^{(s)}) = y_k^{(s)},$$
$$k \in [0:n], \quad s=1, 2, \ldots$$

If

$$t_k^{(s)} \to t_k^*, \quad y_k^{(s)} \to y_k^*, \quad k \in [0:n],$$

as $s \to \infty$, *then* $A_s \to A^*$.

The proof follows directly from (2.3).

4. To end this section, we recall Newton's formula for the interpolating polynomial:

$$L_n(t) = y_0 + \sum_{k=1}^{n} y[t_0, t_1, \ldots, t_k](t-t_0)(t-t_1) \ldots (t-t_{k-1}).$$

Appendix II

CONVEX SETS AND CONVEX FUNCTIONS

§1. CONVEX HULLS. SEPARATION THEOREM

1. As usual, we let E_n denote euclidean n-space, whose elements are vectors $X=(x_1, \ldots, x_n)$. Our notation for the zero of E_n and for an indexed vector will be, respectively,

$$\mathbf{0}=(0, \ldots, 0), \quad X_k=\left(x_1^{(k)}, \ldots, x_n^{(k)}\right).$$

We shall use the notation

$$(X_1, X_2)=\sum_{i=1}^{n} x_i^{(1)} x_i^{(2)}, \qquad \| X \|=\sqrt{(X, X)}.$$

We recall the elementary Cauchy-Bunyakovskii inequality: for any $X_1, X_2 \in E_n$,

$$|(X_1, X_2)| \leqslant \| X_1 \| \cdot \| X_2 \|.$$

By definition, the two limiting relations

$$X_s \xrightarrow[s\to\infty]{} X^* \quad \text{and} \quad \| X_s - X^* \| \xrightarrow[s\to\infty]{} 0$$

are equivalent.

A system of vectors $X_1, \ldots, X_r$ in E_n is said to be l i n e a r l y i n d e p e n d e n t if the equality

$$\sum_{k=1}^{r} \alpha_k X_k = \mathbf{0}$$

is possible if and only if all the coefficients α_k, $k \in [1:r]$, vanish. It is elementary that if $r \geqslant n+1$ any vectors $X_1, \ldots, X_r$ in E_n are linearly dependent, i. e., there exist numbers $\beta_1, \ldots, \beta_r$, $\sum_{k=1}^{r} |\beta_k| > 0$, such that

$$\sum_{k=1}^{r} \beta_k X_k = \mathbf{0}. \tag{1.1}$$

If $r \geqslant n+2$, we may assume that the coefficients satisfy both (1.1) and the additional condition

$$\sum_{k=1}^{r} \beta_k = 0. \tag{1.2}$$

Indeed, considering the vectors

$$\bar{X}_k = (x_1^{(k)}, \ldots, x_n^{(k)}, 1) \in E_{n+1}, \quad k \in [1:r], \quad r \geqslant n+2,$$

we can find numbers β_k, $\sum_{k=1}^{r} |\beta_k| > 0$, such that

$$\sum_{k=1}^{r} \beta_k \bar{X}_k = 0. \tag{1.3}$$

It is easily seen that (1.3) is equivalent to the combination of (1.1) and (1.2).

2. We let $S_\delta(X_0)$ denote the δ-neighborhood of the point X_0:

$$S_\delta(X_0) = \{X \in E_n \mid \| X - X_0 \| < \delta\}, \quad \delta > 0.$$

A set $G \subset E_n$ is said to be open if for any $X_0 \in G$ there exists $\delta > 0$ such that

$$S_\delta(X_0) \subset G.$$

A set $F \subset E_n$ is said to be closed if it follows from

$$X_s \xrightarrow[s\to\infty]{} X^*, \quad X_s \in F, \quad s = 1, 2, \ldots,$$

that $X^* \in F$.

The closure of an arbitrary set $H \subset E_n$ is the set of points X such that

$$X = \lim_{s\to\infty} X_s; \quad X_s \in H, \quad s = 1, 2, \ldots$$

A point X_0 is called an interior point of a set H if for some $\delta > 0$

$$S_\delta(X_0) \subset H.$$

A point X_0 is a boundary point of H if for any $\delta > 0$ its δ-neighborhood $S_\delta(X_0)$ contains at least one point not in H and at least one point in H other than X_0.

A set L is said to be convex if together with any two points X_1, $X_2 \in L$ it contains the entire segment joining them:

$$\alpha X_1 + (1-\alpha) X_2 \in L, \quad 0 \leqslant \alpha \leqslant 1.$$

A closed convex set L is said to be strictly convex if, for any points $X_1, X_2 \in L$ and any $\alpha \in (0, 1)$, the point $\alpha X_1 + (1-\alpha) X_2$ is an interior point of L.

A set H is bounded if there exists a number $M > 0$ such that for all $X \in H$

$$\| X \| \leqslant M.$$

3. The convex hull of an arbitrary set $G \subset E_n$, denoted by co G (or co(G)), is defined by

$$\text{co}\, G = \left\{ X = \sum_{k=1}^{r} \alpha_k X_k \mid X_k \in G;\ \alpha_k \geqslant 0,\ \sum_{k=1}^{r} \alpha_k = 1,\ r = 1, 2, \ldots \right\}. \tag{1.4}$$

Lemma 1.1. Any vector $X \in$ co G may be expressed as a convex combination of at most $n+1$ vectors in G.

Proof. Suppose that some point $X \in$ co G has no representation of type (1.4) with less than $n+2$ terms (after terms with $\alpha_k = 0$ have been discarded), i. e., $r \geqslant n+2$ in any representation

$$X = \sum_{k=1}^{r} \alpha_k X_k;\quad X_k \in G;\quad \alpha_k > 0,\quad \sum_{k=1}^{r} \alpha_k = 1.$$

By the remark in subsection 1, there exist numbers β_k, $\sum_{k=1}^{r} |\beta_k| > 0$, such that

$$\sum_{k=1}^{r} \beta_k X_k = 0,\quad \sum_{k=1}^{r} \beta_k = 0.$$

Set $\varepsilon = \min_{\{k \mid \beta_k > 0\}} \frac{\alpha_k}{\beta_k} > 0$ and $\bar{\alpha}_k = \alpha_k - \varepsilon\beta_k$, $k \in [1:r]$. Obviously,

$$X = \sum_{k=1}^{r} \bar{\alpha}_k X_k,\quad \sum_{k=1}^{r} \bar{\alpha}_k = 1.$$

Considering positive and nonpositive β_k separately, we see that $\bar{\alpha}_k \geqslant 0$, $k \in [1:r]$, and at least one $\bar{\alpha}_k$ must vanish.

Thus X is a convex combination of at most $r-1$ vectors of G. Iterating this procedure sufficiently many times, we finally reach a representation with $r \leqslant n+1$, as required.

Lemma 1.2. If $G \subset E_n$ is a bounded closed set, co G *is a bounded closed convex set.*

Proof. If all the components of the vectors of G are bounded in absolute value by a constant M, all components of the vectors of co G are bounded in absolute value by the same constant M. Indeed, for all $\nu \in [1:n]$ it follows from (1.4) that

$$|x_\nu| = \left| \sum_{k=1}^{r} \alpha_k x_\nu^{(k)} \right| \leqslant \sum_{k=1}^{r} \alpha_k |x_\nu^{(k)}| \leqslant M \sum_{k=1}^{r} \alpha_k = M.$$

Thus co G is bounded.

We now show that co G is convex. Let $X \in$ co G and $Y \in$ co G, with

$$X=\sum_{k=1}^{r_1-1}\alpha_k X_k,\qquad Y=\sum_{k=r_1}^{r_2}\alpha_k X_k;\qquad X_k\in G;$$
$$\sum_{k=1}^{r_1-1}\alpha_k=\sum_{k=r_1}^{r_2}\alpha_k=1;\quad \alpha_k\geqslant 0,\qquad k\in[1:r_2].$$

We shall show that $\alpha X+(1-\alpha)Y\in\operatorname{co}G$ for all $\alpha\in[0,1]$. We have

$$\alpha X+(1-\alpha)Y=\sum_{k=1}^{r_1-1}\alpha\alpha_k X_k+\sum_{k=r_1}^{r_2}(1-\alpha)\alpha_k X_k=\sum_{k=1}^{r_2}\bar{\alpha}_k X_k,$$

where

$$\bar{\alpha}_k=\begin{cases}\alpha\alpha_k, & \text{if}\quad k\in[1:r_1-1],\\ (1-\alpha)\alpha_k, & \text{if}\quad k\in[r_1:r_2].\end{cases}$$

Clearly, $\bar{\alpha}_k\geqslant 0,\ k\in[1:r_2]$, and

$$\sum_{k=1}^{r_2}\bar{\alpha}_k=\alpha\sum_{k=1}^{r_1-1}\alpha_k+(1-\alpha)\sum_{k=r_1}^{r_2}\alpha_k=1.$$

Thus, the point $\alpha X+(1-\alpha)Y$ is in co G for all $\alpha\in[0,1]$, and so co G is convex.

It remains to prove that co G is closed.

Let $X_s\xrightarrow[s\to\infty]{}X^*$, $X_s\in\operatorname{co}G$, $s=1,2,\ldots$. We must show that $X^*\in\operatorname{co}G$. Express each X_s as a convex combination of at most $n+1$ vectors of G:

$$X_s=\sum_{k=1}^{r_s}\alpha_{sk}X_{sk};\quad \alpha_{sk}\geqslant 0,\quad \sum_{k=1}^{r_s}\alpha_{sk}=1; \tag{1.5}$$
$$X_{sk}\in G,\quad k\in[1:r_s],\quad s=1,2,\ldots$$

By adding arbitrary vectors of G with zero coefficients (if necessary), we may assure that $r_s=n+1$, $s=1,2,\ldots$.

Using the fact that G is a bounded closed set and $0\leqslant\alpha_{sk}\leqslant 1$, $k\in[1:n+1]$, $s=1,2,\ldots$, we can find a sequence $\{s_i\}$ such that for all $k\in[1:n+1]$

$$X_{s_ik}\xrightarrow[s_i\to\infty]{}X_{0k},$$
$$\alpha_{s_ik}\xrightarrow[s_i\to\infty]{}\alpha_{0k}.$$

By (1.5), we have

$$X^*=\sum_{k=1}^{n+1}\alpha_{0k}X_{0k},$$

where

$$X_{0k}\in G,\quad \alpha_{0k}\geqslant 0,\quad \sum_{k=1}^{n+1}\alpha_{0k}=1.$$

Thus $X^*\in\operatorname{co}G$ and the lemma is proved.

It is readily shown that *if G is a convex set, then*

$$\text{co}\, G = G.$$

Indeed, by the definition of the convex hull, $G \subset \text{co}\, G$. We need therefore only prove that $\text{co}\, G \subset G$. Let $X \in \text{co}\, G$, $X = \sum_{k=1}^{r} \alpha_k X_k$; $X_k \in G$, $\alpha_k > 0$, $\sum_{k=1}^{r} \alpha_k = 1$. If $r = 1$ or $r = 2$, it is clear that $X \in G$.

In the general case, one uses the definition of a convex set and the relation

$$X = \alpha_r X_r + (1 - \alpha_r) \left\{ \sum_{k=1}^{r-1} \frac{\alpha_k}{1 - \alpha_r} X_k \right\}, \quad \sum_{k=1}^{r-1} \frac{\alpha_k}{1 - \alpha_r} = 1.$$

4. *Lemma 1.3. Let $\Omega \subset E_n$ be a closed convex set. Then there exists a point $X^* \in \Omega$ such that for all $X \in \Omega$*

$$(X, X^*) \geqslant (X^*, X^*).$$

Proof. Consider the function $F(X) = (X, X)$. It is readily shown that $F(X)$ assumes its minimum on Ω at some point $X^* \in \Omega$:

$$\min_{X \in \Omega} F(X) = F(X^*) \geqslant 0. \tag{1.6}$$

We shall show that for any $X \in \Omega$

$$(X, X^*) \geqslant (X^*, X^*). \tag{1.7}$$

Indeed, by (1.6) and the convexity of Ω, for $0 < \alpha < 1$,

$$(X^* + \alpha (X - X^*), X^* + \alpha (X - X^*)) \geqslant (X^*, X^*).$$

Hence

$$2 (X - X^*, X^*) + \alpha (X - X^*, X - X^*) \geqslant 0.$$

It is now easy to see, reasoning indirectly, that

$$(X - X^*, X^*) \geqslant 0.$$

This is equivalent to (1.7).

Remark. By (1.6), the point X^* has the minimum norm with respect to all points $X \in \Omega$. It is therefore known as the point of Ω nearest the origin.

Let Ω_1 and Ω_2 be two convex sets, at least one of which is bounded. Set

$$\rho = \inf_{X \in \Omega_1,\, Y \in \Omega_2} \| X - Y \|.$$

Lemma 1.4. There exist points $X_0 \in \Omega_1$ and $Y_0 \in \Omega_2$ such that

$$\| X_0 - Y_0 \| = \rho.$$

These points have the additional property that for any $X \in \Omega_1$ and $Y \in \Omega_2$

$$(X - Y, X_0 - Y_0) \geqslant (X_0 - Y_0, X_0 - Y_0).$$

Proof. Define a set $\Omega = \Omega_1 - \Omega_2$ by

$$\Omega = \{X - Y \mid X \in \Omega_1, Y \in \Omega_2\}.$$

It is easy to see that Ω is closed and convex. It now remains to use Lemma 1.3 and the subsequent remark.

Theorem 1.1 (separation theorem). Let $\Omega \subset E_n$ be a closed convex set and $X_0 \notin \Omega$. Then there exist a vector $g_0 \in E_n$, $\| g_0 \| = 1$, and a number $a > 0$ such that for any $X \in \Omega$

$$(X - X_0, g_0) \leqslant -a.$$

Proof. Consider the set Ω_0 defined by

$$\Omega_0 = \{X - X_0 \mid X \in \Omega\}.$$

It is easy to see that Ω_0 is closed and convex, and $\mathbf{0} \notin \Omega_0$. By Lemma 1.3, there exists a point $Z^* \in \Omega_0 (Z^* \neq \mathbf{0})$ such that for all $Z \in \Omega_0$

$$(Z, Z^*) \geqslant (Z^*, Z^*).$$

Set $g_0 = -Z^*/\| Z^* \|$, $a = \| Z^* \| > 0$. Then for any $Z \in \Omega_0$

$$(Z, g_0) = -\frac{1}{\|Z^*\|}(Z, Z^*) \leqslant -\frac{1}{\|Z^*\|}(Z^*, Z^*) = -a,$$

which is equivalent to our assertion.

Corollary. If $\Omega \subset E_n$ is a closed convex set and X_0 a boundary point of Ω, there exists a vector $g_0 \in E_n$, $\|g_0\| = 1$, such that

$$(X - X_0, g_0) \leqslant 0$$

for all $X \in \Omega$.

Proof. Since X_0 is a boundary point of Ω, there exists a sequence of points $\{X_i\}$ such that

$$X_i \notin \Omega, \ \| X_0 - X_i \| \xrightarrow[i \to \infty]{} 0.$$

Since $X_i \notin \Omega$, it follows from Theorem 1.1 that for all $X \in \Omega$

$$(X - X_i, g_i) \leqslant -a_i, \tag{1.8}$$

where

$$a_i \overset{df}{=} \min_{Z \in \Omega} \| Z - X_i \| = \| Z_i^* - X_i \| > 0,$$

$$g_i = -\frac{Z_i^* - X_i}{\| Z_i^* - X_i \|}, \qquad \| g_i \| = 1.$$

Now, by the definition of a_i,

$$0 < a_i \leqslant \| X_0 - X_i \|.$$

Hence it follows that

$$a_i \xrightarrow[i \to \infty]{} 0.$$

We may assume without loss of generality that the entire sequence $\{g_i\}$ converges to some vector g_0, $\| g_0 \| = 1$. Letting $i \to \infty$ in (1.8), we see that for all $X \in \Omega$

$$(X - X_0, g_0) \leqslant 0,$$

Q. E. D.

§2. CONVEX CONES

1. A set $\Gamma \subset E_n$ is called a c o n e if together with any vector X it contains all vectors λX, $\lambda \geqslant 0$. Trivial examples of cones are $\{0\}$ and E_n.

Let $\Omega \subset E_n$ be a closed convex set, $X^* \in \Omega$. Consider the cone

$$\Gamma(X^*) = \{V = \lambda (X - X^*) \mid \lambda > 0, X \in \Omega\}.$$

It is immediate that $0 \in \Gamma(X^*)$. The closure of the cone $\Gamma(X^*)$ is known as the c o n e o f f e a s i b l e d i r e c t i o n s of Ω at X^*, denoted by $\bar{\Gamma}(X^*)$. By the definition of the closure (see §1.2), a point V is in $\bar{\Gamma}(X^*)$ if and only if

$$V = \lim_{s \to \infty} \lambda_s (X_s - X^*), \tag{2.1}$$

where $\lambda_s > 0$, $X_s \in \Omega$, $s = 1, 2, \ldots$.

L e m m a 1.2. If Ω is a closed convex set and $X^* \in \Omega$, then $\bar{\Gamma}(X^*)$ is a closed convex cone.

P r o o f. It is evident from (2.1) that $\bar{\Gamma}(X^*)$ is a cone and it is moreover closed. To prove that it is convex, it will suffice to show that $\Gamma(X^*)$ is convex.

Let $V_1, V_2 \in \Gamma(X^*)$, $V_1 = \lambda_1 (X_1 - X^*)$, $V_2 = \lambda_2 (X_2 - X^*)$, where $\lambda_1, \lambda_2 > 0$, $X_1, X_2 \in \Omega$. We claim that for $0 \leqslant \alpha \leqslant 1$

$$\alpha V_1 + (1 - \alpha) V_2 \in \Gamma(X^*).$$

Set $\bar{\lambda} = \alpha\lambda_1 + (1-\alpha)\lambda_2 > 0$,

$$\alpha_0 = \frac{\alpha\lambda_1}{\bar{\lambda}}, \quad 0 \leqslant \alpha_0 \leqslant 1.$$

Since Ω is convex,

$$X_0 \stackrel{df}{=} \alpha_0 X_1 + (1-\alpha_0) X_2 \in \Omega.$$

We now have

$$\begin{aligned} \alpha V_1 + (1-\alpha) V_2 &= \alpha\lambda_1 (X_1 - X^*) + (1-\alpha)\lambda_2 (X_2 - X^*) = \\ &= \bar{\lambda}\,[\alpha_0 (X_1 - X^*) + (1-\alpha_0)(X_2 - X^*)] = \\ &= \bar{\lambda}\,(X_0 - X^*) \in \Gamma(X^*). \end{aligned}$$

This proves the lemma.

2. Let $\Gamma \subset E_n$ be a cone. We define the *dual cone* Γ^+ to be

$$\Gamma^+ = \{W \in E_n \mid (W, V) \geqslant 0 \quad \text{for all} \quad V \in \Gamma\}.$$

Note that if $\Gamma = E_n$ then $\Gamma^+ = \{0\}$. Indeed, $\mathbf{0} \in \Gamma^+$. On the other hand, if $W_0 \in \Gamma^+$, $\| W_0 \| > 0$, we can set $V = -W_0 \in \Gamma$ to obtain $(W_0, V) = -\| W_0 \|^2 < 0$, contradicting our assumption $W_0 \in \Gamma^+$.

As another example, let

$$\Gamma = \{X = (x_1, \ldots, x_n) \mid x_i \geqslant 0, \ i \in [1:\ n]\}.$$

It is easy to see that in this case $\Gamma^+ = \Gamma$.

Lemma 2.2. If $\Gamma \subset E_n$ is a cone, then $\Gamma^+ \subset E_n$ is a closed convex cone.

The proof follows easily from the definition of the dual cone.

Theorem 2.1. Let $\Gamma \subset E_n$ be a closed convex cone, $G \subset E_n$ a bounded closed convex set. Then Γ and G are disjoint,

$$\Gamma \cap G = \varnothing, \tag{2.2}$$

if and only if there is a vector $W_0 \in \Gamma^+$ such that

$$\max_{X \in G} (W_0, X) < 0. \tag{2.3}$$

Proof. Sufficiency. Since $W_0 \in \Gamma^+$, it follows that for any $V \in \Gamma$

$$(W_0, V) \geqslant 0.$$

On the other hand, by (2.3), for any $X \in G$

$$(W_0, X) < 0.$$

Hence Γ and G cannot have common points.

Necessity. Set $\Omega = \Gamma - G$:

$$\Omega = \{V - X \mid V \in \Gamma,\ X \in G\}.$$

It is readily seen that Ω is a closed convex set not containing the origin $\mathbf{0}$. By Lemma 1.3, there is a vector $W_0 \in \Omega$, $W_0 \neq \mathbf{0}$, such that for all $W \in \Omega$

$$(W, W) \geqslant (W_0, W_0), \tag{2.4}$$

$$(W, W_0) \geqslant (W_0, W_0). \tag{2.5}$$

Let $W_0 = V_0 - X_0$, $V_0 \in \Gamma$, $X_0 \in G$. We claim that

$$(W_0, V_0) = 0. \tag{2.6}$$

Since $V_0 \in \Gamma$, we have $\lambda V_0 \in \Gamma$ for all $\lambda > 0$, and so

$$W_\lambda \overset{df}{=} \lambda V_0 - X_0 = (\lambda - 1) V_0 + W_0 \in \Omega.$$

We now have

$$(W_\lambda, W_\lambda) = ((\lambda - 1) V_0 + W_0, (\lambda - 1) V_0 + W_0) =$$
$$= (\lambda - 1)[(\lambda - 1)(V_0, V_0) + 2(W_0, V_0)] + (W_0, W_0).$$

Our assumption that $(W_0, V_0) \neq 0$ implies that for some positive λ (greater than unity if $(W_0, V_0) < 0$, less than unity if $(W_0, V_0) > 0$)

$$(W_\lambda, W_\lambda) < (W_0, W_0).$$

But this contradicts (2.4), proving (2.6).

Now substitute $W = V - X_0$, $V \in \Gamma$ into (2.5):

$$(V - X_0, W_0) \geqslant (W_0, V_0 - X_0).$$

Hence, in view of (2.6), we see that for all $V \in \Gamma$

$$(W_0, V) \geqslant 0,$$

so that $W_0 \in \Gamma^+$. Now, substituting $W = V_0 - X$, $X \in G$, into (2.5) we get

$$(V_0 - X, W_0) \geqslant (W_0, W_0).$$

Using (2.6), we obtain, for all $X \in G$,

$$(W_0, X) \leqslant -(W_0, W_0).$$

In particular,

$$\max_{X \in G} (W_0, X) < 0.$$

Lemma 2.3. If $\Gamma \subset E_n$ is a closed convex cone, then

$$\Gamma^{++} = \Gamma. \tag{2.7}$$

Proof. Let $V_0 \in \Gamma$. Then for any $W \in \Gamma^+$

$$(W, V_0) \geqslant 0.$$

Thus $V_0 \in \Gamma^{++}$ and so $\Gamma \subset \Gamma^{++}$.

The converse inclusion is proved indirectly: suppose that $V_0 \in \Gamma^{++}$ but $V_0 \notin \Gamma$. Denote $G = \{V_0\}$. By Theorem 2.1, there exists a vector $W_0 \in \Gamma^+$ such that $(W_0, V_0) < 0$. Thus $V_0 \notin \Gamma^{++}$, contrary to assumption.

Corollary. If $\Gamma \subset E_n$ is a convex cone and $\overline{\Gamma}$ is its closure, then

$$\Gamma^{++} = \overline{\Gamma}. \tag{2.8}$$

Indeed, $\overline{\Gamma}$ is a closed convex cone, and therefore, by (2.7),

$$\overline{\Gamma}^{++} = \overline{\Gamma}.$$

It remains to observe that $\overline{\Gamma}^{+} = \Gamma^+$.

Let $\Gamma_i \subset E_n$, $i \in [1:s]$ be arbitrary cones and

$$\sum_{i=1}^{s} \Gamma_i \overset{df}{=} \left\{ V = \sum_{i=1}^{s} V_i \,\middle|\, V_i \in \Gamma_i, \quad i \in [1:s] \right\}.$$

We have

$$\left(\sum_{i=1}^{s} \Gamma_i \right)^+ = \bigcap_{i=1}^{s} \Gamma_i^+. \tag{2.9}$$

This equality is easily proved using the definition of the dual cone and the fact that $\mathbf{0} \in \Gamma_i$, $i \in [1:s]$.

It follows from (2.7), (2.8), and (2.9) that if $\Gamma_i \subset E_n$, $i \in [1:s]$ are closed convex cones, then

$$\left(\bigcap_{i=1}^{s} \Gamma_i \right)^+ = \overline{\sum_{i=1}^{s} \Gamma_i^+}.$$

Indeed,

$$\left(\bigcap_{i=1}^{s} \Gamma_i \right)^+ = \left(\bigcap_{i=1}^{s} (\Gamma_i^+)^+ \right)^+ = \left(\sum_{i=1}^{s} \Gamma_i^+ \right)^{++} = \overline{\sum_{i=1}^{s} \Gamma_i^+}.$$

3. Let G be an arbitrary subset of E_n. Let $K(G)$ denote the convex cone spanned by G (known as the convex conical hull of G):

$$K(G) = \left\{ v = \sum_{k=1}^{r} \alpha_k V_k \,\middle|\, V_k \in G,\ \alpha_k \geqslant 0,\ k \in [1:r], \quad r = 1, 2, \ldots \right\}.$$

Lemma 2.4. Any point $V \in K(G)$ may be expressed as

$$V = \sum_{k=1}^{r} \alpha_k V_k; \quad V_k \in G, \quad \alpha_k > 0, \quad k \in [1:r],$$

where $1 \leqslant r \leqslant n$ and the vectors V_k, $k \in [1:r]$, are linearly independent (see § 1.1).

Proof. Let $V \in K(G)$. Of all representations of V in the form (2.10), select that with the smallest r. We claim that then the vectors $V_1, \ldots, V_r$ are linearly independent. Indeed, otherwise there exist numbers β_k, $\sum_{k=1}^{r} |\beta_k| > 0$, such that

$$\sum_{k=1}^{r} \beta_k V_k = 0.$$

Set

$$\varepsilon = \min_{\{k \mid \beta_k > 0\}} \frac{\alpha_k}{\beta_k}, \quad \bar{\alpha}_k = \alpha_k - \varepsilon\beta_k, \quad k \in [1:r].$$

Obviously,

$$V = \sum_{k=1}^{r} \bar{\alpha}_k V_k, \tag{2.11}$$

where the coefficients $\bar{\alpha}_k$ are nonnegative and at least one of them is zero. Discarding terms with $\bar{\alpha}_k = 0$, we get a representation of type (2.10) for V in which the number of terms is less than r. But this contradicts the assumption, and so $V_1, \ldots, V_r$ are linearly independent.

That $1 \leqslant r \leqslant n$ follows from the fact that no system of linearly independent vectors in E_n can contain more than n numbers.

Lemma 2.5. Let $G \subset E_n$ be a bounded closed set whose convex hull does not contain the origin: G: $0 \notin$ co G. Then $K(G)$ is a closed convex cone.

Proof. It is obvious that $K(G)$ is a convex cone, and we need only prove that it is closed.

Let

$$V_i \xrightarrow[i \to \infty]{} V^*; \quad V_i \in K(G), \quad i = 1, 2, \ldots \tag{2.12}$$

We have to show that $V^* \in K(G)$.

It is readily shown that $V \in K(G)$ if and only if $V = \lambda\bar{V}$, where $\lambda \geqslant 0$ and $\bar{V} \in \operatorname{co} G$.

We may thus rewrite (2.12) as

$$\lambda_i \bar{V}_i \xrightarrow[i \to \infty]{} V^*; \quad \lambda_i \geqslant 0, \quad \bar{V}_i \in \operatorname{co} G. \tag{2.13}$$

Since $0 \notin \operatorname{co} G$, there exists $c > 0$ such that

$$\|\bar{V}_i\| \geqslant c$$

uniformly in i. Hence, via (2.13), it follows that the sequence $\{\lambda_i\}$ is bounded.

Now let $\{i_s\}$ be an index sequence such that

$$\lambda_{i_s} \xrightarrow[i_s \to \infty]{} \bar{\lambda}, \tag{2.14}$$

$$\bar{V}_{i_s} \xrightarrow[i_s \to \infty]{} \bar{V}. \tag{2.15}$$

By Lemma 1.2, the set co G is closed, and so $\bar{V} \in \operatorname{co} G$. In view of (2.13), (2.14) and (2.15), we obtain

$$V^* = \bar{\lambda}\bar{V}; \quad \bar{\lambda} \geqslant 0, \quad \bar{V} \in \operatorname{co} G.$$

Thus, $V^* \in K(G)$.

4. The assumption that $0 \notin \operatorname{co} G$ in Lemma 2.5 is essential, as evidenced by the following example.

Let $G = \{X = (x_1, x_2) | (x_1 - 1)^2 + x_2^2 \leqslant 1\}$. Clearly, $0 \in \operatorname{co} G$. The cone $K(G)$ is

$$K(G) = \{0\} \cup \{X = (x_1, x_2) | x_1 > 0\}.$$

It is clear that $K(G)$ is not a closed set.

However, if G is a finite set, the condition $0 \notin \operatorname{co} G$ in Lemma 2.5 may be dropped. Indeed:

Lemma 2.6. If G is a finite set, then K(G) is a closed convex cone.

Proof. Let

$$V_s \xrightarrow[s \to \infty]{} V^*; \quad V_s \in K(G), \quad s = 1, 2, \ldots \tag{2.16}$$

We must show that $V^* \in K(G)$.

By Lemma 2.4, we have

$$V_s = \sum_{k=1}^{r_s} \lambda_{ks} V_{ks}; \quad \lambda_{ks} > 0, \quad V_{ks} \in G,$$

where the vectors $V_{1s}, \ldots, V_{r_s s}$ are linearly independent. Consider all possible sequences $\{V_{ks}\}_{k=1}^{r_s}$, $s = 1, 2, \ldots$; since G is finite, at least one of these sequences must occur infinitely many times. Without loss of generality, we may therefore write

$$V_s = \sum_{k=1}^{r} \lambda_{ks} V'_k, \quad \lambda_{ks} > 0, \quad V'_k \in G. \tag{2.17}$$

It is essential that the vectors $V'_1, \ldots, V'_r$ are linearly independent.

We claim that all the sequences $\{\lambda_{ks}\}$, $k \in [1:r]$, are bounded. If this is not true, there is some index sequence $\{s_i\}$ such that

$$\gamma_{s_i} \overset{df}{=} \sum_{k=1}^{r} |\lambda_{ks_i}| \xrightarrow[s_i \to \infty]{} \infty. \tag{2.18}$$

By (2.17), we have

$$\sum_{k=1}^{r} \lambda'_{ks_i} V'_k = V_{s_i}/\gamma_{s_i}, \tag{2.19}$$

where $\lambda'_{ks_i} = \lambda_{ks_i}/\gamma_{s_i}$. It is clear that for all s_i

$$\sum_{k=1}^{r} |\lambda'_{ks_i}| = 1. \tag{2.20}$$

We may assume without loss of generality that all the sequences $\{\lambda'_{ks_i}\}$, $k \in [1:r]$, are convergent:

$$\lambda'_{ks_i} \xrightarrow[s_i \to \infty]{} \lambda'_{k0}.$$

Letting $s_i \to \infty$ in (2.19) and (2.20), we see from (2.16) and (2.18) that

$$\sum_{k=1}^{r} \lambda'_{k0} V'_k = 0, \quad \sum_{k=1}^{r} |\lambda'_{k0}| = 1,$$

contradicting the linear independence of $V'_1, \ldots, V'_r$. Thus all the sequences $\{\lambda_{ks}\}$, $k \in [1:r]$, in (2.17) are bounded.

Clearly, there exists a sequence $\{\bar{s}_i\}$ such that

$$\lambda_{k\bar{s}_i} \xrightarrow[\bar{s}_i \to \infty]{} \lambda_{k0}, \quad k \in [1:r].$$

Going to the limit in (2.17) with the indices confined to the sequence $\{\bar{s}_i\}$, we get

$$V^* = \sum_{k=1}^{r} \lambda_{k0} V'_k, \quad \lambda_{k0} \geqslant 0, \quad V'_k \in G,$$

whence it follows that $V^* \in K(G)$.

§3. CONVEX FUNCTIONS

1. A function $F(X)$ defined on a convex set $\Omega \subset E_n$ is said to be convex there if for any X_1 and X_2 in Ω and any $\alpha \in [0, 1]$

$$F(\alpha X_1 + (1-\alpha) X_2) \leqslant \alpha F(X_1) + (1-\alpha) F(X_2). \tag{3.1}$$

If inequality (3.1) is strict for $X_1 \neq X_2$ and any $\alpha \in (0, 1)$, the function $F(X)$ is said to be strictly convex on Ω. Inequality (3.1) may be rewritten in the equivalent form

$$F(X_2 + \alpha(X_1 - X_2)) \leqslant F(X_2) + \alpha[F(X_1) - F(X_2)].$$

A function $F(X)$ defined on a convex set $\Omega \subset E_n$ is said to be concave there if for any X_1 and X_2 in Ω and any $\alpha \in [0, 1]$

$$F(\alpha X_1 + (1-\alpha) X_2) \geqslant \alpha F(X_1) + (1-\alpha) F(X_2).$$

It follows from the definitions that $F(X)$ is convex on Ω if and only if $-F(X)$ is concave on Ω.

Examples.

I. The function $F_1(X) = (X, X)$ is convex on E_n. Indeed, for any $\alpha \in [0, 1]$,

$$\begin{aligned} F_1(\alpha X_1 + (1-\alpha) X_2) &\leqslant (\alpha X_1 + (1-\alpha) X_2, \alpha X_1 + (1-\alpha) X_2) + \\ &+ \alpha(1-\alpha)(X_1 - X_2, X_1 - X_2) = \\ &= \alpha(X_1, X_1) + (1-\alpha)(X_2, X_2) = \\ &= \alpha F(X_1) + (1-\alpha) F_1(X_2), \end{aligned}$$

as required. It is readily seen that $F_1(X)$ is strictly convex on E_n.

II. The function $F_2(X) = (A, X)^2$ is convex on E_n. Indeed, for any $\alpha \in [0, 1]$,

$$\begin{aligned} F_2(\alpha X_1 + (1-\alpha) X_2) &\leqslant (A, \alpha X_1 + (1-\alpha) X_2)^2 + \\ &+ \alpha(1-\alpha)(A, X_1 - X_2)^2 = \alpha(A, X_1)^2 + (1-\alpha)(A, X_2)^2 = \\ &= \alpha F_2(X_1) + (1-\alpha) F_2(X_2), \end{aligned}$$

as required.

2. Two properties of convex functions, used in the main text of the book, are worthy of mention.

Lemma 3.1. Let $\Omega \subset E_n$ be a convex set and $F_i(X)$, $i \in [0:N]$, convex functions on Ω. Then the functions

$$\varphi(X) = \max_{i \in [0:N]} F_i(X)$$

and

$$\Phi(X) = \sum_{i=0}^{N} \lambda_i F_i(X),$$

where all λ_i, $i \in [0:N]$, are nonnegative, are convex on Ω.

Proof. We give the proof for $\varphi(X)$. By the definition of a convex function, we have for $X_1, X_2 \in \Omega$ and $\alpha \in [0, 1]$

$$F_i(\alpha X_1 + (1-\alpha) X_2) \leqslant \alpha F_i(X_1) + (1-\alpha) F_i(X_2) \leqslant \\ \leqslant \alpha \max_{i \in [0:N]} F_i(X_1) + (1-\alpha) \max_{i \in [0:N]} F_i(X_2).$$

Hence

$$\max_{i \in [0:N]} F_i(\alpha X_1 + (1-\alpha) X_2) \leqslant \\ \leqslant \alpha \max_{i \in [0:N]} F_i(X_1) + (1-\alpha) \max_{i \in [0:N]} F_i(X_2),$$

or, equivalently,

$$\varphi(\alpha X_1 + (1-\alpha) X_2) \leqslant \alpha\varphi(X_1) + (1-\alpha)\varphi(X_2).$$

This completes the proof.

Lemma 3.2. Let $F(X)$ be a continuously differentiable function on an open set $\Omega' \subset E_n$. If $F(X)$ is also convex on some convex subset Ω of Ω', then for all $X_1, X_2 \in \Omega$

$$\left(\frac{\partial F(X_1)}{\partial X},\ X_2 - X_1\right) \leqslant F(X_2) - F(X_1),$$

where

$$\frac{\partial F(X)}{\partial X} = \left(\frac{\partial F(X)}{\partial x_1}, \ldots, \frac{\partial F(X)}{\partial x_n}\right).$$

Proof. It is readily shown that

$$\left(\frac{\partial F(X_1)}{\partial X},\ X_2 - X_1\right) = \lim_{\alpha \to +0} \frac{F(X_1 + \alpha(X_2 - X_1)) - F(X_1)}{\alpha}. \tag{3.2}$$

Since $F(X)$ is convex on Ω, it follows that for all $0 < \alpha < 1$,

$$F(X_1 + \alpha(X_2 - X_1)) - F(X_1) \leqslant \alpha[F(X_2) - F(X_1)].$$

Dividing both sides of this inequality by $\alpha > 0$ and letting $\alpha \to +0$, we see from (3.2) that

$$\left(\frac{\partial F(X_1)}{\partial X},\ X_2 - X_1\right) \leqslant F(X_2) - F(X_1).$$

Remark. Under the assumptions of the lemma, if $F(X)$ is concave on Ω, then

$$\left(\frac{\partial F(X_1)}{\partial X}, X_2 - X_1\right) \geqslant F(X_2) - F(X_1)$$

for all $X_1, X_2 \in \Omega$.

Appendix III

CONTINUOUS AND CONTINUOUSLY DIFFERENTIABLE FUNCTIONS

§1. CONTINUOUS FUNCTIONS

A function $F(X)$ defined on a set $G \subset E_n$ is said to be continuous at a point $X_0 \in G$ if for any $\varepsilon > 0$ there exists $\delta > 0$ such that for any $X \in G$ with $\|X - X_0\| < \delta$

$$|F(X) - F(X_0)| < \varepsilon.$$

$F(X)$ is said to be continuous on G if it is continuous at every point of the set.

If $G \subset E_n$ is a bounded closed set, any function $F(X)$ continuous on G achieves its minimum and maximum values on G, i. e., there exist points $X_1, X_2 \in G$ such that

$$F(X_1) = \sup_{X \in G} F(X), \qquad F(X_2) = \inf_{X \in G} F(X).$$

A function $F(X)$ continuous on a bounded closed set $G \subset F_n$ is said to be uniformly continuous on G. This means that for any $\varepsilon > 0$ there exists $\delta > 0$ such that for all $X_1, X_2 \in G$ with $\|X_1 - X_2\| < \delta$

$$|F(X_1) - F(X_2)| < \varepsilon.$$

A vector function $\Phi(X) = (F_1(X), \ldots, F_r(X))$ is said to be continuous on a set G if every component $F_i(X)$, $i \in [1:r]$, is continuous on G.

Consider a function $F(X, Y)$ of two variables,

$$X = (x_1, \ldots, x_n) \in \Omega, \qquad Y = (y_1, \ldots, y_m) \in G.$$

We denote

$$\Omega \times G = \{W = (x_1, \ldots, x_n, y_1, \ldots, y_m) \in \\ \in E_{n+m} \mid X = (x_1, \ldots, x_n) \in \Omega, \quad Y = (y_1, \ldots, y_m) \in G\}.$$

We shall say that $F(X, Y)$, $X \in \Omega$, $Y \in G$, is jointly continuous if $F(W)$ is continuous on $\Omega \times G$.

Note that if $\Omega \subset E_n$, $G \subset E_m$ are bounded closed sets, the set $\Omega \times G \subset E_{n+m}$ is also bounded and closed.

§2. SOME EQUALITIES AND INEQUALITIES FOR CONTINUOUS FUNCTIONS

1. In this subsection all functions will be assumed continuous on a bounded closed set $G \subset E_n$.

The following propositions are easily proved.

I.

$$\max_{X \in G} F(X) = -\min_{X \in G}(-F(X)), \tag{2.1}$$

$$\min_{X \in G} F(X) = -\max_{X \in G}(-F(X)). \tag{2.2}$$

Proof. Let $\overline{X} \in G$ be a point at which $F(X)$ assumes its maximum on G, and $X_0 \in G$ a point at which $-F(X)$ assumes its minimum on G. Then

$$\max_{X \in G} F(X) = F(\overline{X}) = -(-F(\overline{X})) \leqslant -\min_{X \in G}(-F(X)). \tag{2.3}$$

On the other hand,

$$-\min_{X \in G}(-F(X)) = -(-F(X_0)) = F(X_0) \leqslant \max_{X \in G} F(X). \tag{2.4}$$

Combining (2.3) and (2.4), we get (2.1).

The proof of (2.2) is analogous.

II.

$$\max_{X \in G} \lambda F(X) = \lambda \max_{X \in G} F(X), \tag{2.5}$$

$$\min_{X \in G} \lambda F(X) = \lambda \min_{X \in G} F(X). \tag{2.6}$$

Here $\lambda \geqslant 0$ *is a constant.*

Proof. For $\lambda \geqslant 0$, $Z \in G$, we have

$$\lambda F(Z) \leqslant \lambda \max_{X \in G} F(X),$$

and so

$$\max_{Z \in G} \lambda F(Z) = \max_{X \in G} \lambda F(X) \leqslant \lambda \max_{X \in G} F(X). \tag{2.7}$$

On the other hand,

$$\lambda \max_{X \in G} F(X) = \lambda F(\overline{X}) \leqslant \max_{X \in G} \lambda F(X). \tag{2.8}$$

Combining (2.7) and (2.8), we get (2.5).

The proof of (2.6) is analogous.

III.

$$\max_{X\in G}[F(X)+C]=\max_{X\in G}F(X)+C,$$
$$\min_{X\in G}[F(X)+C]=\min_{X\in G}F(X)+C.$$

Here C is any constant.

The proof is left to the reader.

IV.

$$|\max_{X\in G}F(X)|\leqslant\max_{X\in G}|F(X)|, \tag{2.9}$$

$$|\min_{X\in G}F(X)|\leqslant\max_{X\in G}|F(X)|. \tag{2.10}$$

Proof. We have

$$|\max_{X\in G}F(X)|=|F(\bar{X})|\leqslant\max_{X\in G}|F(X)|.$$

This proves (2.9). Inequality (2.10) is a consequence of (2.9) and (2.2). Indeed,

$$|\min_{X\in G}F(X)|=|-\max_{X\in G}(-F(X))|\leqslant\max_{X\in G}|-F(X)|=$$
$$=\max_{X\in G}|F(X)|.$$

V.

$$\max_{X\in G}[F_1(X)+F_2(X)]\leqslant\max_{X\in G}F_1(X)+\max_{X\in G}F_2(X), \tag{2.11}$$

$$\max_{X\in G}[F_1(X)+F_2(X)]\geqslant\max_{X\in G}F_1(X)+\max_{X\in Q}F_2(X), \tag{2.12}$$

where

$$Q=\{X\in G\,|\,F_1(X)=\max_{Z\in G}F_1(Z)\}.$$

Proof. Inequality (2.11) is obvious; only (2.12) requires proof. To prove it, we observe that $F_1(X)$ is constant on Q and use III. We obtain

$$\max_{X\in G}[F_1(X)+F_2(X)]\geqslant\max_{X\in Q}[F_1(X)+F_2(X)]=$$
$$=\max_{X\in G}F_1(X)+\max_{X\in Q}F_2(X).$$

VI.

$$\min_{X\in G}[F_1(X)+F_2(X)]\geqslant\min_{X\in G}F_1(X)+\min_{X\in G}F_2(X),$$
$$\min_{X\in G}[F_1(X)+F_2(X)]\leqslant\min_{X\in G}F_1(X)+\min_{X\in Q_1}F_2(X),$$

where $Q_1=\{X\in G\,|\,F_1(X)=\min_{Z\in G}F_1(Z)\}$.

The proof is left to the reader.

VII.

$$|\max_{X\in G} F_1(X) - \max_{X\in G} F_2(X)| \leqslant \max_{X\in G} |F_1(X) - F_2(X)|, \tag{2.13}$$

$$|\min_{X\in G} F_1(X) - \min_{X\in G} F_2(X)| \leqslant \max_{X\in G} |F_1(X) - F_2(X)|. \tag{2.14}$$

Proof. By (2.11), we have

$$\max_{X\in G} F_1(X) \leqslant \max_{X\in G} F_2(X) + \max_{X\in G} |F_1(X) - F_2(X)|,$$
$$\max_{X\in G} F_2(X) \leqslant \max_{X\in G} F_1(X) + \max_{X\in G} |F_2(X) - F_1(X)|.$$

This clearly implies (2.13).

To prove inequality (2.14), we use (2.13) and (2.2).

VIII.

$$|\min_{X\in G} \max_{i\in[0:N]} F_i(X) - \min_{X\in G} \max_{i\in[0:N]} \Phi_i(X)| \leqslant$$
$$\leqslant \max_{X\in G} \max_{i\in[0:N]} |F_i(X) - \Phi_i(X)|.$$

Proof. It is readily seen that the functions

$$\varphi_1(X) = \max_{i\in[0:N]} F_i(X) \quad \text{and} \quad \varphi_2(X) = \max_{i\in[0:N]} \Phi_i(X)$$

are continuous on G. Using (2.14) and (2.13), we get

$$|\min_{X\in G} \max_{i\in[0:N]} F_i(X) - \min_{X\in G} \max_{i\in[0:N]} \Phi_i(X)| \leqslant$$
$$\leqslant \max_{X\in G} |\varphi_1(X) - \varphi_2(X)| \leqslant \max_{X\in G} \max_{i\in[0:N]} |F_i(X) - \Phi_i(X)|,$$

as required.

IX.

$$(\max_{X\in G} |F(X)|)^2 = \max_{X\in G} (F(X))^2, \tag{2.15}$$

$$(\min_{X\in G} |F(X)|)^2 = \min_{X\in G} (F(X))^2. \tag{2.16}$$

Proof. We have

$$(\max_{X\in G} |F(X)|)^2 = (|F(X')|)^2 \leqslant \max_{X\in G} (F(X))^2. \tag{2.17}$$

On the other hand, for all $X\in G$,

$$(F(X))^2 = (|F(X)|)^2 \leqslant (\max_{Z\in G} |F(Z)|)^2.$$

Hence it follows that

$$\max_{X \in G} (F(X))^2 \leqslant (\max_{X \in G} |F(X)|)^2. \tag{2.18}$$

Combining (2.17) and (2.18), we get (2.15). The proof of (2.16) is analogous.

Note, in particular, that by (2.15)

$$\max_{X \in G} |F(X)| = \sqrt{\max_{X \in G} (F(X))^2}.$$

X. *If* $G = \bigcup_{k=1}^{r} G_k$, *then*

$$\max_{X \in G} F(X) = \max_{k \in [1:r]} \sup_{X \in G_k} F(X). \tag{2.19}$$

$$\min_{X \in G} F(X) = \min_{k \in [1:r]} \inf_{X \in G_k} F(X). \tag{2.20}$$

Proof. For all $k \in [1:r]$, we have

$$\max_{X \in G} F(x) \geqslant \sup_{X \in G_k} F(X),$$

and thus

$$\max_{X \in G} F(X) \geqslant \max_{k \in [1:r]} \sup_{X \in G_k} F(X). \tag{2.21}$$

On the other hand, if $F(X)$ achieves its maximum on G at a point $\overline{X}$, the latter must be in one of the sets G_k, say G_1. Therefore,

$$\max_{X \in G} F(X) = F(\overline{X}) \leqslant \sup_{X \in G_1} F(X) \leqslant \max_{k \in [1:r]} \sup_{X \in G_k} F(X). \tag{2.22}$$

Combining (2.21) and (2.22), we get (2.19).

The proof of (2.20) is analogous.

2. Now let $F(t)$ be continuous on an interval $[c, d]$.

XI.

$$\max_{c \leqslant t \leqslant d} F(t) = \max_{-d \leqslant z \leqslant -c} F(-z),$$

$$\max_{c \leqslant t \leqslant d} F(t) = \max_{\alpha \leqslant z \leqslant \beta} F\left(c + \frac{d-c}{\beta - \alpha}(z - \alpha)\right). \tag{2.23}$$

We give the proof for (2.23). Fix some $t_0 \in [c, d]$ and set

$$z_0 = \alpha + \frac{\beta - \alpha}{d - c}(t_0 - c).$$

Clearly, $z_0 \in [\alpha, \beta]$. We now have

$$F(t_0) = F\left(c + \frac{d-c}{\beta-\alpha}(z_0 - \alpha)\right) \leqslant \max_{\alpha \leqslant z \leqslant \beta} F\left(c + \frac{d-c}{\beta-\alpha}(z - \alpha)\right).$$

Since this is true for any $t_0 \in [c, d]$, it follows that

$$\max_{c \leqslant t \leqslant d} F(t) \leqslant \max_{\alpha \leqslant z \leqslant \beta} F\left(c + \frac{d-c}{\beta-\alpha}(z-\alpha)\right). \tag{2.24}$$

To prove the inverse inequality, we let $\bar{z} \in [\alpha, \beta]$. Set

$$\bar{t} = c + \frac{d-c}{\beta-\alpha}(\bar{z}-\alpha).$$

Clearly, $\bar{t} \in [c, d]$. Now

$$F\left(c + \frac{d-c}{\beta-\alpha}(\bar{z}-\alpha)\right) = F(\bar{t}) \leqslant \max_{c \leqslant t \leqslant d} F(t).$$

Since this is true for any $\bar{z} \in [\alpha, \beta]$, it follows that

$$\max_{\alpha \leqslant z \leqslant \beta} F\left(c + \frac{d-c}{\beta-\alpha}(z-\alpha)\right) \leqslant \max_{c \leqslant t \leqslant d} F(t). \tag{2.25}$$

Combining (2.24) and (2.25), we get (2.23).

§3. CONTINUOUSLY DIFFERENTIABLE FUNCTIONS

1. *Lemma 3.1. Let $F(X)$ be continuous, with continuous first partial derivatives, on an open set $\Omega' \subset E_n$. Then for any $X \in \Omega'$ and $g \in E_n$, $\|g\| = 1$, for sufficiently small α,*

$$F(X + \alpha g) = F(X) + \alpha\left(\frac{\partial F(X)}{\partial X}, g\right) + o(X, g; \alpha),$$

where

$$\frac{\partial F(X)}{\partial X} = \left(\frac{\partial F(X)}{\partial x_1}, \ldots, \frac{\partial F(X)}{\partial x_n}\right)$$

and $o(X, g; \alpha)$ has the property that, uniformly in g, $\|g\| = 1$,

$$\frac{o(X, g; \alpha)}{\alpha} \xrightarrow[\alpha \to 0]{} 0.$$

Proof. For fixed X and g, we set

$$h(\alpha) = F(X + \alpha g). \tag{3.1}$$

The function $h(\alpha)$ has a continuous derivative in the neighborhood of $\alpha = 0$, and, as is easily verified,

$$h'(\alpha) = \left(\frac{\partial F(X + \alpha g)}{\partial X}, g\right). \tag{3.2}$$

By elementary calculus,

$$h(\alpha)=h(0)+\int_0^{\alpha} h'(t)\,dt. \tag{3.3}$$

This may also be written

$$h(\alpha)=h(0)+\alpha h'(0)+\int_0^{\alpha} [h'(t)-h'(0)]\,dt. \tag{3.4}$$

Using (3.4), (3.1) and (3.2), we obtain

$$F(X+\alpha g)=F(X)+\alpha\left(\frac{\partial F(X)}{\partial X},\ g\right)+ \\ +\int_0^{\alpha}\left(\frac{\partial F(X+tg)}{\partial X}-\frac{\partial F(X)}{\partial X},\ g\right)dt.$$

We denote

$$o(X,\ g;\ \alpha)=\int_0^{\alpha}\left(\frac{\partial F(X+tg)}{\partial X}-\frac{\partial F(X)}{\partial X},\ g\right)dt.$$

To complete the proof, it is now sufficient to show that

$$\frac{o(X,\ g;\ \alpha)}{\alpha}\xrightarrow[\alpha\to 0]{} 0.$$

uniformly in g, $\|g\|=1$. But this follows directly from the inequality

$$|o(X,\ g;\ \alpha)|\leqslant|\alpha|\max_{t\in[-|\alpha|,\ |\alpha|]}\left\|\frac{\partial F(X+tg)}{\partial X}-\frac{\partial F(X)}{\partial X}\right\|$$

and from the continuity of the vector function $\partial F(X)/\partial X$ in the neighborhood of X.

Remark 1. Let $X\in\Omega'$ and $X+tg\in\Omega'$ for $t\in[0,\ \alpha]$, $\alpha>0$. By the mean-value theorem of integral calculus, it follows from (3.3) that for some τ, $0<\tau<\alpha$,

$$h(\alpha)=h(0)+\alpha h'(\tau). \tag{3.5}$$

In view of (3.1), (3.2) and (3.5), we obtain

$$F(X+\alpha g)=F(X)+\alpha\left(\frac{\partial F(X+\tau g)}{\partial X},\ g\right),$$

where $0<\tau<\alpha$.

Remark 2. If $M\subset\Omega'$ is a bounded closed set, there exists $\alpha_0>0$ such that for all $\alpha\in[0,\ \alpha_0]$

$$F(X+\alpha g)=F(X)+\alpha\left(\frac{\partial F(X)}{\partial X}, g\right)+o(X, g; \alpha),$$

where $\frac{o(X, g; \alpha)}{\alpha} \xrightarrow[\alpha\to 0]{} 0$ uniformly in $X \in M$ and $g \in E_n$, $\| g \| = 1$.

2. Let $F(X)$ be defined on some open set $\Omega' \subset E_n$.

We shall say that $F(X)$ is differentiable at a point $X \in \Omega'$ if there exists a vector $V(X)$ such that for all $g \in E_n$, $\| g \| = 1$,

$$F(X+\alpha g)=F(X)+\alpha(g, V(X))+o(X, g; \alpha), \tag{3.6}$$

where $\frac{o(X, g; \alpha)}{\alpha} \xrightarrow[\alpha\to 0]{} 0$ uniformly in g.

The function $F(X)$ is said to be continuously differentiable on Ω' if it is differentiable at every point of this set and the vector function $V(X)$ figuring in (3.6) is continuous on Ω'.

In Lemma 3.1 we in effect established that if $F(X)$ is continuous and has continuous first partial derivatives on an open set Ω', then it is continuously differentiable on Ω'. The next lemma gives another sufficient condition for continuous differentiability.

Lemma 3.2. Suppose that for all $X \in \Omega'$ *and* $g \in E_n$, $\| g \| = 1$,

$$\frac{\partial F(X)}{\partial g} \overset{\mathrm{df}}{=} \lim_{\alpha\to+0} \frac{F(X+\alpha g)-F(X)}{\alpha}=(g, V(X)), \tag{3.7}$$

where the vector function $V(X)$ *is continuous on* Ω'. *Then* $F(X)$ *is continuously differentiable on* Ω' *and*

$$\frac{\partial F(X)}{\partial X}=V(X).$$

Proof. Fix g, $\| g \| = 1$, and $X_0 \in \Omega'$, and consider the function

$$h(\alpha)=F(X_0+\alpha g). \tag{3.8}$$

Since $X_0 \in \Omega'$, there exists $\delta > 0$ such that

$$X_0+\alpha g \in \Omega' \quad \text{for} \quad \alpha \in (-\delta, \delta).$$

We claim that $h(\alpha)$ has a continuous derivative on $(-\delta, \delta)$. Indeed, by (3.7),

$$\begin{aligned}\lim_{\beta\to+0} \frac{1}{\beta}[h(\alpha+\beta)-h(\alpha)] &= \\ =\lim_{\beta\to+0} \frac{1}{\beta}[F((X_0+\alpha g)+\beta g)-F(X_0+\alpha g)] &= \\ =(g, V(X_0+\alpha g)).\end{aligned} \tag{3.9}$$

On the other hand,

$$\lim_{\beta\to -0}\frac{h(\alpha+\beta)-h(\alpha)}{\beta}=$$
$$=-\lim_{\beta\to -0}\frac{F((X_0+\alpha g)+(-\beta)(-g))-F(X_0+\alpha g)}{-\beta}=$$
$$=-\lim_{\gamma\to +0}\frac{F((X_0+\alpha g)+\gamma(-g))-F(X_0+\alpha g)}{\gamma}=$$
$$=-(-g,\ V(X_0+\alpha g))=(g,\ V(X_0+\alpha g)). \tag{3.10}$$

Thus $h(\alpha)$ has left and right derivatives at every point of $(-\delta,\ \delta)$, and these one-sided derivatives are equal. Hence $h(\alpha)$ has a derivative at every point of $(-\delta,\ \delta)$, and by (3.9) and (3.10),

$$h'(\alpha)=(g,\ V(X_0+\alpha g)). \tag{3.11}$$

Since $V(X)$ is continuous on Ω', we see that $h'(\alpha)$ is continuous on $(-\delta,\ \delta)$.

By elementary calculus,

$$h(\alpha)=h(0)+\int_0^\alpha h'(t)\,dt=$$
$$=h(0)+\alpha h'(0)+\int_0^\alpha [h'(t)-h'(0)]\,dt.$$

By (3.8) and (3.11), this formula may be written

$$F(X_0+\alpha g)=F(X_0)+\alpha(g,\ V(X_0))+o(X_0,\ g;\ \alpha), \tag{3.12}$$

where

$$o(X_0,\ g,\ \alpha)=\int_0^\alpha (g,\ V(X_0+tg)-V(X_0))\,dt.$$

It is readily seen that

$$\frac{o(X_0,\ g;\ \alpha)}{\alpha}\xrightarrow[\alpha\to 0]{}0$$

uniformly in g, $\|g\|=1$. Hence $F(X)$ is differentiable at any point $X_0\in\Omega'$. Using the continuity of the function $V(X)$ and formula (3.12), we conclude that $F(X)$ is continuously differentiable on Ω'.

The formula

$$\frac{\partial F(X_0)}{\partial X}=V(X_0)$$

follows from (3.12), as is seen by equating g to all the directions of the coordinate axes in turn.

3. *Lemma 3.3. Let $F(X)$ be continuous with continuous partial derivatives of order up to and including l on an open set $\Omega' \subset E_n$. Then, for fixed $X = (x_1, \ldots, x_n) \in \Omega'$ and $g = (g_1, \ldots, g_n)$, $\|g\| = 1$, for sufficiently small α,*

$$F(X + \alpha g) = F(X) + \sum_{k=1}^{l} \frac{\alpha^k}{k!} \frac{\partial^k F(X)}{\partial g^k} + o(X, g; \alpha^l),$$

where

$$\frac{\partial^k F(X)}{\partial g^k} = \sum_{j_1, \ldots, j_k = 1}^{n} \frac{\partial^k F(X)}{\partial x_{j_1} \ldots \partial x_{j_k}} g_{j_1} \ldots g_{j_k}$$

and $o(X, g; \alpha^l)$ has the property that, uniformly in g, $\|g\| = 1$,

$$\frac{o(X, g; \alpha^l)}{\alpha^l} \xrightarrow[\alpha \to 0]{} 0.$$

Proof. The function

$$h(\alpha) = F(X + \alpha g) \tag{3.13}$$

has continuous derivatives of order up to and including l in the neighborhood of $\alpha = 0$, and therefore, for sufficiently small α, it satisfies Maclaurin's formula:

$$h(\alpha) = h(0) + \sum_{k=1}^{l-1} \frac{\alpha^k}{k!} h^{(k)}(0) + R_l(\alpha), \tag{3.14}$$

where the remainder term $R_l(\alpha)$ may be written

$$R_l(\alpha) = \frac{\alpha^l}{l!} h^{(l)}(\tau), \qquad 0 < |\tau| < |\alpha|, \tag{3.15}$$

or

$$R_l(\alpha) = \frac{1}{(l-1)!} \int_0^{\alpha} (\alpha - t)^{l-1} h^{(l)}(t)\, dt.$$

Setting

$$R_l(\alpha) = \frac{\alpha^l}{l!} h^{(l)}(0) + o(X, g; \alpha^l),$$

where

$$o(X, g; \alpha^l) = \frac{1}{(l-1)!} \int_0^{\alpha} (\alpha - \tau)^{l-1} \left[h^{(l)}(t) - h^{(l)}(0)\right] dt, \tag{3.16}$$

we infer from (3.14) that

$$h(\alpha)=h(0)+\sum_{k=1}^{l}\frac{\alpha^k}{k!}h^{(k)}(0)+o(X,\ g;\ \alpha^l). \tag{3.17}$$

We now observe that, by (3.13),

$$h'(0)=\sum_{j=1}^{n}\frac{\partial F(X)}{\partial x_j}g_j=\left(\frac{\partial F(X)}{\partial X},\ g\right)\overset{df}{=}\frac{\partial F(X)}{\partial g},$$

$$h''(0)=\sum_{j_1,\ j_2=1}^{n}\frac{\partial^2 F(X)}{\partial x_{j_1}\partial x_{j_2}}g_{j_1}g_{j_2}\overset{df}{=}\frac{\partial^2 F(X)}{\partial g^2},$$

. . . .

$$h^{(l)}(0)=\sum_{j_1,\ \dots,\ j_l=1}^{n}\frac{\partial^l F(x)}{\partial x_{j_1}\dots\partial x_{j_l}}g_{j_1}\dots g_{j_l}\overset{df}{=}\frac{\partial^l F(X)}{\partial g^l}.$$

Substituting these formulas into (3.17), we obtain

$$F(X+\alpha g)=F(X)+\sum_{k=1}^{l}\frac{\alpha^k}{k!}\frac{\partial^k F(X)}{\partial g^k}+o(X,\ g;\ \alpha^l).$$

To complete the proof of the lemma, it remains to show that

$$\frac{o(X,\ g;\ \alpha^l)}{\alpha^l}\xrightarrow[\alpha\to 0]{}0 \tag{3.18}$$

uniformly in g, $\|g\|=1$.

Indeed, by (3.16),

$$|o(X,\ g;\ \alpha^l)|\leqslant\frac{|\alpha|^l}{l!}\max_{t\in[-|\alpha|,\ |\alpha|]}\sum_{j_1,\ \dots,\ j_l=1}^{n}\left|\frac{\partial^l F(X+tg)}{\partial x_{j_1}\dots\partial x_{j_l}}-\frac{\partial^l F(X)}{\partial x_{j_1}\dots\partial x_{j_l}}\right|.$$

Formula (3.18) now follows from the continuity of the l-th derivatives of $F(X)$ on Ω'.

Remark. Let us denote

$$\frac{\partial^2 F(X)}{\partial X^2}=\begin{pmatrix}\frac{\partial^2 F(X)}{\partial x_1\partial x_1} & \dots & \frac{\partial^2 F(X)}{\partial x_1\partial x_n}\\ \dots & \dots & \dots\\ \frac{\partial^2 F(X)}{\partial x_n\partial x_1} & \dots & \frac{\partial^2 F(X)}{\partial x_n\partial x_n}\end{pmatrix}.$$

Then

$$\frac{\partial^2 F(X)}{\partial g^2}\overset{df}{=}\sum_{j_1,\ j_2=1}^{n}\frac{\partial^2 F(X)}{\partial x_{j_1}\partial x_{j_2}}g_{j_1}g_{j_2}=\left(\frac{\partial^2 F(X)}{\partial X^2}g,\ g\right). \tag{3.19}$$

Let $X \in \Omega'$ and $X + \gamma g \in \Omega'$ for $\gamma \in [0, \alpha]$. Setting $l = 2$ in (3.14), we see via (3.15) and (3.19) that

$$F(X + \alpha g) = \\ = F(X) + \alpha \left(\frac{\partial F(X)}{\partial X}, g \right) + \frac{\alpha^2}{2} \left(\frac{\partial^2 F(X + \tau g)}{\partial X^2} g, g \right),$$

where $0 < \tau < \alpha$.

Appendix IV

DETERMINATION OF THE POINT NEAREST THE ORIGIN ON A POLYHEDRON. ITERATIVE METHODS

1. Consider some finite set of points $H=\{Z_i\}_{i=1}^{s}$ in euclidean n-space E_n. Let L denote the convex hull of H:

$$L=\left\{Z=\sum_{i=1}^{s}\alpha_i Z_i \,\middle|\, \alpha_i\geqslant 0,\ \sum_{i=1}^{s}\alpha_i=1\right\}.$$

L is of course a bounded closed convex set.

Let Z^* denote the point of L nearest the origin:

$$(Z^*, Z^*)=\min_{Z\in L}(Z, Z).$$

It is readily shown that $Z^*\in L$ exists and is unique. In addition, for any $Z\in L$ (see Appendix II, Lemma 1.3)

$$(Z, Z^*)\geqslant(Z^*, Z^*). \tag{1}$$

We are going to describe two successive approximation procedures to determine the point Z^*.

2. *Lemma 1. For any* V, $Z\in L$,

$$(V, Z)\geqslant\min_{i\in[1:s]}(Z_i, Z). \tag{2}$$

Proof. Let

$$V=\sum_{i=1}^{s}\alpha_i' Z_i;\quad \alpha_i'\geqslant 0,\quad \sum_{i=1}^{s}\alpha_i'=1.$$

Then

$$(V, Z)=\sum_{i=1}^{s}\alpha_i'(Z_i, Z)\geqslant\left[\min_{i\in[1:s]}(Z_i, Z)\right]\sum_{i=1}^{s}\alpha_i'=\min_{i\in[1:s]}(Z_i, Z).$$

This proves the lemma.

For any point $Z \in L$, we define $\delta(Z)$:

$$\delta(Z) = (Z, Z) - \min_{i \in [1:s]} (Z_i, Z).$$

By virtue of (2),

$$\delta(Z) \geqslant 0.$$

Lemma 2. For any $Z \in L$,

$$\| Z - Z^* \| \leqslant \min \{ \sqrt{\delta(Z)}, \| Z \| \}.$$

Proof. The inequality

$$\| Z - Z^* \| \leqslant \sqrt{\delta(Z)} \tag{3}$$

is equivalent to

$$(Z - Z^*, Z - Z^*) - (Z, Z) + \min_{i \in [1:s]} (Z_i, Z) \leqslant 0,$$

or, after simple rearrangements,

$$-2(Z^*, Z) + (Z^*, Z^*) + \min_{i \in [1:s]} (Z_i, Z) \leqslant 0.$$

This inequality is readily verified using (1) and (2). Indeed,

$$\begin{aligned} -2(Z^*, Z) + (Z^*, Z^*) + \min_{i \in [1:s]} (Z_i, Z) &= \\ = \{(Z^*, Z^*) - (Z, Z^*)\} + \{ \min_{i \in [1:s]} (Z_i, Z) - (Z^*, Z)\} &\leqslant 0. \end{aligned}$$

Inequality (3) is "best possible"; in the situation illustrated in Figure 35 it becomes an equality.

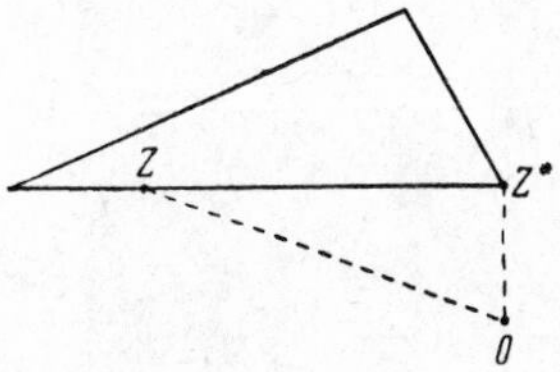

FIGURE 35.

We now show that

$$\| Z - Z^* \| \leqslant \| Z \|. \tag{4}$$

Indeed,

$$2(Z, Z^*) \geqslant 2(Z^*, Z^*) \geqslant (Z^*, Z^*).$$

Hence

$$(Z, Z) - 2(Z, Z^*) + (Z^*, Z^*) \leqslant (Z, Z),$$

which is equivalent to (4). Inequality (4) is also best possible: it becomes an equality when $Z^* = \mathbf{0}$, i. e., L contains the origin. Combining (3) and (4), we complete the proof.

Corollary 1. If $\{V_k\}, V_k \in L$, $k = 0, 1, 2, \ldots$, *is a sequence of points such that* $\delta(V_k) \xrightarrow[k\to\infty]{} 0$, *then*

$$V_k \xrightarrow[k\to\infty]{} Z^*.$$

This follows in an obvious manner from (3).

Corollary 2. If $\{V_k\}$, $V_k \in L$, $k = 0, 1, 2, \ldots$, *is a sequence of points such that*

$$\| V_{k+1} \| \leqslant \| V_k \|,$$

and there exists a subsequence $\{V_{k_j}\}$ *such that*

$$\delta(V_{k_j}) \xrightarrow[k_j\to\infty]{} 0,$$

then

$$V_k \xrightarrow[k\to\infty]{} Z^*.$$

Proof. Since $\{\| V_k \|\}$ is a nonincreasing sequence bounded below, it has a limit:

$$\| V_k \| \xrightarrow[k\to\infty]{} \mu.$$

On the other hand, by (3),

$$\| V_{k_j} \| \xrightarrow[k_j\to\infty]{} \| Z^* \|.$$

Thus, $\| V_k \| \xrightarrow[k\to\infty]{} \| Z^* \|$.

By the uniqueness of Z^* (if $\bar{Z} \in L$ and $\| \bar{Z} \| = \| Z^* \|$ then $\bar{Z} = Z^*$), any convergent subsequence of $\{V_k\}$ necessarily converges to Z^*. Thus the entire sequence $\{V_k\}$ converges to Z^*, and the proof is complete.

Theorem 1. A point $\bar{Z} \in L$ *is the point of* L *nearest the origin if and only if*

$$\delta(\bar{Z}) = 0.$$

Proof. Sufficiency follows from (3).

Necessity. Let $\bar{Z}=Z^*$. Then, first,

$$\delta(Z^*)\geqslant 0. \tag{5}$$

On the other hand, by (1), it follows that for any $i\in[1:s]$

$$(Z_i,\ Z^*)\geqslant(Z^*,\ Z^*),$$

because $Z_i\in H\subset L$.

Hence

$$\min_{i\in[1:s]}(Z_i,\ Z^*)\geqslant(Z^*,\ Z^*),$$

or, what is the same,

$$\delta(Z^*)\leqslant 0. \tag{6}$$

Combining (5) and (6), we get the required equality.

3. First method of successive approximations. The initial approximation V_0 will be any point of L (a reasonable choice is the point Z_{i_0} in H for which

$$(Z_{i_0},\ Z_{i_0})=\min_{i\in[1:s]}(Z_i,\ Z_i)).$$

Suppose we have already found the k-th approximation $V_k\in L$. We now determine V_{k+1}.

Let $\bar{Z}_k$ denote the point of H for which

$$(\bar{Z}_k,\ V_k)=\min_{i\in[1:s]}(Z_i,\ V_k)$$

(if there are several such points, we select any one).

Then

$$\delta(V_k)=(V_k-\bar{Z}_k,\ V_k). \tag{7}$$

Consider the segment

$$V_k(t)=V_k+t(\bar{Z}_k-V_k),\quad 0\leqslant t\leqslant 1. \tag{8}$$

Let t_k, $0\leqslant t_k\leqslant 1$, be defined by the condition

$$(V_k(t_k),\ V_k(t_k))=\min_{t\in[0,\ 1]}(V_k(t),\ V_k(t)).$$

Set $V_{k+1}=V_k(t_k)$ (Figure 36). Clearly, $V_{k+1}\in L$.

Continuing this way, we obtain a sequence $V_k\in L$, $k=0,\ 1,\ 2,\ \ldots$, with

$$\|V_{k+1}\|\leqslant\|V_k\|. \tag{9}$$

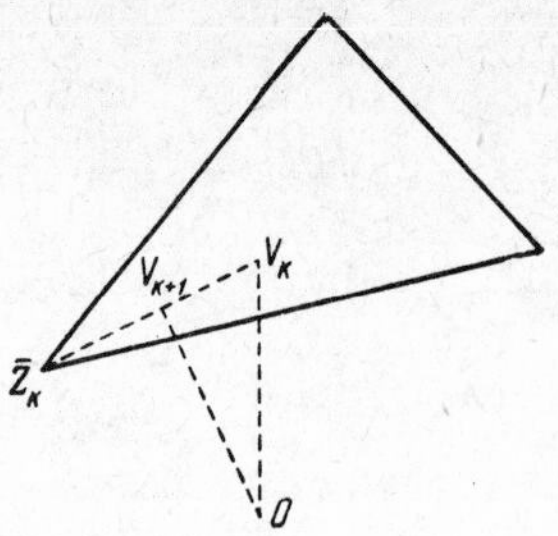

FIGURE 36.

Lemma 3. $\lim_{k\to\infty} \delta(V_k)=0.$

Proof. We first observe that, by (7) and (8),

$$(V_k(t), V_k(t))=(V_k, V_k)-2t\delta(V_k)+t^2\|\bar{Z}_k-V_k\|^2. \qquad (10)$$

Let us denote $\delta_k=\delta(V_k)$.

Suppose that the assertion is false, so that there exists a subsequence $\{V_{k_j}\}$ for which

$$\delta_{k_j}\geqslant\varepsilon>0.$$

By (10), we have

$$(V_{k_j}(t), V_{k_j}(t))\leqslant(V_{k_j}, V_{k_j})-2t\varepsilon+t^2d^2,$$

where

$$d=\max_{Z, V\in L}\|Z-V\|>0.$$

Hence it follows that for $t_0=\min\left\{\frac{\varepsilon}{d^2}, 1\right\}$ (it is trivial that $0<t_0\leqslant 1$)

$$(V_{k_j}(t_0), V_{k_j}(t_0))\leqslant(V_{k_j}, V_{k_j})-t_0\varepsilon.$$

Since by definition we have $(V_{k_j+1}, V_{k_j+1})\leqslant(V_{k_j}(t_0), V_{k_j}(t_0))$, it follows that

$$(V_{k_j+1}, V_{k_j+1})\leqslant(V_{k_j}, V_{k_j})-t_0\varepsilon \qquad (11)$$

uniformly in k_j.

This immediately yields a contradiction. Indeed, by (9), the sequence $\{\|V_k\|^2\}$ is nonincreasing and bounded below, so that it has a limit:

$$\|V_k\|^2\xrightarrow[k\to\infty]{}\mu,$$

with

$$\|V_k\|^2\geqslant\mu, \quad k=0, 1, 2, \ldots \qquad (12)$$

Choose $\bar{k}_j$ so large that

$$\|V_{\bar{k}_j}\|^2 \leqslant \mu + \frac{1}{2} t_0\varepsilon.$$

Then, by (11),

$$\|V_{\bar{k}_j+1}\|^2 \leqslant \|V_{\bar{k}_j}\|^2 - t_0\varepsilon \leqslant \mu - \frac{1}{2} t_0\varepsilon,$$

contradicting (12).

Theorem 2. The above sequence $\{V_k\}$ converges to the point Z^*.

Proof. Follows in an obvious manner from Lemma 3 and Corollary 1 to Lemma 2.

Remark. Suppose that $\delta(V_k)=0$, i. e., $V_k = Z^*$, for some k; then $\delta(V_{k+j})=0$ (so that $V_{k+j}=Z^*$) for all $j=1, 2, \ldots$.

This follows from (10).

4. Let Ξ be the set of all vectors A such that

$$A=(\alpha_1, \ldots, \alpha_s); \quad \alpha_i \geqslant 0, \quad \sum_{i=1}^{s} \alpha_i = 1.$$

Set

$$Z(A) = \sum_{i=1}^{s} \alpha_i Z_i; \qquad Z(A) \in L.$$

We define a function

$$\Delta(A) = \max_{\{i \mid \alpha_i > 0\}} (Z_i, Z(A)) - \min_{i \in [1:s]} (Z_i, Z(A)). \tag{13}$$

Let $i' = i'(A)$ denote an index corresponding to the maximum on the right of (13) (if there are several, we choose any one).

Thus, $\alpha_{i'} > 0$ and

$$(Z_{i'}, Z(A)) = \max_{\{i \mid \alpha_i > 0\}} (Z_i, Z(A)).$$

Lemma 4. For any vector $Z(A)$, $A \in \Xi$,

$$\alpha_{i'}\Delta(A) \leqslant \delta(Z(A)) \leqslant \Delta(A).$$

Proof. We first note that

$$(Z(A), Z(A)) = \sum_{i=1}^{s} \alpha_i (Z_i, Z(A)) \leqslant \max_{\{i \mid \alpha_i > 0\}} (Z_i, Z(A)).$$

This implies the inequality $\delta(Z(A)) \leqslant \Delta(A)$.

Now let $Z_{i''}$, $i'' = i''(A)$, denote a point of H such that

$$(Z_{i''}, Z(A)) = \min_{i \in [1:s]} (Z_i, Z(A)).$$

Then

$$\Delta(A) = (Z_{i'} - Z_{i''}, Z(A)). \tag{14}$$

Set $\bar{A} = (\bar{\alpha}_1, \ldots, \bar{\alpha}_s) \in \Xi$, where

$$\bar{\alpha}_i = \begin{cases} \alpha_i, & i \neq i', i''; \\ 0, & i = i'; \\ \alpha_{i'} + \alpha_{i''}, & i = i''. \end{cases}$$

Obviously,

$$Z(\bar{A}) = Z(A) + \alpha_{i'}(Z_{i''} - Z_{i'}). \tag{15}$$

Since $Z(\bar{A}) \in L$, we have from (2) that

$$(Z(\bar{A}), Z(A)) \geqslant \min_{i \in [1:s]} (Z_i, Z(A)). \tag{16}$$

In view of (16), (15) and (14), we obtain

$$\delta(Z(A)) \geqslant \alpha_{i'} \Delta(A).$$

This proves the lemma.

Theorem 3. $Z(A)$, $A \in \Xi$, is the point of L nearest the origin if and only if $\Delta(A) = 0$.

Proof. Follows in an obvious manner from Lemma 4 and Theorem 1.

5. Second method of successive approximations. Select a vector $A_0 \in \Xi$ arbitrarily, and set

$$V_0 = Z(A_0).$$

Suppose we have already found the k-th approximation $V_k \in L$,

$$V_k = Z(A_k), \quad A_k = (\alpha_1^{(k)}, \ldots, \alpha_s^{(k)}) \in \Xi.$$

In order to determine V_{k+1}, we first find vectors $Z_{i'_k}$ and $Z_{i''_k}$ in H such that

$$\left(Z_{i'_k}, V_k\right) = \max_{\{i \mid \alpha_i^{(k)} > 0\}} (Z_i, Z(A_k)),$$
$$\left(Z_{i''_k}, V_k\right) = \min_{i \in [1:s]} (Z_i, Z(A_k)).$$

Then

$$\Delta_k \stackrel{df}{=} \Delta(A_k) = \left(Z_{i'_k} - Z_{i''_k}, V_k\right). \tag{17}$$

Consider the segment

$$V_k(t) = V_k + t\alpha_{i'_k}^{(k)} \left(Z_{i''_k} - Z_{i'_k}\right), \quad 0 \leqslant t \leqslant 1. \tag{18}$$

Let t_k, $0 \leqslant t_k \leqslant 1$, be defined by the condition

$$(V_k(t_k), V_k(t_k)) = \\ = \min_{t \in [0, 1]} (V_k(t), V_k(t)).$$

Set $V_{k+1} = V_k(t_k)$ (Figure 37). It is readily checked that $V_{k+1} = Z(A_{k+1})$, where

$$A_{k+1} = (\alpha_1^{(k+1)}, \ldots, \alpha_s^{(k+1)}) \in \Xi,$$

$$\alpha_i^{(k+1)} = \begin{cases} \alpha_i^{(k)}, & i \neq i_k', i_k''; \\ \alpha_{i_k''}^{(k)} - t_k \alpha_{i_k'}^{(k)}, & i = i_k'; \\ \alpha_{i_k''}^{(k)} + t_k \alpha_{i_k'}^{(k)}, & i = i_k''. \end{cases}$$

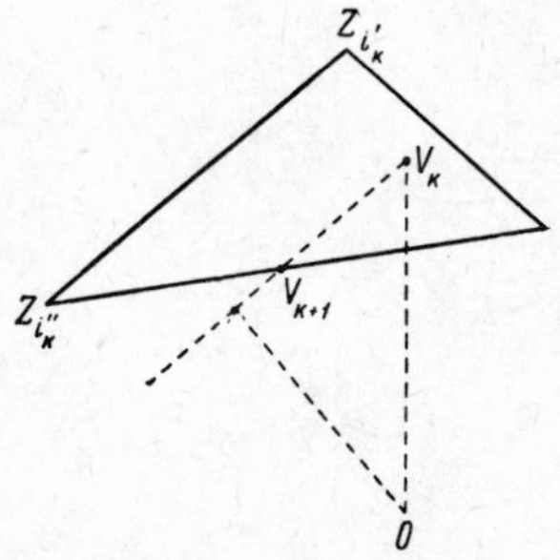

FIGURE 37.

To simplify the notation, from now on we set

$$\alpha_k' = \alpha_{i_k'}^{(k)}, \quad Z_k' = Z_{i_k'}, \quad \bar{Z}_k = Z_{i_k''}.$$

Continuing in this way, we obtain a sequence $\{V_k\}$, $V_k \in L$, $k = 0, 1, 2, \ldots$ where

$$\| V_{k+1} \| \leqslant \| V_k \|. \tag{19}$$

Lemma 5.

$$\lim_{k \to \infty} \alpha_k' \Delta_k = 0. \tag{20}$$

Proof. We first observe that, by (17) and (18),

$$(V_k(t), V_k(t)) = (V_k, V_k) - 2t\alpha_k'\Delta_k + t^2 (\alpha_k' \| \bar{Z}_k - Z_k' \|)^2. \tag{21}$$

Suppose the assertion false. Then there is a subsequence $\{V_{k_j}\}$ for which

$$\alpha'_{k_j}\Delta_{k_j} \geqslant \varepsilon > 0.$$

By (21), we have

$$(V_{k_j}(t),\ V_{k_j}(t)) \leqslant (V_{k_j},\ V_{k_j}) - 2t\varepsilon + t^2 d^2,$$

for all $t \in [0, 1]$, uniformly in k_j, where

$$d = \max_{l,\ p \in [1:s]} \| Z_l - Z_p \| > 0.$$

Together with (19), this last inequality immediately yields a contradiction (see the proof of Lemma 3), proving the lemma.

Lemma 6.

$$\lim_{k\to\infty} \Delta_k = 0. \tag{22}$$

Proof. Suppose the contrary:

$$\overline{\lim_{k\to\infty}}\, \Delta_k = \Delta' > 0. \tag{23}$$

Then, for sufficiently large $k > K_1$,

$$\Delta_k \geqslant \frac{\Delta'}{2}. \tag{24}$$

In view of (20), we see that

$$\alpha'_k \xrightarrow[k\to\infty]{} 0. \tag{25}$$

Let $\bar{t}_k$ denote a point at which $(V_k(t),\ V_k(t))$ assumes its global minimum. Clearly (see (21))

$$\bar{t}_k = \frac{\Delta_k}{\alpha'_k \| \bar{Z}_k - Z'_k \|^2}.$$

By (24) and (25),

$$\bar{t}_k \xrightarrow[k\to\infty]{} \infty.$$

Hence it follows, via (21), that for sufficiently large $k > K \geqslant K_1$ the minimum of $(V_k(t),\ V_k(t))$ on the segment $0 \leqslant t \leqslant 1$ is achieved at $t_k = 1$, so that for such values k

$$V_{k+1} = V_k + \alpha'_k (\bar{Z}_k - Z'_k). \tag{26}$$

Let $\{V_{k_j}\}$ be a subsequence such that

$$\Delta_{k_j} \xrightarrow[k_j\to\infty]{} \Delta'. \tag{27}$$

By discarding terms (if necessary) from the sequence $\{V_{k_j}\}$, we may assure that the following conditions are fulfilled:

I. $$\alpha_i^{(k_j)} \xrightarrow[k_j\to\infty]{} \alpha_i^*, \quad i\in[1:s].$$

This implies

$$V_{k_j} \xrightarrow[k_j\to\infty]{} V^* \overset{df}{=} \sum_{i=1}^{s} \alpha_i^* Z_i.$$

II. $$\bar{Z}_{k_j} = \bar{Z} \in H.$$

This means that $\min_{i\in[1:s]}(Z_i, V_{k_j})$ is achieved for all k_j at the same value of i:

$$(\bar{Z}, V_{k_j}) = \min_{i\in[1:s]}(Z_i, V_{k_j}).$$

III. $$Z'_{k_j} = Z' \in H.$$

Thus,

$$(Z', V_{k_j}) = \max_{\{i \mid \alpha_i^{(k_j)}>0\}} (Z_i, Z(A_{k_j})).$$

Using (27) and conditions I — III of the sequence $\{V_{k_j}\}$ we readily show that

$$(Z' - \bar{Z}, V^*) = \Delta'. \tag{28}$$

Now introduce the notation

$$\rho^* = \min_{i\in[1:s]}(Z_i, V^*),$$
$$H_1^* = \{Z_i \in H \mid (Z_i, V^*) = \rho^*\}, \quad H_2^* = H \setminus H_1^*.$$

$Z_i \in H_2^*$ if and only if

$$(Z_i, V^*) > \rho^*.$$

Note that $\bar{Z} \in H_1^*$ and $Z' \in H_2^*$; this follows from II and (28).

Set

$$\rho' = \min_{Z_i\in H_2^*}(Z_i, V^*), \quad \rho' > \rho^*;$$
$$\tau = \min\{\Delta', \rho' - \rho^*\}.$$

Note that for $k_i \in H_2^*$

$$(Z_i, V^*) \geqslant \rho^* + \tau. \tag{29}$$

Choose $\delta_0 > 0$ such that, whenever $\|V - V^*\| < \delta_0$,

$$\max_{i \in [1:s]} |(Z_i, V) - (Z_i, V^*)| < \frac{\tau}{4}. \tag{30}$$

Let k_j be such that $\|V_{k_j} - V^*\| < \delta_0$. Then

$$H_1(V_{k_j}) \subset H_1^*, \tag{31}$$

where

$$H_1(V_{k_j}) = \{Z_i \in H \mid (Z_i, V_{k_j}) = \min_{l \in [1:s]} (Z_l, V_{k_j})\}.$$

Indeed, let $Z_i \in H_1(V_{k_j})$. Then, by (30),

$$\begin{aligned}(Z_i, V^*) &= (Z_i, V_{k_j}) + (Z_i, V^* - V_{k_j}) = \\ &= \min_{l \in [1:s]} (Z_l, V_{k_j}) + (Z_i, V^* - V_{k_j}) = \rho^* + (Z_i, V^* - V_{k_j}) + \\ &+ [\min_{l \in [1:s]} (Z_l, V_{k_j}) - \min_{l \in [1:s]} (Z_l, V^*)] \leqslant \rho^* + (Z_i, V^* - V_{k_j}) + \\ &\qquad + \max_{l \in [1:s]} |(Z_l, V_{k_j}) - (Z_l, V^*)| \leqslant \rho^* + \frac{\tau}{2}.\end{aligned}$$

In view of (29), we conclude that $Z_i \in H_1^*$, and this proves (31).
Now let $k_j > K$ be such that the following conditions hold:

(a) $$\Delta_k \geqslant \Delta' - \frac{\tau}{4} \quad \text{for} \quad k > k_j;$$

(b) $$\|V_{k_j} - V^*\| < \frac{\delta_0}{2};$$

(c) $$\sum_{l=1}^{s} \alpha'_{k_j+l-1} < \frac{\delta_0}{2d},$$

where

$$d = \max_{l,\, p \in [1:s]} \|Z_l - Z_p\| > 0.$$

Inequality (c) may be guaranteed on the basis of (25).
In view of (26), we write for $p \in [1:s]$

$$V_{k_j+p} = V_{k_j} + \sum_{l=1}^{p} \alpha'_{k_j+l-1} (\overline{Z}_{k_j+l-1} - Z'_{k_j+l-1}).$$

Note that, by (c),

$$\|V_{k_j+p} - V_{k_j}\| \leqslant d \sum_{l=1}^{p} \alpha'_{k_j+l-1} < \frac{\delta_0}{2},$$

so that by (b), for $p \in [1:s]$,

$$\|V_{k_j+p} - V^*\| < \delta_0.$$

Hence it follows that for these values p

$$H_1\left(V_{k_j+p}\right)\subset H_1^*. \tag{32}$$

(The proof is analogous to that of (31).

Next, for all $Z_i\in H_1^*$ and $p\in[1:s]$, we have

$$\left(Z_i,\ V_{k_j+p}\right)\leqslant\Delta'+\rho^*-\frac{3\tau}{4}. \tag{33}$$

Indeed,

$$\left(Z_i,\ V_{k_j+p}\right)=\left(Z_i,\ V^*\right)+\left(Z_i,\ V_{k_j+p}-V^*\right)\leqslant$$
$$\leqslant\rho^*+\frac{\tau}{4}\leqslant\Delta'+\rho^*-\frac{3\tau}{4}.$$

It is now easy to prove that for some vector V_{k_j+p}, $p\in[1:s]$,

$$\max_{\left\{i\mid\alpha_i^{(k_j+p)}>0\right\}}\left(Z_i,\ V_{k_j+p}\right)\leqslant\Delta'+\rho^*-\frac{3\tau}{4}. \tag{34}$$

Indeed, if this inequality fails to hold for some p, it follows from (33) that

$$Z'_{k_j+p}\in H_2^*.$$

On the basis of (26) and (32), we conclude that the representation of V_{k_j+p+1} involves the vector Z'_{k_j+p} with zero coefficient, while the newly introduced vector $\overline{Z}_{k_j+p}$ satisfies inequality (33). Since the set H_2^* contains at most $s-1$ vectors and inequality (33) holds for all the vectors appearing in their representations, there exists $p'\in[1:s]$ for which (34) is fulfilled.

On the other hand,

$$\min_{i\in[1:s]}\left(Z_i,\ V_{k_j+p'}\right)\geqslant\rho^*-\frac{\tau}{4}. \tag{35}$$

Thus, by (34) and (35), we get

$$\Delta_{k_j+p'}\overset{\text{df}}{=}\max_{\left\{i\mid\alpha_i^{(k_j+p')}>0\right\}}\left(Z_i,\ V_{k_j+p'}\right)-\min_{i\in[1:s]}\left(Z_i,\ V_{k_j+p'}\right)\leqslant\Delta'-\frac{\tau}{2},$$

contradicting (a) and proving the lemma.

Theorem 4. The sequence $\{V_k\}$ converges to Z^*.

Proof. Follows in an obvious manner from Lemmas 4 and 6 and Corollary 2 to Lemma 2.

Remark. If $\Delta_k=0$, i.e., $V_k=Z^*$, for some k, then $\Delta_{k+j}=0$ (so that $V_{k+j}=Z^*$) for all $j=1, 2, \ldots$

This follows from (21).

6. Some special features of the second procedure are worthy of note. Consider the hyperplane G defined by

$$G=\{Z\mid(Z,\ Z^*)=(Z^*,\ Z^*)\}.$$

Theorem 5. If $Z^* \neq 0$, i.e., L does not contain the origin, then, for sufficiently large k,

$$V_k \in G.$$

Proof. By Theorem 1,

$$\min_{i \in [1:s]} (Z_i, Z^*) = (Z^*, Z^*).$$

We set

$$H_1 = \{Z_i \in H \mid (Z_i, Z^*) = (Z^*, Z^*)\},$$
$$H_2 = H \setminus H_1 = \{Z_i \in H \mid (Z_i, Z^*) > (Z^*, Z^*)\}.$$

If H_2 is empty, then $V_k \in G$ for all $k = 0, 1, 2, \ldots$ We may therefore assume henceforth that H_2 is a nonempty set. We denote

$$\tau = \min_{Z_i \in H_2} (Z_i, Z^*) - (Z^*, Z^*) > 0.$$

Since $V_k \xrightarrow[k \to \infty]{} Z^*$, it follows that for sufficiently large $k > K_1$

$$\max_{i \in [1:s]} |(Z_i, V_k) - (Z_i, Z^*)| < \frac{\tau}{4}.$$

It is readily shown that for these values $k > K_1$

$$H_1(V_k) \subset H_1, \tag{36}$$
$$(Z_i, V_k) \leqslant (Z^*, Z^*) + \frac{\tau}{4}, \quad \text{if} \quad Z_i \in H_1,$$
$$(Z_i, V_k) \geqslant (Z^*, Z^*) + \frac{3\tau}{4}, \quad \text{if} \quad Z_i \in H_2.$$

Hence it follows that if the expression for V_k, $k > K_1$, involves $Z_i \in H_2$ with a nonzero coefficient, then

$$\Delta_k \geqslant \frac{\tau}{2}. \tag{37}$$

Let

$$V_k = \sum_{\{i \mid Z_i \in H_1\}} \alpha_i^{(k)} Z_i + \sum_{\{i \mid Z_i \in H_2\}} \alpha_i^{(k)} Z_i.$$

By the definition of H_1 and τ,

$$(V_k - Z^*, Z^*) = \sum_{\{i \mid Z_i \in H_2\}} \alpha_i^{(k)} (Z_i - Z^*, Z^*) \geqslant \tau \sum_{\{i \mid Z_i \in H_2\}} \alpha_i^{(k)}.$$

Since the left-hand side of this inequality tends to zero as $k \to \infty$, we have

$$\sum_{\{i \mid Z_i \in H_2\}} \alpha_i^{(k)} \xrightarrow[k \to \infty]{} 0.$$

Choose $K > K_1$ so large that for $k \geqslant K$

$$\sum_{\{i \mid Z_i \in H_2\}} \alpha_i^{(k)} \leqslant \frac{\tau}{2d^2}, \tag{38}$$

where

$$d = \max_{Z_i \in H_1,\ Z_j \in H_2} \| Z_i - Z_j \| > 0.$$

Let $\bar{t}_k$ be a point at which $(V_k(t), V_k(t))$ assumes a global minimum. If the representation of V_k, $k \geqslant K$, involves $Z_i \in H_2$ with a nonzero coefficient, then by (37) and (38)

$$\bar{t}_k = \frac{\Delta_k}{\alpha_k' \|\bar{Z}_k - Z_k'\|^2} \geqslant \frac{\tau}{2\left(\sum_{\{i \mid Z_i \in H_2\}} \alpha_i^{(k)}\right) d^2} \geqslant 1.$$

Hence it follows that

$$V_{k+1} = V_k + \alpha_k' (\bar{Z}_k - Z_k').$$

It is now easy to see that the representation of the vectors V_{K+s} does not involve the vectors $Z_i \in H_2$ (i.e., involves them with zero coefficient; see the proof of Lemma 6). In view of (36), we conclude that for all $k \geqslant K + s$

$$V_k = \sum_{\{i \mid Z_i \in H_1\}} \alpha_i^{(k)} Z_i; \quad \alpha_i^{(k)} \geqslant 0, \quad \sum_{\{i \mid Z_i \in H_1\}} \alpha_i^{(k)} = 1,$$

whence it follows immediately that $(V_k, Z^*) = (Z^*, Z^*)$, i. e., $V_k \in G$.

Theorem 6. As $k \to \infty$, we have

$$\Delta_k \to 0.$$

Proof. If $Z^* = \mathbf{0}$, this follows directly from the definition of Δ_k and the fact that $\| V_k \| \xrightarrow[k \to \infty]{} 0$.

Suppose, then, that $Z^* \neq \mathbf{0}$. By Theorem 5, if $k \geqslant K + s$,

$$V_k = \sum_{\{i \mid Z_i \in H_1\}} \alpha_i^{(k)} Z_i; \quad \alpha_i^{(k)} \geqslant 0, \quad \sum_{\{i \mid Z_i \in H_1\}} \alpha_i^{(k)} = 1.$$

In addition, $H_1(V_k) \subset H_1$. Hence

$$\Delta_k \leqslant \max_{Z_i \in H_1} (Z_i, V_k) - \min_{Z_i \in H_1} (Z_i, V_k).$$

By the definition of H_1, the right-hand side of this inequality tends to zero as $k \to \infty$. Thus also $\Delta_k \xrightarrow[k \to \infty]{} 0$, since $\Delta_k \geqslant 0$.

This completes the proof.

We now set

$$\tilde{\Delta}_k = \frac{\Delta_k}{\|V_k\|^2}, \quad k = 0, 1, 2, \ldots$$

If $\| V_k \| = 0$, we set $\tilde{\Delta}_k = \infty$ by definition.

Theorem 7. The origin is a point of the set L if and only if, for all $k = 0, 1, 2, \ldots,$

$$\tilde{\Delta}_k \geqslant 1. \tag{39}$$

Proof. Necessity. By assumption, $Z^* = \mathbf{0}$. Since $V_k \in L$, we infer from Lemmas 2 and 4 that

$$\tilde{\Delta}_k = \frac{\Delta_k}{\|V_k\|^2} \geqslant \frac{\delta(V_k)}{\|V_k\|^2} \geqslant 1.$$

Sufficiency. Let $Z^* \neq \mathbf{0}$. We have

$$\tilde{\Delta}_k \leqslant \frac{\Delta_k}{\|Z^*\|^2}.$$

By Theorem 6, $\tilde{\Delta}_k \xrightarrow[k\to\infty]{} 0$, and this is inconsistent with (39), completing the proof.

Thus, if $Z^* = \mathbf{0}$, then $\|V_k\| \xrightarrow[k\to\infty]{} 0$, and for all $k=0, 1, 2, \ldots$ we have the inequality $\tilde{\Delta}_k \geqslant 1$. But if $Z^* \neq \mathbf{0}$, then $\|V_k\| \geqslant \|Z^*\|$ and $\tilde{\Delta}_k \xrightarrow[k\to\infty]{} 0$.

If $\tilde{\Delta}_k < 1$ for some k, it follows from Theorem 7 that the origin is not in L. In addition, it is readily shown that then the hyperplane

$$(V_k, Z) - (V_k, \bar{Z}_k) = 0$$

strictly separates the origin from the set L.

7. Some further features of the second successive approximation procedure will now be examined.

Let $V_k = Z(A_k)$, $A_k \in \Xi$, and set

$$J_k = \{i \mid \alpha_i^{(k)} > 0\}.$$

We define a set B_k by

$$B_k = \left\{ Z = Z_{i_0} + \sum_{\substack{i \in J_k \\ i \neq i_0}} \alpha_i (Z_i - Z_{i_0}) \,\middle|\, \alpha_i \in (-\infty, \infty) \right\}.$$

Here i_0 is an arbitrary but fixed index in J_k.

B_k is clearly a convex set. Let $\tilde{V}_k$ be a vector in B_k with minimal norm:

$$\|\tilde{V}_k\| = \min_{Z \in B_k} \|Z\|.$$

Note that the point $\tilde{V}_k \in B_k$ is unique, though it may have more than one representation in the form

$$\tilde{V}_k = Z_{i_0} + \sum_{\substack{i \in J_k \\ i \neq i_0}} \tilde{\alpha}_i (Z_i - Z_{i_0}).$$

It is readily seen that the numbers $\tilde{\alpha}_i$, $i \in J_k$, $i \neq i_0$, constitute a solution of the linear system

$$\left(Z_{i_0} + \sum_{\substack{i \in J_k \\ i \neq i_0}} \alpha_i (Z_i - Z_{i_0}),\ Z_j - Z_{i_0}\right) = 0,\ j \in J_k,\ j \neq i_0. \tag{40}$$

Theorem 8. There exists an infinite subsequence $\{\tilde{V}_{k_j}\}$ *such that for all* k_j

$$\tilde{V}_{k_j} = Z^*.$$

Proof. We shall assume that $Z^* \neq \mathbf{0}$ (if $Z^* = \mathbf{0}$ the proof is only simplified). We first determine a subsequence $\{V_{k_j}\}$ with the following properties:

I. $\alpha_i^{(k_j)} \xrightarrow[k_j \to \infty]{} \alpha_i^*$, $i \in [1:s]$; thus

$$V_{k_j} \xrightarrow[k_j \to \infty]{} Z^* = \sum_{i=1}^{s} \alpha_i^* Z_i.$$

II. $V_{k_j} \in G$ (see Theorem 5).

Set

$$J^* = \{i \mid \alpha_i^* > 0\}.$$

It is clear that for sufficiently large k_j

$$J^* \subset J_{k_j}. \tag{41}$$

Throughout the sequel we shall confine attention to such values k_j.

Let L^* denote the convex hull of the points Z_i, $i \in J^*$; clearly, $Z^* \in L^*$.

Let L_{k_j} denote the convex hull of the points Z_i, $i \in J_{k_j}$. By (41), II and the definition of B_{k_j}, we have

$$L^* \subset L_{k_j} \subset B_{k_j} \subset G.$$

Further,

$$\| Z^* \| = \min_{Z \in G} \| Z \| \leqslant \min_{Z \in B_{k_j}} \| Z \|.$$

Since $Z^* \in L^*$, it follows that $Z^* \in B_{k_j}$. Thus,

$$\| Z^* \| = \min_{Z \in B_{k_j}} \| Z \|.$$

Since the point of B_{k_j} with minimal norm is unique, we obtain

$$\tilde{V}_{k_j} = Z^*.$$

This proves the theorem.

The import of this theorem is that determination of the point of L nearest the origin reduces to the solution of finitely many systems of linear equations of type (40). Note that these systems are readily solved only for values k such that $\| V_k \|$ or $\tilde{\Delta}_k$ are sufficiently small.

8. Our results hitherto may be generalized.
Consider a vector $V = Z(A)$, $A \in \Xi$, and let

$$\rho(V) = \min_{i \in [1:s]} (Z_i, V).$$

For fixed $c \in [0, 1]$, we introduce the set

$$F_c \overset{df}{=} F_c(Z(A)) = \{Z_i \in H \mid \alpha_i > 0,\ (Z_i, V) - \rho(V) \geqslant c\,\Delta(A)\}.$$

Next, we set

$$V'_c = \sum_{Z_i \in F_c} \frac{\alpha_i}{\gamma_c} Z_i,$$

where

$$\gamma_c = \sum_{Z_i \in F_c} \alpha_i.$$

Define

$$\Delta_c(V) = (V'_c, V) - \min_{i \in [1:s]} (Z_i, V).$$

It follows from the definition that

$$\Delta(V) \geqslant \Delta_c(V) \geqslant c\,\Delta(V).$$

In particular, $\Delta(V) = \Delta_1(V)$.

It is not hard to see that

$$\delta(V) = \Delta_0(V).$$

Indeed, if $c = 0$, then $\gamma_0 = 1$, $V'_0 = V$, so that $\Delta_0(V) = \delta(V)$.

Lemma 7. If $0 \leqslant c_1 \leqslant c_2 \leqslant 1$, then, for any vector $V = Z(A)$, $A \in \Xi$,

$$\Delta_{c_1}(V) \leqslant \Delta_{c_2}(V).$$

Proof. It is obvious that

$$F_{c_1} \supset F_{c_2}, \quad \gamma_{c_1} \geqslant \gamma_{c_2}. \tag{42}$$

By virtue of (42) and the definition of F_{c_2},

$$(V'_{c_2} - V'_{c_1}, V) = \left(\frac{1}{\gamma_{c_2}} - \frac{1}{\gamma_{c_1}}\right) \sum_{\{i \mid Z_i \in F_{c_2}\}} \alpha_i (Z_i, V) -$$

$$- \frac{1}{\gamma_{c_1}} \sum_{\{i \mid Z_i \in F_{c_1} \setminus F_{c_2}\}} \alpha_i (Z_i, V) \geqslant \left(\frac{1}{\gamma_{c_2}} - \frac{1}{\gamma_{c_1}}\right)(c_2 \Delta(V) +$$

$$+\rho(V))\sum_{\{i\mid Z_i\in F_{c_2}\}}\alpha_i-\frac{1}{\gamma_{c_1}}(c_2\Delta(V)+\rho(V))\sum_{\{i\mid Z_i\in F_{c_1}\setminus F_{c_2}\}}\alpha_i=$$
$$=(c_2\Delta(V)+\rho(V))\left(\sum_{\{i\mid Z_i\in F_{c_2}\}}\frac{\alpha_i}{\gamma_{c_2}}-\sum_{\{i\mid Z_i\in F_{c_1}\}}\frac{\alpha_i}{\gamma_{c_1}}\right)=0.$$

Thus,

$$(V'_{c_2}-V'_{c_1},\ V)\geqslant 0,$$

whence the assertion evidently follows.

Theorem 9. In order that $V=Z(A)$, $A\in\Xi$ be the point of L nearest the origin, it is sufficient (necessary) that for some (all) $c\in[0,\ 1]$

$$\Delta_c(V)=0.$$

This follows from the relations

$$\delta(V)=\Delta_0(V)\leqslant\Delta_c(V)\leqslant\Delta_1(V)=\Delta(V)$$

and from Theorems 1 and 3.

9. To end this appendix, we describe yet another method for determining the point of L nearest the origin; this method generalizes our first and second procedures.

Fix some $c\in[0,\ 1]$. Let A_0 be an arbitrary vector in Ξ and set

$$V_0=Z(A_0).$$

Suppose we have already determined the k-th approximation $V_k\in L$:

$$V_k=Z(A_k),\ A_k=(\alpha_1^{(k)},\ \ldots,\ \alpha_s^{(k)})\in\Xi.$$

To construct V_{k+1}, we find a vector $\bar{Z}_k\in H$ such that

$$(\bar{Z}_k,\ V_k)=\min_{i\in[1:s]}(Z_i,\ V_k).$$

(If there are several such vectors, we choose any one.)

Consider the segment

$$V_k(t)=V_k+t\gamma_k(\bar{Z}_k-V'_k),\quad 0\leqslant t\leqslant 1,$$

where

$$\gamma_k=\sum_{\{i\mid Z_i\in F_c(V_k)\}}\alpha_i,\ V'_k=\sum_{\{i\mid Z_i\in F_c(V_k)\}}\frac{\alpha_i}{\gamma_k}Z_i\in L.$$

Observe that

$$\Delta_c(V_k)=(V'_k-\bar{Z}_k,\ V_k).$$

Let t_k, $0 \leqslant t_k \leqslant 1$, denote the number defined by the condition

$$(V_k(t_k), V_k(t_k)) = \min_{0 \leqslant t \leqslant 1} (V_k(t), V_k(t)).$$

Set $V_{k+1} = V_k(t_k)$. It is easy to see that

$$V_{k+1} = Z(A_{k+1}), \text{ where } A_{k+1} = (\alpha_1^{(k+1)}, \ldots, \alpha_s^{(k+1)}) \in \Xi.$$

If $\bar{Z}_k \notin F_c(V_k)$, then

$$\alpha_i^{(k+1)} = \begin{cases} \alpha_i^{(k)}, & \text{if } Z_i \notin F_c(V_k),\ Z_i \neq \bar{Z}_k, \\ \alpha_i^{(k)} - t_k\alpha_i^{(k)}, & \text{if } Z_i \in F_c(V_k), \\ \alpha_i^{(k)} + t_k\gamma_k, & \text{if } Z_i = \bar{Z}_k. \end{cases}$$

But if $\bar{Z}_k \in F_c(V_k)$, then

$$\alpha_i^{(k+1)} = \begin{cases} \alpha_i^{(k)}, & \text{if } Z_i \notin F_c(V_k), \\ \alpha_i^{(k)} - t_k\alpha_i^{(k)}, & \text{if } Z_i \in F_c(V_k), Z_i \neq \bar{Z}_k, \\ \alpha_i^{(k)} - t_k\alpha_i^{(k)} + t_k\gamma_k, & \text{if } Z_i = \bar{Z}_k. \end{cases}$$

Proceeding in this way, we obtain a sequence $\{V_k\}$, $V_k \in L$, $k = 0, 1, 2, \ldots$, with

$$\| V_{k+1} \| \leqslant \| V_k \|.$$

If $c = 0$, then $\gamma_k = 1$, $V'_k = V_k$,

$$V_k(t) = V_k + t(\bar{Z}_k - V_k), \quad 0 \leqslant t \leqslant 1,$$

and this is simply the first successive approximation procedure.

If $c = 1$,

$$F_1(V_k) = \{Z_i \in H \mid \alpha_i^{(k)} > 0,\ (Z_i - \bar{Z}_k, V_k) = \Delta(V_k)\}.$$

Any point of the set $F_1(V_k)$ may be taken as Z'_k (see subsection 5), so that when $c = 1$ we obtain a natural generalization of the second successive approximation procedure.

We state the main features of this method.

Theorem 10. *$V_k \to Z^*$ as $k \to \infty$.*

Theorem 11. *If $Z^* \neq 0$ and $c \in (0, 1]$, then, for sufficiently large k,*

$$V_k \in G,$$

where $G = \{Z \mid (Z, Z^*) = (Z^*, Z^*)\}$.

The proofs are entirely analogous to those of Theorems 4 and 5.

10. The methods of subsections 4 and 5 converge quite slowly if the origin is in the polyhedron L. The following device is recommended to deal with this eventuality. We first solve a linear programming problem in the space of vectors $W=(\alpha_1, \ldots, \alpha_s, u)$: find min u under the constraints

$$\sum_{i=1}^{s} \alpha_i z_k^{(i)} - u \leqslant 0, \quad k \in [1:n],$$

$$-\sum_{i=1}^{s} \alpha_i z_k^{(i)} - u \leqslant 0, \quad k \in [1:n],$$

$$\sum_{i=1}^{s} \alpha_i = 1; \; \alpha_i \geqslant 0, \; i \in [1:s]; \quad u \geqslant 0,$$

where $Z_i = (z_1^{(i)}, \ldots, z_n^{(i)})$ are points of H. Let $W_0 = (\alpha_1^{(0)}, \ldots, \alpha_s^{(0)}, u^{(0)})$ be a solution of this problem. It is readily checked that $\mathbf{0} \in L$ if and only if $u^{(0)} = 0$. If $u^{(0)} = 0$, then $Z^* = \mathbf{0}$. If $u^{(0)} > 0$, Z^* may be determined using the first or second successive approximation procedure, the initial approximation being the point

$$Z_0 = \sum_{i=1}^{s} \alpha_i^{(0)} Z_i,$$

where the components of the vector $(\alpha_1^{(0)}, \ldots, \alpha_s^{(0)})$ are the first s components of W_0.

We note that in many cases it will suffice to find the first term of the sequence $\{V_k\}$ converging to Z^* for which

$$\frac{\min\limits_{Z \in H} (Z, V_k)}{(V_k, V_k)} \geqslant \theta,$$

where $0 < \theta < 1$ (see Chap. III, §9). A point V_k with this property may be found in finitely many trials.

Supplement

ON MANDEL'SHTAM'S PROBLEM

In this supplement we illustrate the numerical solution of a continuous minimax problem.

1. Statement of the problem. Set

$$Z=(z_1, z_2, \ldots, z_{n+1}) \in E_{n+1},$$

$$W_n(Z, t)=\sum_{k=1}^{n+1} \cos(kt+z_k),$$

$$\Phi_n(Z)=\max_{t \in [-\pi, \pi]} |W_n(Z, t)|.$$

Problem: minimize the function $\Phi_n(Z)$ on E_{n+1}.

2. Preliminaries. We first observe that $\Phi_n(z_1, z_2, \ldots, z_{n+1})$ is continuous on E_{n+1} and 2π-periodic in each variable. Since

$$\inf_{Z \in E_{n+1}} \Phi_n(Z)=\inf_{Z \in Q_{n+1}} \Phi_n(Z),$$

where $Q_{n+1}=\{Z=(z_1, z_2, \ldots, z_{n+1}) \mid -\pi \leqslant z_k \leqslant \pi,\ k \in [1:n+1]\}$ is a bounded closed set, the infimum of $\Phi_n(Z)$ on E_{n+1} is attained: there exists a point Z^* such that

$$\Phi_n(Z^*)=\inf_{Z \in E_{n+1}} \Phi_n(Z).$$

It is not difficult to prove the following:

I.
$$\max_{Z \in E_{n+1}} \Phi_n(Z)=\Phi_n(0)=(n+1)^2.$$

II.
$$\Phi_n(-Z)=\Phi_n(Z).$$

III. For any real λ,

$$\Phi_n(Z+\lambda p)=\Phi_n(Z),$$

where $p=(1, 2, \ldots, n+1)$.

We prove only III. Since the maxima of a 2π-periodic function on any two segments of length 2π are equal, we have

$$\Phi_n(Z+\lambda p) = \max_{t\in[-\pi,\pi]} \left|\sum_{k=1}^{n+1} \cos(kt + z_k + \lambda k)\right| =$$
$$= \max_{u\in[-\pi+\lambda,\,\pi+\lambda]} \left|\sum_{k=1}^{n+1} \cos(ku + z_k)\right| = \Phi_n(Z),$$

and this is our assertion.

Lemma. If $Z^{(s)}$ denotes an arbitrary $(n+1)$-vector, with s-th coordinate zero, then for any $s\in[1:n+1]$

$$\min_{\{Z\}} \Phi_n(Z) = \min_{\{Z^{(s)}\}} \Phi_n(\dot{Z}^{(s)}).$$

Proof. Obviously,

$$\min_{\{Z\}} \Phi_n(Z) \leqslant \min_{\{Z^{(s)}\}} \Phi_n(Z^{(s)}). \tag{1}$$

Now, any vector $Z=(z_1, z_2, \ldots, z_{n+1})$ may be expressed as $Z=\left(Z-\frac{z_s}{s}p\right)+ +\frac{z_s}{s}p$. Since the s-th coordinate of $Z-\frac{z_s}{s}p$ is zero, it follows from III that

$$\Phi_n(Z) = \Phi_n\left(Z-\frac{z_s}{s}p\right) \geqslant \min_{\{Z^{(s)}\}} \Phi_n(Z^{(s)}).$$

Hence

$$\min_{\{Z\}} \Phi_n(Z) \geqslant \min_{\{Z^{(s)}\}} \Phi_n(Z^{(s)}). \tag{2}$$

Combining (1) and (2), we complete the proof.

Henceforth we assume that $z_1=0$.

3. Computation results. We introduce the notation

$$X=(x_1, \ldots, x_n)\in E_n,$$
$$V_n(X,t) = \cos t + \sum_{k=1}^{n} \cos((k+1)t + x_k),$$
$$\varphi_n(X) = \max_{t\in[-\pi,\pi]} |V_n(X,t)|,$$
$$R(X) = \{t\in[-\pi,\pi] \mid V_n(X,t) = \varphi_n(X)\},$$

and solve a problem equivalent to the original one: minimize the function $\varphi_n(X)$ on E_n.

Table 4 presents the results of the computations for $n\in[1:5]$ (in the fourth column, $t_i^*\in R(X^*)$).

For $n=1$ and $n=2$, the vector X^* is a global minimum point of $\varphi_n(X)$ on E_n; for $n\in[3:5]$ all we can guarantee is that X^* is a strict local minimum point.

TABLE 4.

n	X^*	$R(X^*)$	$V_n(X^*, t_i^*)$
1	1.570 796	−2.506 726 −0.634 867	−1.760 173 1.760 173
2	2.859 712 2.146 750	−0.840 986 1.399 331 2.707 688	1.979 806 1.979 806 −1.979 806
3	1.396 481 1.072 719 −1.209 531	−1.991 906 −0.083 577 1.860 299 2.769 081	−2.039 103 2.039 103 2.039 103 −2.039 103
4	1.084 832 2.087 535 −0.250 345 −1.277 773	−1.874 150 −0.953 170 0.833 524 1.545 230 2.277 249	−2.343 516 2.343 516 −2.343 516 2.343 516 −2.343 516
5	1.927 212 2.311 026 −2.514 500 2.102 134 1.076 190	−1.881 850 −1.058 158 −0.481 051 0.249 616 0.855 029 1.988 655	−2.549 372 2.549 372 2.549 372 −2.549 372 2.549 372 2.549 372

4. Method of solution. If $n=1$, the exact solution is readily determined: $X^*=\pi/2$.

For $n=2$ and $n=3$, we used different methods to select the initial approximation X_0, which possessed the following property: the ε-strip

$$\{t \in [-\pi, \pi] \mid \varphi_n(X_0) - |V_n(X_0, t)| \leqslant \varepsilon\},$$

where $\varepsilon=0.01$, contains exactly $n+1$ points at which the function $|V_n(X_0, t)|$ has a local maximum. Denote these points by $t_i^{(0)}$, $i \in [1:n+1]$:

$$-\pi < t_1^{(0)} < t_2^{(0)} < \ldots < t_{n+1}^{(0)} \leqslant \pi.$$

For $n \in [4:5]$, a satisfactory initial approximation was obtained by considering a discrete minimax problem:

$$\varphi_{nN}(X) \overset{df}{=} \max_{i \in [0:N]} V_n^2(X, t_i) \to \min_{X \in E_n},$$

where $t_i = -\pi + i\frac{2\pi}{N}$. Setting $N=400$ and $\varepsilon=0.05$, we determined an ε-stationary point X_0 of the function $\varphi_{nN}(X)$. Note that the set

$$\{t_i \mid \varphi_{nN}(X_0) - V_n^2(X_0, t_i) \leqslant \varepsilon; \quad V_n^2(X_0, t_i) \geqslant V_n^2(X_0, t_{i-1});\ V_n^2(X_0, t_i) \geqslant V_n^2(X_0, t_{i+1})\}$$

contains exactly $n+1$ points:

$$-\pi < t_1^{(0)} < t_2^{(0)} < \ldots < t_{n+1}^{(0)} \leqslant \pi.$$

For all $n \in [2:5]$, the second step consisted in equalizing the maxima; this involved solving the following nonlinear system of order $(2n+1)$

$$\left.\begin{aligned} &\sigma_k V_n(X, y_k) - \sigma_{k+1} V_n(X, y_{k+1}) = 0, \quad k \in [1:\ n], \\ &V_n'(X, y_k) = 0, \quad k \in [1:\ n+1]. \end{aligned}\right\} \tag{3}$$

Here $\sigma_k = \operatorname{sign} V_n(X_0, t_k^{(0)})$, and the unknowns are $x_1, \ldots, x_n, y_1, \ldots, y_{n+1}$.

To solve system (3) we resorted to a special modification of Newton's method, due to Mitchell. The initial approximation was the vector whose components are those of X_0 and the numbers $t_1^{(0)}, \ldots, t_{n+1}^{(0)}$.

Denote the solution of system (3) by $x_1^*, \ldots, x_n^*, t_1^*, t_2^*, \ldots, t_{n+1}^*$. Then the vector $X^* = (x_1^*, \ldots, x_n^*)$ is a stationary point of the function $\varphi_n(X)$, and the numbers $t_1^*, t_2^*, \ldots, t_{n+1}^*$ form an extremal basis for $R(X^*)$.

The condition

$$\sum_{i=1}^{n+1} \alpha_i \frac{\partial |V_n(X^*, t_i^*)|}{\partial X} = 0$$

for some $\alpha_i > 0$, $\sum_{i=1}^{n+1} \alpha_i = 1$, guaranteed that X^* was indeed a strict local minimum point of $\varphi_n(X)$.

NOTES

The principle of optimal selection of parameters based on minimization of the maximum deviation was originally suggested by Chebyshev and described in detail in his classical memoir /80/.

Preface

Mandel'shtam's problem as an example of a minimax problem appears in Chebotarev /79/. The theory of matrix games is discussed in detail in Karlin /39/, von Neumann and Morgenstern /57/.

Chapter I

The discrete problem of best approximation by algebraic polynomials was considered by de la Vallée-Poussin /6/. In particular, de la Vallée-Poussin proposed the idea of basing the theory of best approximation on Chebyshev interpolation. He also worked out the algorithm described in §3 for constructing the polynomial of best approximation.

The R-algorithm is proposed in /49/.

§5. The first geometric interpretation of the best approximation problem (Theorem 5.1) may be found in Remez /65/. The reduction of the problem to determination of the extreme point of intersection of an axis with a polyhedron was probably first suggested by Rubinshtein /70/.

Kalashnikov /35/ and Eterman /84/ studied these polynomials (2.9) from different standpoints. In particular, in /84/ these polynomials are used to solve various problems of computational mathematics.

Chapter II

This chapter is devoted to the pivotal question of the transition from the discrete problem to the continuous case. The limit theorems (§3) and Remez' method of successive Chebyshev interpolations (§4) may be found in /65/. Related questions are discussed in /64/ and /4/. Under certain additional assumptions, Theorem 5.1 (convergence of the grid method) was proved in /50/. Concerning Theorem 5.2, see, e.g., /4/.

The concept of an ill-conditioned polynomial was introduced by Samokish, to whom are due Theorem 6.1 and other more delicate results concerning the behavior of the coefficients of polynomials of best approximation /72/.

Chapter III

The central achievement of recent years in minimax theory has been the derivation of a formula for the directional derivative of the maximum function. In some special cases, this formula was established in /10, 61/. The general case was considered in /14, 15, 17–19/. Theorem 2.2 was proved in /20/.

§3. Theorems 3.1 and 3.2 were essentially known to Chebotarev /79/. The expression for the direction of steepest descent of the maximum function (Theorem 3.3) was obtained in /20/.

§4. Some of the estimates presented here may be found in /18/.

§5. The idea of the method of steepest descent goes back to Cauchy /44/ and Hadamard /1/. The details were worked out in the forties by Levenberg /47/, Kantarovich /37/ and Curry /40/.

For continuously differentiable functions, the method of coordinatewise descent and the method of steepest descent both yield a stationary point. In minimax problems, as shown in the text, the situation is less favorable. Our example of "jamming" of steepest descent is taken from /23/.

A good review of publications on the method of steepest descent may be found in Kelley /41/.

§6. The first method of successive approximations was published in /18/. The modification described in subsection 5 is related to the algorithm of Zoutendijk /30—32/ and Zukhovitskii, Polyak and Primak /33/.

§7. The method of subsection 1 was described in /3/ for a special case.

§8. The D-method was proposed in /21, 23/.

§9. Full expositions of the theory of linear programming may be found in /16, 34, 85/. For methods of minimization for functions of one variable, see /75/. In /2/ one can find ALGOL-60 procedures suitable for the solution of minimax problems.

Concerning other methods for the solution of discrete minimax problems, including second order methods, see /86/.

Chapter IV

Dual cones were first applied to necessary conditions for an extremum by Dubovitskii and Milyutin /28, 29/.

Theorem 3.1 is essentially to be found in /61, 62/ (see also /51/).

§5. Some of the estimates may be found in /18/.

Chapter V

This chapter applies the general theory of Chapters III and IV to the generalized problem of mathematical programming.

The theory of mathematical (or optimal, nonlinear) programming has recently undergone rapid development. We are not in a position here to provide a detailed bibliography; the reader may consult the monographs /32, 39, 54, 71/, the papers /12, 48/ and the review of Künzi and Oettli /46/, where an extensive list of literature may be found.

§2. The Slater condition originates from /73/.

§5. The Kuhn-Tucker theorem (for the case $N = 0$) was published in /45/. Various generalizations have been proposed, as in Karlin /39/.

§§6—9 are based on /20, 22/.

§10. The idea of penalty functions appears in /41/. Further results on the method as applied to the solution of minimax problems were obtained in /8, 9/.

§11. Other solution methods for nonlinear programming problems may be found in /67—69, 82, 83, 60/.

Chapter VI

§§ 2, 3 generalize the results of Chapters III — V to the continuous case. On this subject, see /18, 51, 61/.

Theorem 3.3 is a special case of a general theorem of Shnirel'man /81/ (see also /74/.

Necessary conditions for an extremum are considered in /7, 11, 26, 28, 29, 36, 54, 59, 63/.

§4. Theorem 4.1 (convergence of the grid method) was published in /52/.

§5. For the minimax theorem, see, e.g., /39/.

§6 is based on /24/. Concerning other methods for determination of saddle points see /13, 27, 57, 66/. It is well known /16, 57/ that matrix games are equivalent to linear programming problems.

§7. The idea of the proof of Theorem 7.1 (existence of a polynomial of best approximation) is borrowed from /38/. Theorem 7.2 was established in /43, 5, 58/. The convergence of the grid method in the problem of best approximation for functions of several variables was examined in /53/.

§8. The fact that Chebyshev's theorem is a corollary of Theorem 2.2 was already known to Chebotarev /79/.

Appendix I

Proofs of the theorems on algebraic interpolation may be found, e.g., in /56/..

Appendix II

In writing this appendix we used the monographs /39, 76/.

Appendix III

More detailed information on the questions examined here may be found in /77/.

Appendix IV

Concerning the first method of successive approximations, see /78, 25, 42/. The second method was proposed in /55/.

Supplement

This supplement was written by E.E. Voiton, V.N. Malozemov and B.F. Mitchell.

BIBLIOGRAPHY

1. Hadamard, J. Mémoire sur le problème d'analyse relatif à l'équilibre des plaques élastiques encastrées. — Mém. Prés. Acad. Sci. France 2 33, No.4 (1908).
2. Romanovskii, I.V. (editor). ALGOL Procedures. Manual of Methods in Computer Software, No.9. — Vychislitel'nyi Tsentr Leningradskogo Gosudarstvennogo Universiteta (rotaprint), 1971. (Russian)
3. Birzak, B. and B.N. Pshenichnyi. Some minimization problems for nonsmooth functions. — Kibernetika, No.6 (1966), 53—57. (Russian)
4. Burov, V.N. Some effective techniques for solution of Chebyshev's problem of best approximation. — Izv. Vyssh. Uchebn. Zaved. Matematika, No.1 (1957), 67—79). (Russian)
5. Burov, V.N. On approximation of functions by polynomials satisfying nonlinear relations. — Dokl. Akad. Nauk SSSR 138, No.3 (1961), 515—517. (Russian)
6. de la Vallée-Poussin, C. Leçons sur l'approximation des fonctions d'une variable réelle. — Paris, 1919.
7. Germeier, Yu.B. Necessary conditions for max-min. — Zh. Vychisl. Mat. i Mat. Fiz.9, No.2 (1969), 432—438. (Russian)
8. Germeier, Yu.B. Use of penalty functions in approximate reduction of a max-min problem to a maximization problem. — Zh. Vychisl. Mat. i Mat. Fiz.9, No.3 (1969), 730—731. (Russian)
9. Germeier, Yu.B. On the problem of max-min seeking with constraints. — Zh. Vychisl. Mat. i Mat. Fiz. 10, No.1 (1970), 39—55. (Russian)
10. Girsanov, I.V. Differentiability of solutions of mathematical programming problems. — In: Tezisy dokladov Vsesoyuznoi mezhvuzovskoi konferentsii po primeneniyu metodov funktsional'nogo analiza k resheniyu nelineinykh zadach, Baku, 1965, 43—45. (Russian)
11. Girsanov, I.V. Lectures on Mathematical Theory of Extremum Problems. — Moscow, Izdatel'stvo Moskovskogo Universiteta, 1970. (Russian) [English translation: Berlin, Springer, 1972]
12. Gol'shtein, E.G. Convex Programming (Elements of the Theory). — Moscow, Nauka, 1970. (Russian)
13. Danskin, J.M. Fictitious play for continuous games. — Naval Res. Logist. Quart. 1 (1954), 313—320.
14. Danskin, J.M. The theory of max-min, with applications. — SIAM J. Appl. Math. 14 (1966), 641—664.
15. Danskin, J.M. The Theory of Max-min and its Application to Weapons Allocation Problems. — Berlin, Springer, 1967.
16. Dantzig, G.B. Linear Programming and Extensions. — Princeton University Press, 1963.
17. Dem'yanov, V.F. On minimization of the maximum deviation. — Vestnik Leningrad. Gos. Univ., No.7 (1966), 21—28. (Russian)
18. Dem'yanov, V.F. On the solution of some minimax problems, I, II. — Kibernetika, No.6 (1966), 58—66; No.3 (1967), 62—66. (Russian)
19. Dem'yanov, V.F. A method of successive approximations for solution of a certain minimax problem. — Abstracts of Short Scientific Papers, International Congress of Mathematicians, Sec. 14, p.31, Moscow, 1966 (Russian)
20. Dem'yanov, V.F. Algorithms for some minimax problems. — J. Computer and System Sci.2, No.4 (1968), 342—380.
21. Dem'yanov, V.F. On the minimax problem. — Dokl. Akad. Nauk SSSR 187, No.2 (1969), 255—258. (Russian)
22. Dem'yanov, V.F. On seeking the minimax on a bounded set. — Dokl. Akad. Nauk SSSR 191, No.6 (1970), 1216—1219. (Russian)
23. Dem'yanov, V.F. Differentiation of the maximum function. I, III. — Zh. Vychisl. Mat. i Mat. Fiz. 8, No.6 (1968), 1186—1195; 10, No.1 (1970), 26—38. (Russian)
24. Dem'yanov, V.F. Finding saddle points on polyhedra. — Dokl. Akad. Nauk SSSR 192, No.1 (1970), 13—15. (Russian)

25. Dem'yanov, V.F. and A.M. Rubinov. Minimization of a smooth convex functional on a convex set. — Vestnik Leningrad. Gos. Univ., No.19 (1964), 5—17. (Russian)

26. Dem'yanov, V.F. and A.M. Rubinov. Approximation Methods for Solution of Extremum Problems. — Leningrad, Izdatel'stvo Leningradskogo Gosudarstvennogo Universiteta, 1968. (Russian)

27. Dresher, M. and S. Karlin. Solutions of convex games as fixed-points. — In: Contributions to the Theory of Games, Vol.2 (Annals of Mathematics Studies, No.28), 75—86, Princeton University Press, 1953.

28. Dubovitskii, A.Ya. and A.A. Milyutin. The extremum problem in the presence of constraints. — Dokl. Akad. Nauk SSSR 149, No.4 (1963), 759—762. (Russian)

29. Dubovitskii, A.Ya. and A.A. Milyutin. Extremum problems in the presence of constraints. — Zh. Vychisl. Mat. i Mat. Fiz. 5, No.3 (1965), 395—453. (Russian)

30. Zoutendijk, G. Maximizing a function in a convex region. — J. Roy. Statist. Soc. Ser. B21 (1959), 338—355.

31. Zoutendijk, G. Nonlinear programming: A numerical survey. — SIAM J. Control 4 (1966), 194—210.

32. Zoutendijk, G. Methods of Feasible Directions — A Study in Linear and Nonlinear Programming. — Amsterdam, Elsevier, 1960.

33. Zukhovitskii, S.I., R.A. Polyak and M.E. Primak. An algorithm for solution of convex programming problems. — Dokl. Akad. Nauk SSSR 153, No.5 (1963), 991—994. (Russian)

34. Zukhovitskii, S.I. and L.I. Avdeev. Linear and Convex Programming (Second edition). — Moscow, Nauka, 1967. (Russian)

35. Kalashnikov, M.D. On polynomials of best quadratic approximation in a given system of points. — Dokl. Akad. Nauk SSSR 105 (1955), 634—636. (Russian)

36. Kantorovich, L.V. On an effective method for solution of certain extremum problems. — Dokl. Akad. Nauk SSSR 28 (1940), 212—215. (Russian)

37. Kantorovich, L.V. On the method of steepest descent. — Dokl. Akad. Nauk SSSR 56 (1947), 233—236. (Russian)

38. Kantorovich, L.V. and G.P. Akilov. Functional Analysis in Normed Spaces. — Moscow, Fizmatgiz, 1959 (Russian)

39. Karlin, S. Mathematical Methods and Theory in Games, Programming and Economics. — Reading, Mass., Addison-Wesley, 1959.

40. Curry, H.B. The method of steepest descent for nonlinear minimization problems. — Quart. Appl. Math. 2, No.3 (1944), 258—261.

41. Kelley, H.J. Methods of gradients. — In: Optimization Techniques (G. Leitmann, Ed.), New York, Academic Press, 1962.

42. Kozinets, B.N. On a learning algorithm for a linear perception. — In: Vychislitel'naya Tekhnika i Voprosy Programmirovaniya, No.3, 80—83, 1964. (Russian)

43. Kolmogorov, A.N. Remarks on Chebyshev's polynomials of least deviation from a given function. — Uspekhi Mat. Nauk 3, No.1 (1948), 216—221. (Russian)

44. Cauchy, A.L. Méthode générale pour la résolution des systèmes d'équations simultanées. — C.R. Acad. Sci. 25 (1847), 536—538.

45. Kuhn, H.M. and A.W. Tucker. Nonlinear programming. — Proceedings of the Second Berkeley Symposium on Mathematical Statistics and Probability, 481—491, Berkeley, University of California Press, 1951.

46. Künzi, H.P. and W. Oettli. Nichtlineare Optimierung: Neuere Verfahren, Bibliographie. — Berlin, Springer, 1969.

47. Levenberg, K.A. A method for the solution of certain nonlinear problems in least squares. — Quart. Appl. Math. 2 (1944), 164—168.

48. Levitin, E.S. and B.T. Polyak. Methods of minimization in the presence of constraints. — Zh. Vychisl. Mat. i Mat. Fiz. 6, No.5 (1966), 787—823. (Russian)

49. Malozemov, V.N. On finding the algebraic polynomial of best approximation. — Kibernetika 5 (1969), 125—131. (Russian)

50. Malozemov, V.N. On convergence of the grid method in the problem of best approximation by polynomials. — Vestnik Leningrad. Gos. Univ., No.19 (1970), 138—140. (Russian)

51. Malozemov, V.N. On the theory of nonlinear minimax problems. — Kibernetika, No.3 (1970), 121—125. (Russian)

52. Malozemov, V.N. On convergence of the grid method in nonlinear minimax problems. — Vestnik Leningrad. Gos. Univ., No.19 (1971), 35—37. (Russian)

53. Malozemov, V.N. Best uniform approximation of functions of several variables. — Zh. Vychisl. Mat. i Mat. Fiz. 10, No.3 (1970), 575—586. (Russian)

54. Mangasarian, O.L. Nonlinear Programming. — New York, McGraw-Hill, 1970.

55. Mitchell, B.F., V.F. Dem'yanov and V.N. Malozemov. Finding the point of a polyhedron nearest the origin. — Vestnik Leningrad. Gos. Univ., No. 13 (1971), 38—45. (Russian)

56. Mysovskikh, I.P. Lectures on Computing Methods. — Moscow, Fizmatgiz, 1962. (Russian)

57. Von Neumann, J. and O. Morgenstern. Theory of Games and Economic Behavior (Third edition). — New York, Wiley, 1964.

58. Nikol'skii, V.N. A characteristic feature of the least deviating elements in convex sets. — In: Issledovaniya po sovremennym problemam konstruktivnoi teorii funktsii, 80—84, Baku, Izdatel'stvo Akad. Nauk Azerb. SSR, 1965. (Russian)

59. Neustadt, L.W. An abstract variational theory with applications to a broad class of optimization problems, I, II. — SIAM J. Control 4 (1966), 505—525; 5 (1967), 90—137.

60. Polyak, B.T. A general method for solution of extremum problems. — Dokl. Akad. Nauk SSSR 174, No.1 (1967), 33—36. (Russian)

61. Pshenichnyi, B.N. The dual method in extremum problems, I, II. — Kibernetika, No. 3 (1965), 89—95; No.4 (1965), 64—69. (Russian)

62. Pshenichnyi, B.N. Convex programming in a normed space. — Kibernetika, No.5 (1965), 46—54. (Russian)

63. Pshenichnyi, B.N. Necessary Conditions for Extremum. — Moscow, Nauka, 1969. (Russian)

64. Remez, E.Ya. On the method of successive Chebyshev interpolations and various variants of its realization. — Ukrain. Mat. Zh. 12, No.2 (1960), 170—180. (Russian)

65. Remez, E. Ya. Elements of Numerical Methods of Chebyshev Approximation. — Kiev, Naukova Dumka, 1969. (Russian)

66. Robinson, J. An iterative method of solving a game. — Ann. of Math. (2) 54 (1951), 296—301.

67. Rosen, J.B. Nonlinear programming. The gradient projection method. — Bull. Amer. Math. Soc. 63 (1957), 25—26.

68. Rosen, J.B. The gradient projection method for nonlinear programming, Part I. Linear constraints; Part II. Nonlinear constraints. — SIAM J. Appl. Math. 8 (1960), 181—217; 9 (1961), 514—532.

69. Rosen, J.B. Gradient projection as a least squares solution of Kuhn-Tucker conditions. — Technical Report, University of Wisconsin, July 1965.

70. Rubinshtein, G. Sh. The problem of the extreme point of intersection of an axis with a polyhedron and its application to investigation of a finite system of linear inequalities. — Dokl. Akad. Nauk SSSR 100 (1955), 627—630 (Russian)

71. Rubinshtein, G. Sh. Finite-Dimensional Models of Optimization. — Course of Lectures, Novosibirsk University (rotaprint), 1970. (Russian)

72. Samokish, B.A. On behavior of the coefficients of polynomials approximating a regular function on a closed interval. — In: Metody Vychislenii, No.1, 58—65, Izdatel'stvo Leningrad. Gos. Univ., 1963. (Russian)

73. Slater, M. Lagrange Multipliers Revisited. A Contribution to Nonlinear Programming. — RAND Corporation RM-676, Santa Monica, Calif., 1951.

74. Smirnov, V.I. and N.A. Lebedev. Constructive Theory of Functions of a Complex Variable. — Moscow, Nauka, 1964. (Russian)

75. Wilde, D.J. Optimum-Seeking Methods. — Englewood Cliffs, N.J., Prentice-Hall, 1963.

76. Faddeev, D.K. and V.N. Faddeeva. Numerical Methods of Linear Algebra. — Moscow, Fizmatgiz, 1960. (Russian)

77. Fikhtengol'ts, G.M. Course of Differential and Integral Calculus, Vol.1. — Moscow, Fizmatgiz, 1962. (Russian)

78. Frank, M. and P. Wolfe. An algorithm for quadratic programming. — Naval Res. Logist. Quart. 3 (1956), 93—110.

79. Chebotarev, N.G. On a certain minimax criterion. — Dokl. Akad. Nauk SSSR 39 (1943), 373—376 (see also Collected Works, Vol.2, Moscow, Izdatel'stvo Akad. Nauk SSSR, 1949). (Russian)

80. Chebyshev, P.L. Questions concerning the smallest quantities connected with approximate representation of functions. — Collected Works, Vol.II, 151—238, Moscow, Izdatel'stvo Akad. Nauk SSSR, 1947. (Russian)

81. Shnirel'man, L.G. On uniform approximations. — Izv. Akad. Nauk SSSR Ser. Mat. 2, No.1 (1938), 53—59. (Russian)

82. Shor, E.Z. Application of generalized gradient descent in block programming. — Kibernetika, No.3 (1967), 53—55. (Russian)

83. Shor, E.Z. Generalized gradient descent. — In: Trudy pervoi zimnei shkoly po matematicheskomu programmirovaniyu (Drogobych), No.3, 578—585, Moscow, 1969. (Russian)

84. Eterman, I.I. Polynomial Approximation and Solution of Some Problems of Applied Mathematics. — Penza, Izdatel'stvo Penzenskogo Politekhnicheskogo Instituta, 1960. (Russian)

85. Yudin, D.B. and E.G. Gol'shtein. Linear Programming. — Moscow, Fizmatgiz, 1963. (Russian)

86. Dem'yanov, V.F. and V.N. Malozemov. On the theory of nonlinear minimax problems. — Uspekhi Mat. Nauk 26, No.3 (1971), 53—104. (Russian)

SUBJECT INDEX